D0858245

Methods in Enzymology

Volume 273

RNA POLYMERASE AND ASSOCIATED FACTORS

Part A

METHODS IN ENZYMOLOGY

Methods in Enzymology

Volume 273

RNA Polymerase and Associated Factors

Part A

EDITED BY

Sankar Adhya

NATIONAL CANCER INSTITUTE
NATIONAL INSTITUTES OF HEALTH
BETHESDA, MARYLAND

ACADEMIC PRESS
San Diego London Boston New York Sydney Tokyo Toronto

This book is printed on acid-free paper. ♾

Academic Press, Inc.
525 B Street, Suite 1900, San Diego, California 92101-4495, USA
http://www.apnet.com

Academic Press Limited
24-28 Oval Road, London NW1 7DX, UK
http://www.hbuk.co.uk/ap/

International Standard Serial Number: 0076-6879

International Standard Book Number: 0-12-182174-9

PRINTED IN THE UNITED STATES OF AMERICA
96 97 98 99 00 01 EB 9 8 7 6 5 4 3 2 1

Table of Contents

Section I. Promoter Elements and RNA Polymerase Components

Section II. Biochemical Assays of Transcription Initiation

Section III. RNA Polymerase and Its Subunits in Prokaryotes

Section IV. RNA Polymerase and Associated Factors from Eukaryotes

Section V. Genetic Analysis of Transcription and Its Regulation

Contributors to Volume 273

Article numbers are in parentheses following the names of contributors.
Affiliations listed are current.

CORY ABATE-SHEN (6), *Center for Advanced Biotechnology and Medicine and Department of Neuroscience and Cell Biology, Robert Wood Johnson Medical School, University of Medicine and Dentistry of New Jersey, Piscataway, New Jersey 08854*

CHANG BAI (27), *Verna and Marrs McLean Department of Biochemistry, Howard Hughes Medical Institute, Baylor College of Medicine, Houston, Texas 77030*

STEFAN BJORKLUND (15), *Stanford University School of Medicine, Stanford, California 94305*

RICHARD R. BURGESS (12), *McArdle Laboratory for Cancer Research, University of Wisconsin–Madison, Madison, Wisconsin 53706*

DAVID M. CHAO (16), *Whitehead Institute for Biomedical Research, Cambridge, Massachusetts 02142; and Department of Biology, Massachusetts Institute of Technology, Cambridge, Massachusetts 02139*

JIN-LONG CHEN (19), *Department of Biology, Tularik Incorporated, South San Francisco, California 94080*

JOAN WELIKY CONAWAY (18), *Program in Molecular and Cell Biology, Oklahoma Medical Research Foundation, Oklahoma City, Oklahoma 73104*

RONALD C. CONAWAY (18), *Program in Molecular and Cell Biology, Oklahoma Medical Research Foundation, Oklahoma City, Oklahoma 73104*

CHRISTINE CONESA (22), *Service de Biochimie et de Génétique Moléculaire, Commissariat à l'Energie Atomique, Saclay, F91191 Gif sur Yvette Cedex, France*

MICHAEL E. DAHMUS (17), *Section of Molecular and Cellular Biology, Division of Biological Sciences, University of California, Davis, Davis, California 95616*

BENOIT de CROMBRUGGHE (20), *Department of Molecular Genetics, M.D. Anderson Cancer Center, University of Texas, Houston, Texas 77030*

GIORGIO DIECI (22), *Service de Biochimie et de Génétique Moléculaire, Commissariat à l'Energie Atomique, Saclay, F91191 Gif sur Yvette Cedex, France*

ALICIA J. DOMBROSKI (11), *Department of Microbiology and Molecular Genetics, University of Texas Medical School, Houston, Texas 77030*

RICHARD H. EBRIGHT (10), *Department of Chemistry and Waksman Institute, Rutgers University, New Brunswick, New Jersey 08855*

STEPHEN J. ELLEDGE (27), *Verna and Marrs McLean Department of Biochemistry, Howard Hughes Medical Institute, Baylor College of Medicine, Houston, Texas 77030*

HAO FAN (23), *Center for Gene Research, Nagoya University, Nagoya 464-01, Japan*

JAN S. FASSLER (1), *Department of Biological Sciences, University of Iowa, Iowa City, Iowa 52242*

NOBUYUKI FUJITA (9), *Department of Molecular Genetics, National Institute of Genetics, Mishima, Shizuoka 411, Japan*

KARLA PFEIL GARRETT (18), *Program in Molecular and Cell Biology, Oklahoma Medical Research Foundation, Oklahoma City, Oklahoma 73104*

ALEX GOLDFARB (10), *Public Health Research Institute, New York, New York 10016*

JAY D. GRALLA (7), *Department of Chemistry and Biochemistry, University of California, Los Angeles, Los Angeles, California 90095*

MICHAEL R. GREEN (8), *Program in Molecular Medicine, Howard Hughes Medical Institute Research Laboratories, University of Massachusetts Medical Center, Worcester, Massachusetts 01605*

INGRID GRUMMT (21), *Division of Molecular Biology of the Cell II/0120, German Cancer Research Center, D-69120 Heidelberg, Germany*

GARY N. GUSSIN (1, 3), *Department of Biological Sciences, University of Iowa, Iowa City, Iowa 52242*

REGINE HAKENBECK (24), *Max-Planck Institut für Molekulare Genetik, D-14195 Berlin, Germany*

LUCIE S. HEATH (5), *Department of Biological Sciences, University of Alabama, Tuscaloosa, Alabama 35487*

GERALD Z. HERTZ (2), *Department of Molecular, Cellular, and Developmental Biology, University of Colorado, Boulder, Colorado 80309*

LILIAN M. HSU (4), *Program in Biochemistry, Mount Holyoke College, South Hadley, Massachusetts 01075*

JANINE HUET (22), *Service de Biochimie et de Génétique Moléculaire, Commissariat à l'Energie Atomique, Saclay, F91191 Gif sur Yvette Cedex, France*

NANCY ILER (6), *Center for Advanced Biotechnology and Medicine and Graduate Program in Microbiology and Molecular Genetics, Rutgers University, Piscataway, New Jersey 08854*

AKIRA ISHIHAMA (9), *Department of Molecular Genetics, National Institute of Genetics, Mishima, Shizuoka 411, Japan*

VINAY K. JAIN (26), *Baylor University Medical Center, Dallas, Texas 75246*

DING JUN JIN (25), *Developmental Genetics Section, Laboratory of Molecular Biology, Division of Basic Sciences, National Cancer Institute, National Institutes of Health, Bethesda, Maryland 20892*

YOUNG-JOON KIM (15), *Stanford University School of Medicine, Stanford, California 94305*

YOUNGGYU KIM (10), *Department of Chemistry and Waksman Institute, Rutgers University, New Brunswick, New Jersey 08855*

ANTHONY J. KOLESKE (16), *Department of Biology, Massachusetts Institute of Technology, Cambridge, Massachusetts 02139*

ROGER D. KORNBERG (15), *Department of Structural Biology, Stanford University School of Medicine, Stanford, California 94305*

OLIVIER LEFEBVRE (22), *Service de Biochimie et de Génétique Moléculaire, Commissariat à l'Energie Atomique, Saclay, F91191 Gif sur Yvette Cedex, France*

YANG LI (15), *Department of Structural Biology, Stanford University School of Medicine, Stanford, California 94305*

CHONGGUANG LIU (5), *Department of Cell Biology and Physiology, University of Pittsburgh, Pittsburgh, Pennsylvania 15260*

SANKAR N. MAITY (20), *Department of Molecular Genetics, M.D. Anderson Cancer Center, University of Texas, Houston, Texas 77030*

NATHALIE MANAUD (22), *Service de Biochimie et de Génétique Moléculaire, Commissariat à l'Energie Atomique, Saclay, F91191 Gif sur Yvette Cedex, France*

CHARLES P. MORAN, JR. (13), *Department of Microbiology and Immunology, Emory University School of Medicine, Atlanta, Georgia 30322*

GÉRALD PEYROCHE (22), *Service de Biochimie et de Génétique Moléculaire, Commissariat à l'Energie Atomique, Saclay, F91191 Gif sur Yvette Cedex, France*

WADE POWELL (18), *Department of Biochemistry, Emory University School of Medicine, Atlanta, Georgia 30322*

FENGXIA QI (5), *Department of Oral Biology, University of Alabama at Birmingham, Birmingham, Alabama 35294*

DANIEL REINES (18), *Department of Biochemistry, Emory University School of Medicine, Atlanta, Georgia 30322*

MICHEL RIVA (22), *Service de Biochimie et de Génétique Moléculaire, Commissariat à l'Energie Atomique, Saclay, F91191 Gif sur Yvette Cedex, France*

STEFAN G. E. ROBERTS (8), *Wellcome Sciences Institute, Department of Biochemistry, University of Dundee, Dundee DD1 4HN, United Kingdom*

ROBERT G. ROEDER (14), *Laboratory of Biochemistry and Molecular Biology, The Rockefeller University, New York, New York 10021*

ANNY RUET (22), *Service de Biochimie et de Génétique Moléculaire, Commissariat à l'Energie Atomique, Saclay, F91191 Gif sur Yvette Cedex, France*

ANDREAS SCHNAPP (21), *Division of Molecular Biology of the Cell II/0120, German Cancer Research Center, D-69120 Heidelberg, Germany*

ANDRÉ SENTENAC (22), *Service de Biochimie et de Génétique Moléculaire, Commissariat à l'Energie Atomique, Saclay, F91191 Gif sur Yvette Cedex, France*

KONSTANTINE SEVERINOV (10), *The Rockefeller University, New York, New York 10021*

JEFFRY B. STOCK (24), *Department of Molecular Biology, Princeton University, Princeton, New Jersey 08544*

GARY D. STORMO (2), *Department of Molecular, Cellular, and Developmental Biology, University of Colorado, Boulder, Colorado 80309*

MASAHIRO SUGIURA (23), *Center for Gene Research, Nagoya University, Nagoya 464-01, Japan*

HONG TANG (10), *Department of Chemistry and Waksman Institute, Rutgers University, New Brunswick, New Jersey 08855*

KATHLEEN M. TATTI (13), *Department of Microbiology and Immunology, Emory University School of Medicine, Atlanta, Georgia 30322*

ROBERT TJIAN (19), *Department of Molecular and Cell Biology, Howard Hughes Medical Institute, University of California, Berkeley, Berkeley, California 94720*

CHARLES L. TURNBOUGH, JR. (5), *Department of Microbiology, University of Alabama at Birmingham, Birmingham, Alabama 35294*

RICHARD A. YOUNG (16), *Whitehead Institute for Biomedical Research, Cambridge, Massachusetts 02142; and Department of Biology, Massachusetts Institute of Technology, Cambridge, Massachusetts 02139*

HAILAN ZHANG (6), *Center for Advanced Biotechnology and Medicine and Graduate Program in Molecular Genetics and Microbiology, Robert Wood Johnson Medical School, University of Medicine and Dentistry of New Jersey, Piscataway, New Jersey 08854*

YAN NING ZHOU (25), *Developmental Genetics Section, Laboratory of Molecular Biology, Division of Basic Sciences, National Cancer Institute, National Institutes of Health, Bethesda, Maryland 20892*

Preface

One cannot fully understand the biology of a cell without understanding the central role that gene expression and its regulation play. RNA polymerase was discovered in eukaryotes in 1959 and in prokaryotes in 1960, and the subject of transcription regulation was reported in the early 1960s. Although many breakthrough experiments were performed in the 1970s, there has been, unquestionably, an explosion in our knowledge in the field since the 1980s thanks to the rapid development and use of powerful genetic, biochemical, and physical techniques. As a result, many plausible, sometimes unexpected, ideas have been generated. More encouragingly, some of the ideas have been accepted.

Volumes 273 and 274 of *Methods in Enzymology* cover, for the first time, methods and other analytical approaches for the study of transcription and its regulation in prokaryotes and eukaryotes. The chapters in these two volumes describe steps of transcription; component machinery and their specificity; purification, assays, and properties of RNA polymerases and their intrinsic and extrinsic (including regulatory) factors that guide transcription initiation, elongation, and termination; and the assembly of RNA polymerase holoenzymes and many regulatory protein–protein and nucleoprotein complexes, including chromatins. A few chapters dealing with specialized techniques analyzing transcriptional regulation are also included.

These volumes will help further exploration of how transcription controls cellular adaptation, development, and differentiation. We underscore the importance of DNA–protein interactions in studying transcription and its regulation, a subject covered in Volume 208 of this series.

SANKAR ADHYA

METHODS IN ENZYMOLOGY

VOLUME XVI. Fast Reactions
Edited by KENNETH KUSTIN

VOLUME XVII. Metabolism of Amino Acids and Amines (Parts A and B)
Edited by HERBERT TABOR AND CELIA WHITE TABOR

VOLUME XVIII. Vitamins and Coenzymes (Parts A, B, and C)
Edited by DONALD B. MCCORMICK AND LEMUEL D. WRIGHT

VOLUME XIX. Proteolytic Enzymes
Edited by GERTRUDE E. PERLMANN AND LASZLO LORAND

VOLUME XX. Nucleic Acids and Protein Synthesis (Part C)
Edited by KIVIE MOLDAVE AND LAWRENCE GROSSMAN

VOLUME XXI. Nucleic Acids (Part D)
Edited by LAWRENCE GROSSMAN AND KIVIE MOLDAVE

VOLUME XXII. Enzyme Purification and Related Techniques
Edited by WILLIAM B. JAKOBY

VOLUME XXIII. Photosynthesis (Part A)
Edited by ANTHONY SAN PIETRO

VOLUME XXIV. Photosynthesis and Nitrogen Fixation (Part B)
Edited by ANTHONY SAN PIETRO

VOLUME XXV. Enzyme Structure (Part B)
Edited by C. H. W. HIRS AND SERGE N. TIMASHEFF

VOLUME XXVI. Enzyme Structure (Part C)
Edited by C. H. W. HIRS AND SERGE N. TIMASHEFF

VOLUME XXVII. Enzyme Structure (Part D)
Edited by C. H. W. HIRS AND SERGE N. TIMASHEFF

VOLUME XXVIII. Complex Carbohydrates (Part B)
Edited by VICTOR GINSBURG

VOLUME XXIX. Nucleic Acids and Protein Synthesis (Part E)
Edited by LAWRENCE GROSSMAN AND KIVIE MOLDAVE

VOLUME XXX. Nucleic Acids and Protein Synthesis (Part F)
Edited by KIVIE MOLDAVE AND LAWRENCE GROSSMAN

VOLUME XXXI. Biomembranes (Part A)
Edited by SIDNEY FLEISCHER AND LESTER PACKER

VOLUME XXXII. Biomembranes (Part B)
Edited by SIDNEY FLEISCHER AND LESTER PACKER

VOLUME XXXIII. Cumulative Subject Index Volumes I–XXX
Edited by MARTHA G. DENNIS AND EDWARD A. DENNIS

VOLUME XXXIV. Affinity Techniques (Enzyme Purification: Part B)
Edited by WILLIAM B. JAKOBY AND MEIR WILCHEK

Volume XXXV. Lipids (Part B)
Edited by John M. Lowenstein

Volume XXXVI. Hormone Action (Part A: Steroid Hormones)
Edited by Bert W. O'Malley and Joel G. Hardman

Volume XXXVII. Hormone Action (Part B: Peptide Hormones)
Edited by Bert W. O'Malley and Joel G. Hardman

Volume XXXVIII. Hormone Action (Part C: Cyclic Nucleotides)
Edited by Joel G. Hardman and Bert W. O'Malley

Volume XXXIX. Hormone Action (Part D: Isolated Cells, Tissues, and Organ Systems)
Edited by Joel G. Hardman and Bert W. O'Malley

Volume XL. Hormone Action (Part E: Nuclear Structure and Function)
Edited by Bert W. O'Malley and Joel G. Hardman

Volume XLI. Carbohydrate Metabolism (Part B)
Edited by W. A. Wood

Volume XLII. Carbohydrate Metabolism (Part C)
Edited by W. A. Wood

Volume XLIII. Antibiotics
Edited by John H. Hash

Volume XLIV. Immobilized Enzymes
Edited by Klaus Mosbach

Volume XLV. Proteolytic Enzymes (Part B)
Edited by Laszlo Lorand

Volume XLVI. Affinity Labeling
Edited by William B. Jakoby and Meir Wilchek

Volume XLVII. Enzyme Structure (Part E)
Edited by C. H. W. Hirs and Serge N. Timasheff

Volume XLVIII. Enzyme Structure (Part F)
Edited by C. H. W. Hirs and Serge N. Timasheff

Volume XLIX. Enzyme Structure (Part G)
Edited by C. H. W. Hirs and Serge N. Timasheff

Volume L. Complex Carbohydrates (Part C)
Edited by Victor Ginsburg

Volume LI. Purine and Pyrimidine Nucleotide Metabolism
Edited by Patricia A. Hoffee and Mary Ellen Jones

Volume LII. Biomembranes (Part C: Biological Oxidations)
Edited by Sidney Fleischer and Lester Packer

Volume LIII. Biomembranes (Part D: Biological Oxidations)
Edited by Sidney Fleischer and Lester Packer

VOLUME 107. Posttranslational Modifications (Part B)
Edited by FINN WOLD AND KIVIE MOLDAVE

VOLUME 108. Immunochemical Techniques (Part G: Separation and Characterization of Lymphoid Cells)
Edited by GIOVANNI DI SABATO, JOHN J. LANGONE, AND HELEN VAN VUNAKIS

VOLUME 109. Hormone Action (Part I: Peptide Hormones)
Edited by LUTZ BIRNBAUMER AND BERT W. O'MALLEY

VOLUME 110. Steroids and Isoprenoids (Part A)
Edited by JOHN H. LAW AND HANS C. RILLING

VOLUME 111. Steroids and Isoprenoids (Part B)
Edited by JOHN H. LAW AND HANS C. RILLING

VOLUME 112. Drug and Enzyme Targeting (Part A)
Edited by KENNETH J. WIDDER AND RALPH GREEN

VOLUME 113. Glutamate, Glutamine, Glutathione, and Related Compounds
Edited by ALTON MEISTER

VOLUME 114. Diffraction Methods for Biological Macromolecules (Part A)
Edited by HAROLD W. WYCKOFF, C. H. W. HIRS, AND SERGE N. TIMASHEFF

VOLUME 115. Diffraction Methods for Biological Macromolecules (Part B)
Edited by HAROLD W. WYCKOFF, C. H. W. HIRS, AND SERGE N. TIMASHEFF

VOLUME 116. Immunochemical Techniques (Part H: Effectors and Mediators of Lymphoid Cell Functions)
Edited by GIOVANNI DI SABATO, JOHN J. LANGONE, AND HELEN VAN VUNAKIS

VOLUME 117. Enzyme Structure (Part J)
Edited by C. H. W. HIRS AND SERGE N. TIMASHEFF

VOLUME 118. Plant Molecular Biology
Edited by ARTHUR WEISSBACH AND HERBERT WEISSBACH

VOLUME 119. Interferons (Part C)
Edited by SIDNEY PESTKA

VOLUME 120. Cumulative Subject Index Volumes 81–94, 96–101

VOLUME 121. Immunochemical Techniques (Part I: Hybridoma Technology and Monoclonal Antibodies)
Edited by JOHN J. LANGONE AND HELEN VAN VUNAKIS

VOLUME 122. Vitamins and Coenzymes (Part G)
Edited by FRANK CHYTIL AND DONALD B. MCCORMICK

VOLUME 123. Vitamins and Coenzymes (Part H)
Edited by FRANK CHYTIL AND DONALD B. MCCORMICK

VOLUME 124. Hormone Action (Part J: Neuroendocrine Peptides)
Edited by P. MICHAEL CONN

VOLUME 125. Biomembranes (Part M: Transport in Bacteria, Mitochondria, and Chloroplasts: General Approaches and Transport Systems)
Edited by SIDNEY FLEISCHER AND BECCA FLEISCHER

VOLUME 126. Biomembranes (Part N: Transport in Bacteria, Mitochondria, and Chloroplasts: Protonmotive Force)
Edited by SIDNEY FLEISCHER AND BECCA FLEISCHER

VOLUME 127. Biomembranes (Part O: Protons and Water: Structure and Translocation)
Edited by LESTER PACKER

VOLUME 128. Plasma Lipoproteins (Part A: Preparation, Structure, and Molecular Biology)
Edited by JERE P. SEGREST AND JOHN J. ALBERS

VOLUME 129. Plasma Lipoproteins (Part B: Characterization, Cell Biology, and Metabolism)
Edited by JOHN J. ALBERS AND JERE P. SEGREST

VOLUME 130. Enzyme Structure (Part K)
Edited by C. H. W. HIRS AND SERGE N. TIMASHEFF

VOLUME 131. Enzyme Structure (Part L)
Edited by C. H. W. HIRS AND SERGE N. TIMASHEFF

VOLUME 132. Immunochemical Techniques (Part J: Phagocytosis and Cell-Mediated Cytotoxicity)
Edited by GIOVANNI DI SABATO AND JOHANNES EVERSE

VOLUME 133. Bioluminescence and Chemiluminescence (Part B)
Edited by MARLENE DELUCA AND WILLIAM D. MCELROY

VOLUME 134. Structural and Contractile Proteins (Part C: The Contractile Apparatus and the Cytoskeleton)
Edited by RICHARD B. VALLEE

VOLUME 135. Immobilized Enzymes and Cells (Part B)
Edited by KLAUS MOSBACH

VOLUME 136. Immobilized Enzymes and Cells (Part C)
Edited by KLAUS MOSBACH

VOLUME 137. Immobilized Enzymes and Cells (Part D)
Edited by KLAUS MOSBACH

VOLUME 138. Complex Carbohydrates (Part E)
Edited by VICTOR GINSBURG

VOLUME 139. Cellular Regulators (Part A: Calcium- and Calmodulin-Binding Proteins)
Edited by ANTHONY R. MEANS AND P. MICHAEL CONN

VOLUME 140. Cumulative Subject Index Volumes 102–119, 121–134

VOLUME 141. Cellular Regulators (Part B: Calcium and Lipids)
Edited by P. MICHAEL CONN AND ANTHONY R. MEANS

VOLUME 142. Metabolism of Aromatic Amino Acids and Amines
Edited by SEYMOUR KAUFMAN

VOLUME 143. Sulfur and Sulfur Amino Acids
Edited by WILLIAM B. JAKOBY AND OWEN GRIFFITH

VOLUME 144. Structural and Contractile Proteins (Part D: Extracellular Matrix)
Edited by LEON W. CUNNINGHAM

VOLUME 145. Structural and Contractile Proteins (Part E: Extracellular Matrix)
Edited by LEON W. CUNNINGHAM

VOLUME 146. Peptide Growth Factors (Part A)
Edited by DAVID BARNES AND DAVID A. SIRBASKU

VOLUME 147. Peptide Growth Factors (Part B)
Edited by DAVID BARNES AND DAVID A. SIRBASKU

VOLUME 148. Plant Cell Membranes
Edited by LESTER PACKER AND ROLAND DOUCE

VOLUME 149. Drug and Enzyme Targeting (Part B)
Edited by RALPH GREEN AND KENNETH J. WIDDER

VOLUME 150. Immunochemical Techniques (Part K: *In Vitro* Models of B and T Cell Functions and Lymphoid Cell Receptors)
Edited by GIOVANNI DI SABATO

VOLUME 151. Molecular Genetics of Mammalian Cells
Edited by MICHAEL M. GOTTESMAN

VOLUME 152. Guide to Molecular Cloning Techniques
Edited by SHELBY L. BERGER AND ALAN R. KIMMEL

VOLUME 153. Recombinant DNA (Part D)
Edited by RAY WU AND LAWRENCE GROSSMAN

VOLUME 154. Recombinant DNA (Part E)
Edited by RAY WU AND LAWRENCE GROSSMAN

VOLUME 155. Recombinant DNA (Part F)
Edited by RAY WU

VOLUME 156. Biomembranes (Part P: ATP-Driven Pumps and Related Transport: The Na,K-Pump)
Edited by SIDNEY FLEISCHER AND BECCA FLEISCHER

VOLUME 157. Biomembranes (Part Q: ATP-Driven Pumps and Related Transport: Calcium, Proton, and Potassium Pumps)
Edited by SIDNEY FLEISCHER AND BECCA FLEISCHER

VOLUME 158. Metalloproteins (Part A)
Edited by JAMES F. RIORDAN AND BERT L. VALLEE

VOLUME 176. Nuclear Magnetic Resonance (Part A: Spectral Techniques and Dynamics)
Edited by NORMAN J. OPPENHEIMER AND THOMAS L. JAMES

VOLUME 177. Nuclear Magnetic Resonance (Part B: Structure and Mechanism)
Edited by NORMAN J. OPPENHEIMER AND THOMAS L. JAMES

VOLUME 178. Antibodies, Antigens, and Molecular Mimicry
Edited by JOHN J. LANGONE

VOLUME 179. Complex Carbohydrates (Part F)
Edited by VICTOR GINSBURG

VOLUME 180. RNA Processing (Part A: General Methods)
Edited by JAMES E. DAHLBERG AND JOHN N. ABELSON

VOLUME 181. RNA Processing (Part B: Specific Methods)
Edited by JAMES E. DAHLBERG AND JOHN N. ABELSON

VOLUME 182. Guide to Protein Purification
Edited by MURRAY P. DEUTSCHER

VOLUME 183. Molecular Evolution: Computer Analysis of Protein and Nucleic Acid Sequences
Edited by RUSSELL F. DOOLITTLE

VOLUME 184. Avidin–Biotin Technology
Edited by MEIR WILCHEK AND EDWARD A. BAYER

VOLUME 185. Gene Expression Technology
Edited by DAVID V. GOEDDEL

VOLUME 186. Oxygen Radicals in Biological Systems (Part B: Oxygen Radicals and Antioxidants)
Edited by LESTER PACKER AND ALEXANDER N. GLAZER

VOLUME 187. Arachidonate Related Lipid Mediators
Edited by ROBERT C. MURPHY AND FRANK A. FITZPATRICK

VOLUME 188. Hydrocarbons and Methylotrophy
Edited by MARY E. LIDSTROM

VOLUME 189. Retinoids (Part A: Molecular and Metabolic Aspects)
Edited by LESTER PACKER

VOLUME 190. Retinoids (Part B: Cell Differentiation and Clinical Applications)
Edited by LESTER PACKER

VOLUME 191. Biomembranes (Part V: Cellular and Subcellular Transport: Epithelial Cells)
Edited by SIDNEY FLEISCHER AND BECCA FLEISCHER

VOLUME 192. Biomembranes (Part W: Cellular and Subcellular Transport: Epithelial Cells)
Edited by SIDNEY FLEISCHER AND BECCA FLEISCHER

VOLUME 193. Mass Spectrometry
Edited by JAMES A. MCCLOSKEY

VOLUME 194. Guide to Yeast Genetics and Molecular Biology
Edited by CHRISTINE GUTHRIE AND GERALD R. FINK

VOLUME 195. Adenylyl Cyclase, G Proteins, and Guanylyl Cyclase
Edited by ROGER A. JOHNSON AND JACKIE D. CORBIN

VOLUME 196. Molecular Motors and the Cytoskeleton
Edited by RICHARD B. VALLEE

VOLUME 197. Phospholipases
Edited by EDWARD A. DENNIS

VOLUME 198. Peptide Growth Factors (Part C)
Edited by DAVID BARNES, J. P. MATHER, AND GORDON H. SATO

VOLUME 199. Cumulative Subject Index Volumes 168–174, 176–194

VOLUME 200. Protein Phosphorylation (Part A: Protein Kinases: Assays, Purification, Antibodies, Functional Analysis, Cloning, and Expression)
Edited by TONY HUNTER AND BARTHOLOMEW M. SEFTON

VOLUME 201. Protein Phosphorylation (Part B: Analysis of Protein Phosphorylation, Protein Kinase Inhibitors, and Protein Phosphatases)
Edited by TONY HUNTER AND BARTHOLOMEW M. SEFTON

VOLUME 202. Molecular Design and Modeling: Concepts and Applications (Part A: Proteins, Peptides, and Enzymes)
Edited by JOHN J. LANGONE

VOLUME 203. Molecular Design and Modeling: Concepts and Applications (Part B: Antibodies and Antigens, Nucleic Acids, Polysaccharides, and Drugs)
Edited by JOHN J. LANGONE

VOLUME 204. Bacterial Genetic Systems
Edited by JEFFREY H. MILLER

VOLUME 205. Metallobiochemistry (Part B: Metallothionein and Related Molecules)
Edited by JAMES F. RIORDAN AND BERT L. VALLEE

VOLUME 206. Cytochrome P450
Edited by MICHAEL R. WATERMAN AND ERIC F. JOHNSON

VOLUME 207. Ion Channels
Edited by BERNARDO RUDY AND LINDA E. IVERSON

VOLUME 208. Protein–DNA Interactions
Edited by ROBERT T. SAUER

VOLUME 209. Phospholipid Biosynthesis
Edited by EDWARD A. DENNIS AND DENNIS E. VANCE

VOLUME 210. Numerical Computer Methods
Edited by LUDWIG BRAND AND MICHAEL L. JOHNSON

VOLUME 211. DNA Structures (Part A: Synthesis and Physical Analysis of DNA)
Edited by DAVID M. J. LILLEY AND JAMES E. DAHLBERG

VOLUME 212. DNA Structures (Part B: Chemical and Electrophoretic Analysis of DNA)
Edited by DAVID M. J. LILLEY AND JAMES E. DAHLBERG

VOLUME 213. Carotenoids (Part A: Chemistry, Separation, Quantitation, and Antioxidation)
Edited by LESTER PACKER

VOLUME 214. Carotenoids (Part B: Metabolism, Genetics, and Biosynthesis)
Edited by LESTER PACKER

VOLUME 215. Platelets: Receptors, Adhesion, Secretion (Part B)
Edited by JACEK J. HAWIGER

VOLUME 216. Recombinant DNA (Part G)
Edited by RAY WU

VOLUME 217. Recombinant DNA (Part H)
Edited by RAY WU

VOLUME 218. Recombinant DNA (Part I)
Edited by RAY WU

VOLUME 219. Reconstitution of Intracellular Transport
Edited by JAMES E. ROTHMAN

VOLUME 220. Membrane Fusion Techniques (Part A)
Edited by NEJAT DÜZGÜNEŞ

VOLUME 221. Membrane Fusion Techniques (Part B)
Edited by NEJAT DÜZGÜNEŞ

VOLUME 222. Proteolytic Enzymes in Coagulation, Fibrinolysis, and Complement Activation (Part A: Mammalian Blood Coagulation Factors and Inhibitors)
Edited by LASZLO LORAND AND KENNETH G. MANN

VOLUME 223. Proteolytic Enzymes in Coagulation, Fibrinolysis, and Complement Activation (Part B: Complement Activation, Fibrinolysis, and Nonmammalian Blood Coagulation Factors)
Edited by LASZLO LORAND AND KENNETH G. MANN

VOLUME 224. Molecular Evolution: Producing the Biochemical Data
Edited by ELIZABETH ANNE ZIMMER, THOMAS J. WHITE, REBECCA L. CANN, AND ALLAN C. WILSON

VOLUME 225. Guide to Techniques in Mouse Development
Edited by PAUL M. WASSARMAN AND MELVIN L. DEPAMPHILIS

VOLUME 226. Metallobiochemistry (Part C: Spectroscopic and Physical Methods for Probing Metal Ion Environments in Metalloenzymes and Metalloproteins)
Edited by JAMES F. RIORDAN AND BERT L. VALLEE

VOLUME 263. Plasma Lipoproteins (Part C: Quantitation)
Edited by WILLIAM A. BRADLEY, SANDRA H. GIANTURCO, AND JERE P. SEGREST

VOLUME 264. Mitochondrial Biogenesis and Genetics (Part B)
Edited by GIUSEPPE M. ATTARDI AND ANNE CHOMYN

VOLUME 265. Cumulative Subject Index Volumes 228, 230–262 (in preparation)

VOLUME 266. Computer Methods for Macromolecular Sequence Analysis
Edited by RUSSELL F. DOOLITTLE

VOLUME 267. Combinatorial Chemistry
Edited by JOHN N. ABELSON

VOLUME 268. Nitric Oxide (Part A: Sources and Detection of NO; NO Synthase)
Edited by LESTER PACKER

VOLUME 269. Nitric Oxide (Part B: Physiological and Pathological Processes)
Edited by LESTER PACKER

VOLUME 270. High Resolution Separation and Analysis of Biological Macromolecules (Part A: Fundamentals)
Edited by BARRY L. KARGER AND WILLIAM S. HANCOCK

VOLUME 271. High Resolution Separation and Analysis of Biological Macromolecules (Part B: Applications)
Edited by BARRY L. KARGER AND WILLIAM S. HANCOCK

VOLUME 272. Cytochrome P450 (Part B)
Edited by ERIC F. JOHNSON AND MICHAEL R. WATERMAN

VOLUME 273. RNA Polymerase and Associated Factors (Part A)
Edited by SANKAR ADHYA

VOLUME 274. RNA Polymerase and Associated Factors (Part B)
Edited by SANKAR ADHYA

VOLUME 275. Viral Polymerases and Related Proteins (in preparation)
Edited by LAWRENCE C. KUO, DAVID B. OLSEN, AND STEVEN S. CARROLL

VOLUME 276. Macromolecular Crystallography, Part A (in preparation)
Edited by CHARLES W. CARTER, JR., AND ROBERT M. SWEET

Section I

Promoter Elements and RNA Polymerase Components

[1] Promoters and Basal Transcription Machinery in Eubacteria and Eukaryotes: Concepts, Definitions, and Analogies

By JAN S. FASSLER and GARY N. GUSSIN

Eubacterial Promoters

Historical Concepts and Definitions

The earliest experimental definition of the promoter was genetic and, not surprisingly, originated in the context of the *lac* operon of *Escherichia coli.* The concept of a promoter distinct from the other controlling elements of the operon was first enunciated by Jacob *et al.*[1] and the first putative promoter mutants were isolated by Scaife and Beckwith.[2] The idea that the promoter would include the recognition site for RNA polymerase (RNAP) and the site of transcription initiation became confounded by the subsequent discovery that maximal levels of transcription of the *lac* operon required CRP (cAMP receptor protein), also known as CAP (catabolite gene activator protein) and cAMP.[3,4] In fact, the mutations isolated by Scaife and Beckwith were changes in the CRP-binding site, not the RNAP interaction site.

Perhaps as a consequence, the first published sequence of the *lac* control region[5] designated both the CRP-binding site [centered between 61 and 62 base pairs (bp) preceding the *lac* transcription start site] and the RNAP interaction site as components of the promoter. However, because the definition of the promoter must be equally applicable to all transcription units, many of which function perfectly well in the absence of activators, it makes sense to restrict the definition to the interaction site for RNAP holoenzyme. This idea is reinforced by the isolation of *lac* promoter (*lacP*) mutations that make transcription initiation essentially independent of CRP.[6]

[1] F. Jacob, A. Ullman, and J. Monod, *C. R. Acad. Sci.* **258,** 3125 (1964).
[2] J. Scaife and J. R. Beckwith, *Cold Spring Harbor Symp. Quant. Biol.* **31,** 403 (1966).
[3] R. Perlman, B. Chen, B. deCrombrugghe, M. Emmer, M. Gottesman, H. Varmus, and I. Pastan, *Cold Spring Harbor Symp. Quant. Biol.* **35,** 419 (1970).
[4] G. Zubay, D. Schwartz, and J. Beckwith, *Proc. Natl. Acad. Sci. U.S.A.* **66,** 104 (1970).
[5] R. C. Dickson, J. Abelson, W. M. Barnes, and W. S. Reznikoff, *Science* **187,** 27 (1975).
[6] A. E. Silverstone, R. R. Arditti, and B. Magasanik, *Proc. Natl. Acad. Sci. U.S.A.* **66,** 773 (1970).

Current Definition and Experimental Criteria

Genetic Criteria. Promoter mutations cause several characteristic changes in phenotype: (1) they affect expression of all genes in an operon proportionately; (2) their effects are *cis* specific (noncomplementable in *trans*); and (3) they map close to one end of the operon, but can be distinguished from mutations in structural genes because they do not affect the activities of the protein products of these genes (particularly the activity of the product encoded by the promoter-proximal gene). Originally, it was thought that promoter mutations should not quantitatively alter the effects of regulators of transcription initiation, but this cannot be a general property because the binding of RNAP to a promoter can compete with repressor binding to an operator[7,8] or be synergistic with activator binding to an adjacent DNA target site.[9–12]

Biochemical Criteria. Biochemically, the promoter is the set of sequences required for basal (not maximal) transcription. This definition is not problematic for weak prokaryotic promoters, even those that are so weak that transcription initiation cannot be detected by conventional means *in vivo*. One example is the bacteriophage λ P_{RE} promoter, which is stimulated 100- to 1000-fold by *c*II protein. Although P_{RE} transcript cannot be detected *in vivo* in the absence of the activator,[13] initiation at the promoter can be detected *in vitro,* either by addition of high concentrations of RNAP, or by allowing extremely long times for transcriptionally competent open complexes to form prior to addition of substrates.[14] In addition, even for weak promoters, it is possible to generate stronger promoters by mutations that do not change the site of transcription initiation, but make transcription initiation at least partially activator independent.[6,11,15] The point is that a weak promoter is still a promoter—the requirement for an activator does not force us to change our definition.

RNA Polymerase. Operationally, the promoter is the minimal sequence that can be distinguished from random sequences by RNA polymerase *holoenzyme.* The major RNAP of *E. coli* (designated $E\sigma^{70}$) consists of

[7] D. K. Hawley, A. D. Johnson, and W. R. McClure, *J. Biol. Chem.* **260,** 8618 (1985).

[8] M. Lanzer and H. Bujard, *Proc. Natl. Acad. Sci. U.S.A.* **85,** 8973 (1988).

[9] Y.-L. Ren, S. Garges, S. Adhya, and J. S. Krakow, *Proc. Natl. Acad. Sci. U.S.A.* **85,** 4138 (1988).

[10] Y.-S. Ho and M. Rosenberg, *J. Biol. Chem.* **260,** 11838 (1982).

[11] D. K. Hawley and W. R. McClure, *J. Mol. Biol.* **157,** 493 (1982).

[12] J.-J. Hwang and G. N. Gussin, *J. Mol. Biol.* **200,** 735 (1988).

[13] U. Schmeissner, D. Court, H. Shimatake, and M. Rosenberg, *Proc. Natl. Acad. Sci. U.S.A.* **77,** 3191 (1980).

[14] M.-C. Shih and G. N. Gussin, *J. Mol. Biol.* **172,** 489 (1984).

[15] B. J. Meyer, R. Maurer, and M. Ptashne, *J. Mol. Biol.* **139,** 163 (1980).

a core ($\alpha_2\beta\beta'$) plus σ^{70}, which has been demonstrated genetically and biochemically to be the subunit that determines promoter specificity of the enzyme.[16–18] Shortly after formation of the first several phosphodiester bonds, σ dissociates, enabling the core enzyme to move away from the promoter (translocate) and carry out the relatively nonspecific catalytic function of the enzyme.[19,20] When alternative σ factors bind to core RNAP, the minimal promoter sequence changes accordingly,[21] but the definition of the promoter based on sequence recognition by a particular form of the holoenzyme is still appropriate. This is true even for the Ntr (nitrogen source regulation)-specific form of the holoenzyme (Eσ^{54}), which contains σ^{54} instead of σ^{70} and cannot form transcriptionally competent open complexes in the absence of an activator (e.g., NifA or NtrC). Nevertheless, Eσ^{54} is sufficient for promoter recognition and can form sequence-specific stable complexes with the promoter.[22,23]

Certain bacteriophage-encoded RNA polymerases, such as the T7 gene 1 product and N4 RNAP II, have a much simpler (usually monomeric) structure that is similar to the structure of mitochondrial RNAPs and is sufficient for recognition and initiation at a limited number of promoters whose sequences are fairly uniform.[24]

Early work showing that *E. coli* core enzyme alone is unable to form heparin- or rifampicin-resistant (open) complexes[25] led to the suggestion that, in addition to promoter recognition, σ factors mediate unwinding of about 10–12 bp of promoter DNA that overlap the −10 consensus region and the transcription start site.[26] On the basis of examination of amino acid sequences of *E. coli* σ^{70} and several of its homologs, it was proposed that a region rich in aromatic amino acids interacted with single-stranded DNA in RNAP–promoter open complexes.[27] Subsequently, mutations in this region of the σ subunit of

[16] D. A. Siegele, J. C. Hu, W. A. Walter, and C. A. Gross, *J. Mol. Biol.* **206,** 591 (1989).

[17] T. Gardella, H. Moyle, and M. M. Susskind, *J. Mol. Biol.* **206,** 579 (1989).

[18] A. J. Dombroski, W. A. Walter, M. T. Record, D. A. Siegele, and C. A. Gross, *Cell* **70,** 501 (1992).

[19] A. A. Travers and R. R. Burgess, *Nature (London)* **222,** 537 (1969).

[20] W. R. McClure, *Annu. Rev. Biochem.* **54,** 171 (1985).

[21] C. A. Gross, M. Lonetto, and R. Losick, *in* "Transcriptional Regulation" (K. Yamamoto and S. McKnight, eds.), p. 129. Cold Spring Harbor Laboratory, Cold Spring Harbor, New York, 1992.

[22] D. L. Popham, D. Szeto, J. Keener, and S. Kustu, *Science* **243,** 629 (1989).

[23] S. Sasse-Dwight and J. D. Gralla, *Proc. Natl. Acad. Sci. U.S.A.* **85,** 8934 (1988).

[24] W. T. McAllister, *Mol. Microbiol.* **10,** 1 (1993).

[25] M. J. Chamberlin, *in* "RNA Polymerase" (R. Losick and M. J. Chamberlin, eds.), p. 159. Cold Spring Harbor Laboratory, Cold Spring Harbor, New York, 1976.

[26] U. Siebenlist, R. B. Simpson, and W. Gilbert, *Cell* **20,** 269 (1980).

[27] J. D. Helmann and M. J. Chamberlin, *Annu. Rev. Biochem.* **57,** 839 (1988).

Bacillus subtilis RNAP holoenzyme were shown to inhibit DNA unwinding *in vitro*.[28] However, as yet, there is no direct evidence that σ itself provides the unwinding function.

Minimal Promoter for σ^{70}-Containing Holoenzyme. It is well known[20,29] that promoters recognized by the major RNAP holoenzyme in eubacterial systems contain two consensus regions: the −35 region, which usually extends from −30 to −35 (30 to 35 nucleotides preceding the transcription start site) and the −10 region, which usually extends from −7 to −12 (Fig. 1; see also [2] in this volume[29a]). Studies of DNA binding *in vitro*[18] confirm that these sequences are specifically recognized by two different regions of σ^{70}, designated 4.2 and 2.4, respectively.[27] The spacing between the two consensus regions also influences promoter activity, with 17 bp the optimal distance; possibly the spacing affects the rotational geometry of the two regions of σ^{70} essential for recognition of the consensus sequences or for isomerization of the RNAP–promoter complex following binding.[25,30] In addition, two nonuniversal sequence elements can be shown for certain promoters to affect promoter strength, sometimes profoundly: (1) the so-called UP element is an AT-rich region centered at −52 in the *rrnB P1* promoter. Because this region is recognized by the α subunit of RNAP,[31,32] it is likely that such AT-rich sequences also exist in similar regions of promoters recognized by holoenzymes containing alternative σ subunits, although so far none has been identified; (2) the sequence TG is found at −15, −14 in several promoters. Several promoter down-mutations at −14 have been identified and, in a few cases, the T and G have been shown biochemically to be a determinant of promoter strength.[33,34] It is not known which subunit of RNAP specifically contacts these base pairs.

The effects of mutations in the −10 and −35 regions on open complex formation *in vitro* or on promoter activity *in vivo* correlate surprisingly well with the degree to which the mutations increase or decrease agreement with the consensus sequences.[35,36] However, deviations from the expected

[28] Y.-L. Juang and J. D. Helmann, *J. Mol. Biol.* **235,** 1470 (1994).

[29] C. B. Harley and R. P. Reynolds, *Nucleic Acids Res.* **15,** 2343 (1987).

[29a] G. Z. Hertz and G. D. Stormo, *Methods Enzymol.* **273,** Chap. 2, 1996 (this volume).

[30] J. E. Stefano and J. D. Gralla, *Proc. Natl. Acad. Sci. U.S.A.* **79,** 1069 (1982).

[31] L. Rao, W. Ross, J. A. Appleman, T. Gaal, S. Leirmo, P. J. Schlax, M. T. Record, and R. L. Gourse, *J. Mol. Biol.* **235,** 1421 (1994).

[32] W. Ross, K. K. Gosink, J. Salomon, K. Igarashi, C. Zou, A. Ishihama, K. Severinov, and R. L. Gourse, *Science* **262,** 1407 (1993).

[33] S. Kielty and M. Rosenberg, *J. Biol. Chem.* **262,** 6389 (1987).

[34] B. Chan and S. Busby, *Gene* **84,** 227 (1989).

[35] H. Moyle, C. Waldburger, and M. M. Susskind, *J. Bacteriol.* **173,** 1944 (1991).

[36] M. E. Mulligan, D. K. Hawley, R. Entriken, and W. R. McClure, *Nucleic Acids Res.* **12,** 789 (1984).

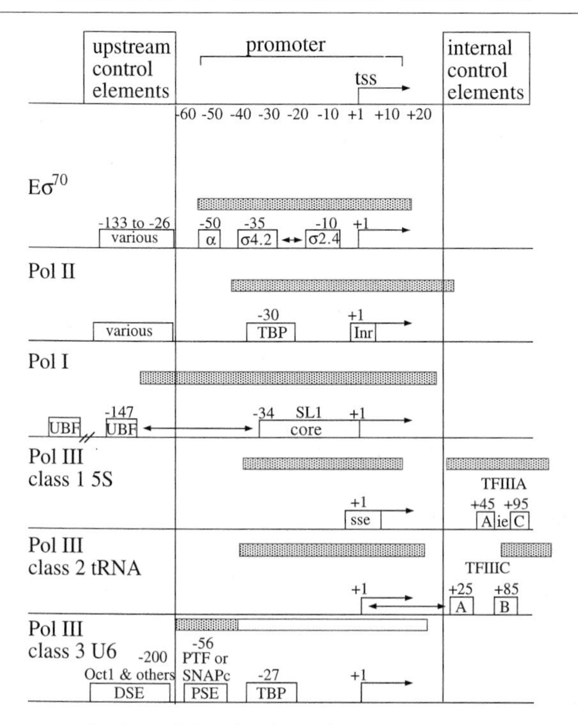

FIG. 1. Structures of eubacterial and eukaryotic promoters. Boxes represent known sequence elements required for transcription initiation or activation. Inside each box is the name of the factor that recognizes the corresponding sequence element, or when there is uncertainty about the DNA-binding protein(s), the name of the site is indicated inside the box and the putative recognition proteins are listed above it. The notations $\sigma 4.2$ and $\sigma 2.4$ indicate the regions of σ^{70} that recognize the -35 and -10 consensus regions, respectively.[21,27] TFIIIA is labeled above its binding site, due to lack of space. Positions of those elements whose location is fixed relative to the start site are also indicated. Double-headed arrows indicate a fixed spatial relationship between two sequence elements. Stippled rectangles denote regions protected by $E\sigma^{70}$ or a eukaryotic basal transcription complex from digestion by DNase I or other cleavage agents.[26,32,143,157,246,247] For Pol III class 1 and 2 promoters, the TFIIIA and TFIIIC footprints are also shown; the footprint for the Pol III U6 TBP complex is inferred from Pol II promoter data. tss, Transcription start site; ie, intermediate element; sse, start site element.

pattern (and differences between results obtained *in vitro* and *in vivo* for near-consensus or consensus promoters)[35,37] implicate sequences centered at -43 (the upstream region, USR)[37a] and between $+1$ and $+20$ (the downstream region, DSR) as determinants of gene expression *in vivo.*

[37] R. Knaus and H. Bujard, *EMBO J.* **7,** 2919 (1988).
[37a] R. Knaus and H. Bujard, *Nucleic Acids and Molec. Biol.* **4,** 110 (1990).

These sequences are not required for promoter recognition per se, but apparently facilitate promoter clearance—translocation of RNAP concomitant with the release of contacts between RNAP in open complexes and nucleotides in consensus regions of the promoter.[37,38]

Transcriptional Activators

Properties of Transcriptional Activators and Their Recognition Sites. Transcriptional activators constitute a class of positive regulators of gene expression. They bind to specific DNA recognition sites that are distinct from the promoters whose activity they regulate. In their absence (due, for example, to mutation), transcription initiation from their target promoters may be reduced 10- to 1000-fold. Mutations in genes encoding transcriptional activators often cannot be distinguished phenotypically from those in genes encoding alternative σ factors. However, biochemically, alternative σ factors and activators can be distinguished by two criteria[39,40]: (1) σ factors can be purified as part of a holoenzyme, but activators cannot; (2) activators do not usually associate with core RNAP under conditions that favor reconstitution of holoenzyme from core plus σ.

The genetic properties of activator-binding sites resemble those of promoters because they are *cis*-acting genetic elements, are often essential for transcription initiation at a rate sufficient to yield a wild-type phenotype, and frequently map close to, or even overlap, the promoters they regulate. However, activator-binding site mutations can usually be distinguished genetically from promoter down-mutations on the basis of their inability to reduce transcription initiation below the level observed with the wild-type template in the absence of activator.[41]

Two properties of particular eubacterial activators are reminiscent of characteristics generally associated with eukaryotic enhancers (see Eukaryotic Promoters and RNA Polymerases). First, activation of Ntr-specific promoters is mediated by a protein whose binding site may be located more than 1000 bp from the promoter, either upstream or downstream.[42] Second, in an artificial construct, the combined effect of CRP/cAMP and λ repressor (cI protein) on transcription initiation at P_{RM} *in vivo* is greater than the sum of the effects of each activator separately.[43]

Activation Mechanisms. The mechanism by which activators stimulate

[38] B. Krummel and M. J. Chamberlin, *Biochemistry* **28,** 7829 (1989).
[39] A. D. Grossman, J. W. Erickson, and C. A. Gross, *Cell* **38,** 383 (1984).
[40] G. Kassavetis and E. P. Geiduschek, *Proc. Natl. Acad. Sci. U.S.A.* **81,** 5101 (1984).
[41] J. D. Hopkins, *J. Mol. Biol.* **87,** 715 (1974).
[42] L. Reitzer and B. Magasanik, *Cell* **45,** 785 (1986).
[43] J. K. Joung, D. M. Koepp, and A. Hochschild, *Science* **265,** 1863 (1994).

transcription initiation can be considered both kinetically and structurally. Kinetically, because promoters subject to activation are rate limited at one or more steps in the pathway to the open complex, the effects of particular activators may be characterized by an increase in the value(s) of kinetic parameter(s) associated with particular steps (these steps are outlined in more detail in [3] in this volume[44]). Thus, CRP/cAMP acting at *lacP* increases K_B, the equilibrium association constant for formation of the initial closed complex, approximately 10-fold[45]; λ *c*I-mediated activation of the P_{RM} promoter is characterized predominantly by a 5- to 10-fold effect on k_f, the phenomenological rate constant for isomerization of closed to open complexes[11]; and λ *c*II protein acting at P_{RE} or P_I causes large increases in both parameters.[14,46] Activation can also be mediated by a change in the rate of promoter clearance.[47,48] The combined effect of activators that stimulate two different slow steps in initiation at a single promoter in theory could be multiplicative.

A complication in understanding the structural basis of activation is the fact that many activators, the prototype being CRP/cAMP, cause DNA bending[49] centered at their binding sites. However, the isolation of activation-defective mutant proteins that bind normally to their recognition sites, and bend DNA normally, demonstrates that DNA bending and protein binding per se are not sufficient for activation.[50,51] What is absolutely required in a large number of cases is contact between a specific region on the solvent-exposed surface of the activator and RNAP.[50,52,53] Initial evidence for such contact was the identification of several point mutations in *rpoA* (the gene encoding the α subunit) that cause defects in the response of RNAP to specific activators *in vivo* or *in vitro*[54,55] and the demonstration that truncation of 73 C-terminal amino acids of the α subunit had the same effect *in vitro*.[56] The distinct effects of different *rpoA* mutations on the

[44] G. N. Gussin, *Methods Enzymol.* **273,** Chap. 3, 1996 (this volume).
[45] T. P. Malan, A. Kolb, H. Buc, and W. R. McClure, *J. Mol. Biol.* **180,** 881 (1984).
[46] B. Hoopes and W. R. McClure, unpublished data, cited by W. R. McClure, *Annu. Rev. Biochem.* **54,** 171 (1985).
[47] W. Kammerer, U. Deuschle, R. Gentz, and H. Bujard, *EMBO J.* **5,** 2995 (1986).
[48] M. Menendez, A. Kolb, and H. Buc, *EMBO J.* **6,** 4227 (1987).
[49] M. R. Gartenberg and D. M. Crothers, *Nature (London)* **333,** 824 (1988).
[50] A. Hochschild, N. Irwin, and M. Ptashne, *Cell* **32,** 219 (1983).
[51] Y. Zhou, X. Zhang, and R. Ebright, *Proc. Natl. Acad. Sci. U.S.A.* **90,** 6081 (1993).
[52] H. Tang, K. Severinov, A. Goldfarb, D. Fenyö, B. Chait, and R. H. Ebright, *Genes Dev.* **8,** 3058 (1994).
[53] Y. Chen, Y. Ebright, and R. Ebright, *Science* **265,** 90 (1994).
[54] S. Busby and R. Ebright, *Cell* **79,** 743 (1994).
[55] F. Russo and T. Silhavy, *J. Biol. Chem.* **267,** 14515 (1992).
[56] K. Igarashi and A. Ishihama, *Cell* **65,** 1015 (1991).

response of RNAP to different activators suggest that α contact points for different activators are not precisely the same. The effects of α C-terminal truncation mutations on activation at other promoters also indicate that different activators may contact separate regions of α or different RNAP subunits altogether. In fact, the search for allele-specific suppressors of λ *c*I mutants defective in activation revealed that the *c*I protein must contact σ, rather than α, to function as an activator at the P_{RM} promoter.[57]

Unlike activation of *lacP* (whose CRP-binding site is centered at −61.5, i.e., between −61 and −62), CRP-mediated activation of *galP1* (CRP site centered at −41.5) is not affected by α truncation,[58] providing evidence for a second CRP contact region on RNAP. When CRP heterodimers are oriented on asymmetric CRP-binding sites, contact with RNAP is established by the promoter-distal monomer at an artificial *galP1*-like promoter (with the CRP site at −41.5), but by the promoter-proximal monomer at *lacP*.[59,60] Remarkably, mutations in the region of monomeric CRP necessary for activation cause defects in activation at both promoters.[61] Thus, one structural subdomain on the surface of an activator protein (the CRP monomer) can, in two distinct orientations, contact different regions of RNAP.

An important consequence of the contact between an activator and RNAP holoenzyme is that the binding of the two proteins is often found to be synergistic.[9–12] In addition, mutations either in an activator-binding site or in the promoter itself could prevent activation by altering the conformation of the bound activator or its contact site on RNAP.[62,63]

DNA Looping, Bending, and Other Conformational Changes. Although reports of DNA looping caused by interactions between proteins bound to DNA sites separated by several hundred base pairs in both prokaryotes and eukaryotes are no longer surprising, evidence for the existence of such interactions—in the *gal* and *ara* operons (see below)—was first reported only a few years ago.[64,65] One role of DNA looping is to facilitate contact between an activator and RNAP, with the DNA tethering the two proteins to each other, thereby increasing the probability of interaction between

[57] M. Li, H. Moyle, and M. M. Susskind, *Science* **263,** 75 (1994).

[58] K. Igarashi, A. Hanamura, K. Makion, H. Aiba, T. Miuno, A. Nakata, and A. Ishihama, *Proc. Natl. Acad. Sci. U.S.A.* **88,** 8958 (1991).

[59] Y. Zhou, S. Busby, and R. H. Ebright, *Cell* **73,** 375 (1993).

[60] Y. Zhou, P. S. Pendergrast, A. Bell, R. Williams, S. Busby, and R. H. Ebright, *EMBO J.* **13,** 4549 (1994).

[61] Y. Zhou, T. J. Merkel, and R. H. Ebright, *J. Mol. Biol.* **243,** 603 (1994).

[62] K. Martin, L. Huo, and R. F. Schleif, *Proc. Natl. Acad. Sci. U.S.A.* **83,** 3654 (1986).

[63] J. Gao and G. N. Gussin, *EMBO J.* **10,** 4137 (1991).

[64] M. H. Irani, L. Orosz, and S. Adhya, *Cell* **32,** 783 (1984).

[65] T. Dunn, S. Hahn, S. Ogden, and R. Schleif, *Proc. Natl. Acad. Sci. U.S.A.* **81,** 5017 (1984).

them.[66] Direct evidence for looping was first obtained electron microscopically for NtrC protein and $E\sigma^{54}$ acting at the *glnA* promoter; in this case, the NtrC-binding site is located more than 100 bp upstream from the transcription start site.[42,67] Similar contact between NifA protein and $E\sigma^{54}$ at the *nifH* promoter and between NtrC and $E\sigma^{54}$ at the *glnH* promoter requires bending of the intervening DNA mediated by the integration host factor (IHF).[68,69] Here again, the role of bending appears to be to facilitate contact between the activator and RNAP. [An alternative method of achieving contact between a DNA-binding activator and RNAP bound at a (T4 late) promoter requires tracking of the activator along the intervening DNA; tracking can initiate at a site upstream or downstream from the promoter, and can span several hundreds base pairs.[70]]

A long-standing question, which has not been conclusively answered, is whether any activator functions exclusively by changing DNA structure at or near the promoter. In an artificial case, DNA bending per se substituted for CRP/cAMP binding to stimulate transcription initiation,[71] and the degree of DNA bending upstream from the plasmid pLS1 P_{ctII} promoter (in the presumed absence of any bound protein) is correlated with up to a threefold increase in promoter activity *in vivo.*[72] It has been suggested[73] that upstream DNA bending in conjunction with an RNAP-induced bend at the promoter[74] facilitates the wrapping of the DNA template around the polymerase.

At the *mer* locus, the binding of MerR protein to a site located remarkably between the −10 and −35 regions of the *merT* promoter (*mer*P_{T}) activates transcription initiation in the presence of mercury[75,76] and causes the DNA to unwind partially.[77] Unwinding (by about 50°) appears to compensate partially for the unusual rotational phasing of the *mer*P_{T} −10 and −35 consensus regions, which are separated by 19 bp instead of the consensus 17 bp. However, the possibility that the distortion of the DNA that

[66] A. Wedel, D. S. Weiss, D. Popham, P. Droge, and S. Kustu, *Science* **248,** 486 (1990).
[67] W. Su, S. Porter, S. Kustu, and H. Echols, *Proc. Natl. Acad. Sci. U.S.A.* **87,** 5504 (1990).
[68] T. R. Hoover, E. Santero, S. Porter, and S. Kustu, *Cell* **63,** 11 (1990).
[69] F. Claverie-Martin and B. Magasanik, *Proc. Natl. Acad. Sci. U.S.A.* **88,** 1631 (1988).
[70] G. M. Sanders, G. A. Kassavetis, and E. P. Geiduschek, *Proc. Natl. Acad. Sci. U.S.A.* **91,** 7703 (1994).
[71] L. Bracco, D. Kotlarz, A. Kolb, S. Diekman, and H. Buc, *EMBO J.* **8,** 4289 (1989).
[72] J. Perez-Martin and M. Espinosa, *J. Mol. Biol.* **241,** 7 (1994).
[73] S. Adhya, M. Gottesman, S. Garges, and A. Oppenheim, *Gene* **132,** 1 (1993).
[74] G. Kuhnke, C. Theres, H.-J. Fritz, and R. Ehring, *EMBO J.* **8,** 1247 (1989).
[75] A. Heltzel, D. Gambill, W. J. Jackson, P. A. Totis, and A. O. Summers, *J. Bacteriol.* **169,** 3379 (1987).
[76] T. V. O'Halloran, B. Frantz, M. K. Shin, D. M. Ralston, and J. G. Wright, *Cell* **56,** 119 (1989).
[77] A. Z. Ansari, J. E. Bradner, and T. V. O'Halloran, *Nature (London)* **374,** 371 (1995).

results in unwinding also functions to establish contact between MerR protein and RNAP has not been excluded. *Escherichia coli* single-stranded binding protein (SSB) facilitates initiation by the N4 virion RNA polymerase by stabilizing a specific stem–loop structure in one of the two complementary strands of the template DNA.[78] Again, the requirement for formation of a specific structure does not preclude an interaction between SSB and N4 RNAP.

Promoters Subject to Activation. Because activation requires that a step in open complex formation be rate limiting, promoters that are subject to activation are generally a poor match to one or more of the consensus sequence elements identified for eubacterial promoters. As mentioned previously, "up"-mutations in several promoters relieve the need for an activator, simply by increasing the rate of open complex formation.[6,11,15] This may have implications for eukaryotic promoters (see Eukaryotic Promoters and RNA Polymerases), most of which require activators, and therefore should be considered "weak" according to criteria used to characterize eubacterial promoters.

Transcriptional Repressors

Properties of Transcriptional Repressors and Operators. Like activators, repressors are site-specific DNA-binding proteins. Mutations that inactivate the repressors themselves or their binding sites (operators) result in constitutive expression of their target genes or operons. Classically, the two types of mutations can be distinguished genetically by their ability or inability to affect gene regulation in *trans.* Operators are also genetically distinguishable from the promoters they regulate,[79] even though the two *cis*-acting sites may be physically inseparable.

Mechanisms of Repression. Initial findings[5,80] that repressor-binding sites overlapped the promoters they regulated led to a straightforward model for repression: the presence of repressor should sterically prevent the binding of RNAP. Detailed kinetic evidence that repressors can indeed block binding of RNAP is now available for λ P_R,[7] the *uvrA* promoter,[81] and *lacP1.*[82] The data for *lacP1* are particularly important because several studies were interpreted to indicate that Lac repressor (LacI) and RNAP could bind DNA simultaneously. Among those frequently cited was an order-of-

[78] L. B. Rothman-Denes, *Methods Enzymol.* **274,** 9 (1996).

[79] F. Jacob and J. Monod, *J. Mol. Biol.* **3,** 318 (1961).

[80] M. Ptashne, K. Backman, M. Z. Humayun, A. Jeffrey, R. Maurer, B. Meyer, and R. T. Sauer, *Science* **194,** 156 (1976).

[81] E. Bertrand-Burggraf, S. Hurstel, M. Daune, and M. Schnarr, *J. Mol. Biol.* **193,** 293 (1987).

[82] P. J. Schlax, M. W. Capp, and M. T. Record, *J. Mol. Biol.* **245,** 331 (1995).

addition experiment performed in 1971[83]; results of this experiment were subsequently contradicted when it became possible to isolate well-defined DNA fragments containing *lacP*.[84] Interpretation of footprinting data seeming to show simultaneous occupancy was confounded by the fact that template DNAs contained both *lacP1* and an upstream overlapping promoter *lacP2*.[85,86]

The fact that repressors block RNAP binding at some promoters does not preclude the possibility that at other promoters, repressors may interfere with steps in initiation that occur subsequent to polymerase binding. At the *gal* promoter, DNA looping between repressor molecules bound to two operators, O_E (centered at −60.5) and O_I (centered at +53.5), prevents initiation even though RNAP and Gal repressor (GalR) can bind simultaneously to the template DNA and neither operator site overlaps the promoter.[74,87] The structure of the loop itself appears to inhibit a step subsequent to RNAP binding, but prior to actual synthesis of phosphodiester bonds.[88,89] Similarly, AraC dimers bound at $araO_2$ (at −276) and $araI_1$ (at −64) interact through DNA looping to repress the $araP_{BAD}$ promoter in the absence of arabinose, while dimers bound at $araO_2$ and $araO_1$ interact to repress $araP_C$.[65,90,91] In neither case is there evidence for RNAP binding under repressing conditions.

In the absence of looping, GalR bound only to the external operator O_E can partially repress open complex formation at *galP1* by interacting with the α subunit of RNAP,[92] again without blocking RNAP binding to the promoter. Finally, acting as a repressor in the absence of Hg(II), MerR can occupy a site that substantially overlaps its target promoter ($merP_T$) without blocking RNAP binding.[77] This example is fair warning, if any is needed, that substantial overlap in two dimensions (as DNA sequences are normally written) does not preclude simultaneous binding.

[83] B. Chen, B. deCrombrugghe, W. B. Anderson, M. E. Gottesman, I. Pastan, and R. L. Perlman, *Nature New Biol.* **233,** 67 (1971).

[84] J. Majors, *Proc. Natl. Acad. Sci. U.S.A.* **72,** 4394 (1975).

[85] A. Schmitz and D. J. Galas, *Nucleic Acids Res.* **6,** 111 (1979).

[86] S. B. Straney and D. M. Crothers, *Cell* **51,** 699 (1987).

[87] A. Majumdar and S. Adhya, unpublished data, cited by S. Adhya and A. Majumdar, *in* "RNA Polymerase and Regulation of Transcription" (W. Reznikoff, R. Burgess, J. Dahlberg, C. Gross, T. Record, and M. Wickens, eds.), pp. 129–135. Elsevier, New York, 1987.

[88] H. E. Choy and S. Adhya, *Proc. Natl. Acad. Sci. U.S.A.* **89,** 11264 (1992).

[89] H. Choy, S.-W. Park, P. Parrack, and S. Adhya, *Proc. Natl. Acad. Sci. U.S.A.* **92,** p. 7327 (1995).

[90] N. Lee, C. Francklyn, and E. P. Hamilton, *Proc. Natl. Acad. Sci. U.S.A.* **84,** 8814 (1987).

[91] R. B. Lobell and R. F. Schleif, *Science* **250,** 528 (1990).

[92] H. E. Choy, S.-W. Park, T. Aki, P. Parrack, N. Fujita, A. Ishihama, and S. Adhya, *EMBO J.*, **14,** 4523 (1995).

Eukaryotic Promoters and RNA Polymerases

Polymerase Core Enzymes

Specialization of eukaryotic RNA polymerases (RNA polymerase I, II, and III or Pol I, II, and III) is one alternative to the use in eubacteria of different sigma factors to transcribe specific gene sets (see also [14] in this volume[92a]). Genes encoding ribosomal RNA precursors are transcribed by Pol I, protein-coding genes and some small nuclear RNA (snRNA) genes are transcribed by Pol II, and 5S and tRNA genes are transcribed by Pol III. Each eukaryotic polymerase is a multisubunit enzyme whose two largest polypeptides are similar in sequence to the β' and β subunits of eubacterial RNAP, possibly reflecting fundamental similarities in active site structure and function.[93–95] In addition, *E. coli* RNAP α-subunit sequences important for its interaction with the remainder of RNAP core enzyme are conserved in the third largest subunit of Pol II as well as in a shared Pol I/Pol III subunit.[96] Several of the smaller (yeast) subunits, which do not have eubacterial counterparts, are shared by two or more polymerases, possibly reflecting a common function in transcription and providing the opportunity to coordinate the activities of multiple polymerases.[97–99]

Promoters and Recognition Proteins for RNA Polymerase II

Like promoters recognized by *E. coli* σ^{70}, a eukaryotic promoter can be defined as the set of sequences that are sufficient to permit the formation of a preinitiation complex and specify the point of transcription initiation. (Here the preinitiation complex will be defined as the fully assembled complex capable of initiating transcription on addition of ribonucleoside triphosphates.) The ability of Pol I, II, and III to initiate transcription at specific promoters depends on additional (general) transcription factors. Thus, physical separation of the nonspecific catalytic function and the specific promoter-recognition function is a general strategy that allows these two dichotomous functions to be performed by a single enzyme.

Minimal RNA Polymerase II Promoter. The eukaryotic promoters that

[92a] R. G. Roeder, *Methods Enzymol.* **273,** Chap. 14, 1996 (this volume).

[93] L. A. Allison, M. Moyle, M. Shales, and C. J. Ingles, *Cell* **42,** 599 (1985).

[94] B. Breant, J. Huet, A. Sentenac, and P. Fromageot, *J. Biol. Chem.* **258,** 11968 (1983).

[95] J. Huet, A. Sentenac, and P. Fromageot, *J. Biol. Chem.* **257,** 2613 (1982).

[96] A. Sentenac, M. Riva, P. Thurizux, J.-M. Buhler, I. Treich, C. Carles, M. Werner, A. Ruet, J. Huet, C. Mann, N. Chiannilkulchai, S. Stettler, and S. Mariotte, *in* "Transcriptional Regulation" (S. L. McKnight and K. R. Yamamoto, eds.), p. 27. Cold Spring Harbor Laboratory Press, Cold Spring Harbor, New York, 1992.

[97] N. A. Woychik, S. M. Liao, P. A. Koldziej, and R. A. Young, *Genes Dev.* **4,** 313 (1990).

[98] C. Carles, I. Treich, F. Bouet, M. Riva, and A. Sentenac, *J. Biol. Chem.* **266,** 24092 (1991).

[99] I. Treich, M. Riva, and A. Sentenac, *Gene Exp.* **2,** 31 (1992).

most resemble those of eubacteria are the TATA-containing promoters recognized by Pol II, which minimally consist of a TATA box (consensus TATAAA) located near −30 and, in some cases, an initiator (Inr) element that includes the site of transcription initiation. Eukaryotic templates containing only the minimal promoter are virtually silent *in vivo,* but have been shown for a number of promoters to direct low (basal) levels of initiation at specific sites *in vitro.*[100,101] Because these basal levels can be stimulated by transcriptional activators, the minimal promoter can be thought of as a weak promoter, analogous to activatable bacterial promoters. Similar features of promoters for Pol I and Pol III are considered below.

Recognition of Minimal RNA Polymerase II Promoter and Assembly of Preinitiation Complex. Recognition of the minimal eukaryotic Pol II promoter is mediated by the protein factor TFIID, a multisubunit complex consisting of TATA-binding protein (TBP) and at least seven or eight TBP-associated (protein) factors (TAFs).[102–105] Although the TAFs are required for activated transcription, they are dispensable for basal transcription as long as the promoter has a TATA element.[102,106] The C-terminal 180 amino acids of TBP constitute a DNA-binding domain that interacts through minor groove contacts[107,108] and causes a 90° bend in the DNA.[109,110] The TBP-induced change in promoter conformation has been proposed to reflect stable TBP–DNA complex formation[111]; however, its significance in preinitiation complex formation remains unclear. The relative absence of opportunities in the minor groove for hydrogen bonding between residues of TBP and specific bases of the TATA element is consistent with the relatively small ratio (less than 10^3) of the affinity of TBP for the promoter to its nonspecific affinity.[111–113] This, together with evidence for TBP diffusion along the DNA, has led to the suggestion that TBP, like other DNA-

[100] J. Corden, B. Wasylyk, A. Buchwalder, P. Sassone-Corsi, C. Kedinger, and P. Chambon, *Science* **209,** 1406 (1980).
[101] R. Breathnach and P. Chambon, *Annu. Rev. Biochem.* **50,** 349 (1981).
[102] B. F. Pugh and R. Tjian, *Genes Dev.* **5,** 1935 (1991).
[103] Q. Zhou, P. M. Lieberman, T. G. Boyer, and A. J. Berk, *Genes Dev.* **6,** 1964 (1992).
[104] B. D. Dynlacht, T. Hoey, and R. Tjian, *Cell* **66,** 563 (1991).
[105] N. Tanese, B. F. Pugh, and R. Tjian, *Genes Dev.* **5,** 2212 (1991).
[106] S. Buratowski, S. Hahn, P. A. Sharp, and L. Guarente, *Nature* (*London*) **334,** 37 (1988).
[107] D. K. Lee, M. Horkoshi, and R. G. Roeder, *Cell* **67,** 1241 (1991).
[108] D. B. Starr and D. K. Hawley, *Cell* **67,** 1231 (1991).
[109] J. D. Griffith, A. Makhov, L. Zawel, and D. Reinberg, *J. Mol. Biol.* **246,** 576 (1995).
[110] M. Horikoshi, C. Bertuccioli, R. Takada, J. Wang, T. Yamamoto, and R. Roeder, *Proc. Natl. Acad. Sci. U.S.A.* **89,** 1060 (1992).
[111] R. A. Coleman and B. F. Pugh, *J. Biol. Chem.* **270,** 13850 (1995).
[112] S. Hahn, S. Buratowski, P. A. Sharp, and L. Guarente, *Proc. Natl. Acad. Sci. U.S.A.* **86,** 5718 (1989).
[113] V. L. Singer, C. R. Wobbe, and K. Struhl, *Genes Dev.* **4,** 636 (1990).

binding proteins, may associate nonspecifically with DNA, later finding the promoter by sliding.[111,114]

TATA-binding protein association with the promoter is a regulated step in preinitiation complex formation. Several transcriptional activators are known to contact TBP or one of the TBP-associated factors. In addition, negative cofactors have been isolated that affect TBP–DNA complex formation. TFIIA is proposed to stimulate preinitiation complex formation by competing with inhibitory factors (see below) for binding to the basic repeat domain of TBP. This is consistent with the requirement for TFIIA in basal transcription catalyzed by unfractionated extracts but not in a purified system. On the basis of its contacts with certain activators and coactivators, an additional role for TFIIA in activation has been proposed,[115–119] but the possibility that this may be accounted for by the earlier requirement for TFIIA in preinitiation complex formation has not been ruled out.

The association of RNA polymerase with TBP–promoter complexes is mediated by the general transcription factors TFIIB and possibly TFIIF. TFIIB consists of separate domains capable of interacting with Pol II, TBP, and the TBP-associated factor, $TAF_{II}40$.[120,121] Although TFIIB does not bind promoter sequences directly, association of TFIIB with the TBP–promoter complex protects the initiation site from DNase I digestion. TFIIB may play a role in start site selection because *sua7* mutations in the gene encoding yeast TFIIB cause altered patterns of initiation.[122]

TFIIF consists of two subunits, RAP30 and RAP74. RAP30 has a small region of weak homology to one RNAP interaction domain of *E. coli* σ^{70} and can, in fact, bind to *E. coli* core RNA polymerase as well as to Pol II core enzyme in the absence of other factors or DNA.[123,124] A second σ^{70}-like feature of RAP30 is its ability to prevent formation of nonspecific Pol II–DNA complexes and together with RAP74 cause their dissociation.[125] RAP30 also interacts with TFIIB.[120]

[114] P. H. von Hippel and O. G. Berg, *J. Biol. Chem.* **264,** 675 (1989).

[115] K. Yokomori, A. Admon, J. A. Goodrich, J. L. Chen, and R. Tjian, *Genes Dev.* **7,** 2235 (1993).

[116] W. Wang, J. D. Gralla, and M. Carey, *Genes Dev.* **6,** 1716 (1992).

[117] P. M. Lieberman and A. J. Berk, *Genes Dev.* **8,** 995 (1994).

[118] H. Ge and R. G. Roeder, *Cell* **78,** 513 (1994).

[119] T. Chi and M. Carey, *Mol. Cell. Biol.* **13,** 7045 (1993).

[120] I. Ha, S. Roberts, E. Maldonado, X. Sun, L.-U. Kim, M. Green, and D. Reinberg, *Genes Dev.* **7,** 1021 (1993).

[121] J. A. Goodrich, T. Hoey, C. J. Thut, A. Admon, and R. Tjian, *Cell* **75,** 519 (1993).

[122] I. Pinto, D. E. Ware, and M. Hampsey, *Cell* **68,** 977 (1992).

[123] S. McCracken and J. Greenblatt, *Science* **253,** 900 (1991).

[124] O. Flores, H. Lu, M. Killeen, J. Greenblatt, Z. F. Burton, and D. Reinberg, *Proc. Natl. Acad. Sci. U.S.A.* **88,** 9999 (1991).

[125] R. C. Conaway and J. W. Conaway, *Annu. Rev. Biochem.* **62,** 161 (1993).

The largest Pol II subunit, corresponding to β', is uniquely characterized by a carboxy terminal domain (CTD) consisting of tandem heptapeptide repeats whose number is species specific.[126,127] Pol II containing unphosphorylated CTD is preferentially recruited into preinitiation complexes[128,129] whereas the CTD isolated from elongation complexes is highly phosphorylated; however, it is now clear that CTD phosphorylation is dispensable for normal basal and activated transcription *in vitro*.[130,131]

Analysis of Pol II promoters lacking TATA boxes led to the identification of the initiator (Inr), an element of the promoter with a highly variable sequence that includes the transcription start site and is capable of independently mediating assembly of Pol II-specific transcription complexes.[132] Initiator elements can be recognized specifically by Inr-binding proteins or in some cases weakly by Pol II itself.[133,134] Candidate initiator-binding proteins also include one or more TAFs.[135,136] Transcription complexes assembled in the absence of a TATA box contain the same subunits, including even the TATA-binding protein, TBP, as do complexes formed at TATA-containing promoters.[102,137] The involvement of TBP in transcription from TATA-less promoters is analogous to the involvement of TBP in initiation by Pol I and Pol III (see below).

The postulated requirement for two additional factors, TFIIE and TFIIH, in the formation of preinitiation complexes[138–141] has been challenged by abortive initiation experiments in which the factors were found to be dispensable. The data indicate that TFIIE and TFIIH are required

[126] J. L. Corden, D. L. Cadena, J. M. Ahearn, and M. E. Dahmus, *Proc. Natl. Acad. Sci. U.S.A.* **82,** 7934 (1985).
[127] L. A. Allison, J. K. Wong, V. D. Fitzpatrick, M. Moyle, and C. J. Ingles, *Mol. Cell. Biol.* **8,** 321 (1988).
[128] H. Lu, O. Flores, R. Weinmann, and D. Reinberg, *Proc. Natl. Acad. Sci. U.S.A.* **88,** 10004 (1991).
[129] B. Bartholomew, M. E. Dahmus, and C. F. Meares, *J. Biol. Chem.* **261,** 14226 (1986).
[130] T. P. Makela, J. D. Parvin, J. Kim, L. J. Huber, P. A. Sharp, and R. A. Weinberg, *Proc. Natl. Acad. Sci. U.S.A.* **92,** 5174 (1995).
[131] H. Serizawa, J. W. Conaway, and R. C. Conaway, *Nature (London)* **363,** 371 (1993).
[132] S. T. Smale and D. Baltimore, *Cell* **57,** 103 (1989).
[133] G. Gill, *Curr. Biol.* **4,** 374 (1994).
[134] L. Weis and D. Reinberg, *FASEB J.* **6,** 3300 (1992).
[135] J. C. Wang and M. W. Van Dyke, *Biochim. Biophys. Acta* **1216,** 73 (1993).
[136] B. A. Purnell and D. S. Gilmour, *Mol. Cell. Biol.* **13,** 2593 (1993).
[137] J. Carcamo, L. Buckbinder, and D. Reinberg, *Proc. Natl. Acad. Sci. U.S.A.* **88,** 1526 (1991).
[138] O. Flores, H. Lu, and D. Reinberg, *J. Biol. Chem.* **267,** 2786 (1992).
[139] J. D. Parvin and P. A. Sharp, *Cell* **73,** 533 (1993).
[140] J. D. Parvin, H. T. M. Timmers, and P. A. Sharp, *Cell* **68,** 1135 (1992).
[141] C. M. Tyree, C. P. George, L. M. Kira-DeVito, S. L. Wampler, M. E. Dahmus, L. Zawel, and J. T. Kadonaga, *Genes Dev.* **7,** 1254 (1993).

for promoter clearance and not for preinitiation complex formation or for open complex formation as previously suggested.[142]

RNA Polymerase II Holoenzyme. Although *in vitro* assays performed with purified chromatographic fractions and recombinant proteins revealed an ordered addition of general transcription factors and Pol II to the TFIID–DNA complex,[143–145] the observation that some general transcription factors are capable of associating with Pol II in the absence of DNA led to speculation that a preformed "holoenzyme" might bind directly to DNA.[146,147] The postulated holoenzyme, isolated only recently,[148,149] may have escaped detection earlier because of the low abundance of Pol II in eukaryotic cells and the apparently fragile nature of Pol II–initiation factor interactions. The holoenzyme contains Pol II and the general transcription factor TFIIF and, in some forms, TFIIB and TFIIH, as well as a large number of accessory proteins.[148,149] The isolation in different laboratories of complexes containing distinct but overlapping sets of proteins suggests that there may exist in the cell an even larger aggregate of general transcription and accessory factors that simultaneously interact with the promoter. However, the eukaryotic holoenzyme, as currently defined, differs significantly from the bacterial holoenzyme in that the complex lacks the ability to recognize a promoter.

RNA Polymerase III Promoter Structural Elements and Recognition Proteins

Recognition and Preinitiation Complex Formation at RNA Polymerase III Promoters. RNA polymerase III-specific genes fall into three main classes: class 1 consists of 5S genes containing a tripartite internal control region or ICR, class 2 consists of tRNA genes containing internal block A and B elements, and class 3 consists of Pol III genes such as U6 of higher eukaryotes that employ only upstream elements. In addition, some Pol III genes have both downstream and upstream elements.

Transcription of all Pol III-specific templates depends on the general

[142] J. A. Goodrich and R. Tjian, *Cell* **77,** 145 (1994).

[143] S. Buratowski, S. Hahn, P. A. Sharp, and L. Guarente, *Cell* **56,** 549 (1989).

[144] K. S. Kim and L. Guarente, *Nature* (*London*) **342,** 200 (1989).

[145] L. Guarente, *in* "Transcriptional Regulation" (S. L. McKnight and K. R. Yamamoto, eds.), p. 1007. Cold Spring Harbor Laboratory Press, Cold Spring Harbor, New York, 1992.

[146] M. Sopta, R. W. Carthew, and J. Greenblatt, *J. Biol. Chem.* **260,** 10353 (1985).

[147] S. Y. Tsai, P. Dicker, P. Fang, M.-J. Tsai, and B. W. O'Malley, *J. Biol. Chem.* **259,** 11587 (1984).

[148] Y. J. Kim, S. Bjorklund, Y. Li, M. H. Sayre, and R. D. Kornberg, *Cell* **77,** 599 (1994).

[149] A. J. Koleske and R. A. Young, *Nature* (*London*) **368,** 466 (1994).

transcription factor, TFIIIB, which in yeast contains at least three subunits, including TBP[150] and a TFIIB homolog known as BRF (B-related factor).[151–153] The general involvement of TBP in transcription of Pol III-specific genes including those lacking TATA elements (class 1 and class 2) has been demonstrated both *in vitro*[154] and *in vivo*.[155,156]

TFIIIB binding to class 1 and class 2 promoters requires an interaction with transcription factor TFIIIC. In the case of class 1 genes, a third (5S-specific) factor, TFIIIA, is required to allow TFIIIC to bind to the promoter.[157] The association of TFIIIB with TFIIIC is mediated by the BRF component of TFIIIB,[158] which shifts TFIIIB to an alternative state with higher affinity for DNA.[150,159,160] Although its initial affinity for DNA lacking a TATA element is weak, once associated with the template, TFIIIB is resistant to high salt and heparin, treatments that suffice to remove TFIIIC.[157] Preformed preinitiation complexes that have lost TFIIIC and TFIIIA can, on addition of Pol III, accurately and efficiently recruit the polymerase and mediate multiple rounds of transcription initiation.[161] The demonstration that Pol III can bind directly to immobilized BRF is consistent with the idea that binding of TFIIIB is sufficient for binding of Pol III and subsequent initiation.[158]

Class 1 Promoter Elements. The three elements of the internal control region (ICR) of class 1 (5S) promoters extend from +50 to +64 (A), +67 to +72 (ie), and +80 to +97 (C); the ICR is recognized by the zinc finger transcription factor, TFIIIA. The stability of the TFIIIA–ICR complex is increased by the subsequent binding of TFIIIC to TFIIIA,[162,163] although no specific DNA-binding site for TFIIIC has been identified as class 1

[150] G. A. Kassavetis, B. Bartholomew, J. A. Blanco, T. E. Johnson, and E. P. Geiduschek, *Proc. Natl. Acad. Sci. U.S.A.* **88,** 7308 (1991).
[151] A. Lopez-de-Leon, M. Librizzi, K. Tuglia, and I. Willis, *Cell* **71,** 211 (1992).
[152] S. Buratowski and H. Zhou, *Cell* **71,** 221 (1992).
[153] T. Colbert and S. Hahn, *Genes Dev.* **6,** 1940 (1992).
[154] R. J. White, S. P. Jackson, and P. W. Rigby, *Proc. Natl. Acad. Sci. U.S.A.* **89,** 1949 (1992).
[155] B. P. Cormack and K. Struhl, *Cell* **69,** 685 (1992).
[156] M. C. Schultz, R. H. Reeder, and S. Hahn, *Cell* **69,** 697 (1992).
[157] E. P. Geiduschek and G. A. Kassavetis, *in* "Transcriptional Regulation" (S. L. McKnight and K. R. Yamamoto, eds.), p. 247. Cold Spring Harbor Laboratory Press, Cold Spring Harbor, New York, 1992.
[158] B. Khoo, B. Brophy, and S. P. Jackson, *Genes Dev.* **8,** 2879 (1994).
[159] G. A. Kassavetis, D. L. Riggs, R. Negri, L. H. Nguyen, and E. P. Geiduschek, *Mol. Cell. Biol.* **9,** 2551 (1991).
[160] B. Bartholomew, G. A. Kassavetis, and E. P. Geiduschek, *Mol. Cell. Biol.* **11,** 5181 (1991).
[161] G. A. Kassavetis, B. R. Braun, L. H. Nguyen, and E. P. Geiduschek, *Cell* **60,** 235 (1990).
[162] A. B. Lassar, P. L. Martin, and R. G. Roeder, *Science* **222,** 740 (1983).
[163] D. F. Bogenhagen, W. M. Wormington, and D. D. Brown, *Cell* **28,** 413 (1982).

promoters.[164] Strong effects of sequence substitutions near the transcription start site of class 1 (and class 2) promoters suggest possible additional promoter elements.[157]

Class 2 Promoter Elements. The promoter-specific sequence elements of tRNA (class 2) genes consist of the internal block A region located approximately at +10 to +20 and the block B region located 30 to 60 bases away. These elements are recognized directly by TFIIIC, which in yeast is thought to consist of six nonidentical polypeptides that form a highly stable complex.[165–167] Human TFIIIC is at least as complex and separates during chromatography into two subcomplexes known as C1 and C2, both of which are required for activity.[168] Which component(s) of this heterogeneous complex are responsible for recognition of blocks A and B remains unclear.[157,169] At least four proteins in the TFIIIC–DNA complex form photocross-links extending over the region from −25 to +75.[160,165] A consensus start site element has been derived from analysis of tRNA gene sequences.[157]

Class 3 Promoter Elements. In most systems, class 3 promoter sequences are located entirely upstream from the transcription start site and consist of a TATA element at −30 and the proximal sequence element (PSE) at −60. In spite of the presence of a TATA element, the PSE fixes the location of the transcription start site,[170] a function performed by the TATA box of TATA-containing Pol II promoters. However, the composition of the complex that binds the PSE is unclear because two nonidentical protein complexes called SNAPc and PTF that bind to PSE have been identified.[171,172]

All class 3 promoters depend on the TATA-binding protein, TBP, which is a constituent of TFIIIB. Conceivably TBP in TFIIIB interacts directly with the TATA element. However, because the PSE is required for initiation, it is likely that PSE-binding factor interactions with the TBP component of TFIIIB are also necessary.

[164] B. R. Braun, D. L. Riggs, G. A. Kassavetis, and E. P. Geiduschek, *Proc. Natl. Acad. Sci. U.S.A.* **86,** 2530 (1989).

[165] B. Bartholomew, G. A. Kassavetis, B. R. Braun, and E. P. Geiduschek, *EMBO J.* **9,** 2197 (1990).

[166] M. C. Parsons and P. A. Weil, *J. Biol. Chem.* **265,** 5095 (1990).

[167] O. S. Gabrielsen, N. Marzouki, A. Ruet, A. Sentenac, and P. Fromgeot, *J. Biol. Chem.* **264,** 7505 (1989).

[168] N. Dean and A. J. Berk, *Nucleic Acids Res.* **15,** 9895 (1987).

[169] B. S. Shastry, *Mol. Cell. Biochem.* **124,** 85 (1993).

[170] N. Hernanadez, *in* "Transcriptional Regulation" (S. L. McKnight and K. R. Yamamoto, eds.), p. 281. Cold Spring Harbor Laboratory Press, Cold Spring Harbor, New York, 1992.

[171] C. L. Sadowski, R. W. Henry, S. M. Lobo, and N. Hernandez, *Genes Dev.* **7,** 1535 (1993).

[172] J.-B. Yoon, S. Murphy, L. Bai, Z. Wang, and R. G. Roeder, *Mol. Cell. Biol.* **15,** 2019 (1995).

Oddly, the mammalian Pol III U6 promoter differs from the mammalian Pol II U2 promoter only in the absence of a U2 TATA element.[173,174] In plants, both promoters contain TATA elements, and a 10-bp change in the spacing between the TATA sequence and an upstream element can convert the U6 promoter from a Pol III-specific to a Pol II-specific promoter.[175] Thus, although numerous details of promoter structure influence polymerase specificity, the mechanism(s) by which Pol II and Pol III promoters are distinguished remain unclear.

RNA Polymerase I Promoter Structural Elements and Recognition Proteins

Pol I-specific promoters contain two core elements extending from −75 to −50 and from −30 to +1[176,177] The core promoter is bound weakly, at best, by a TBP-containing complex (SL1, in human and rat) that also contains three species-specific and Pol I-specific TAFs.[176] Despite the presence of TBP in the SL1 complex, Pol I preinitiation complex assembly does not depend on a specific TBP–DNA interaction; indeed, Pol I-specific promoters lack a TATA element. Among the SL1 TAFS, TAF63 contains a zinc finger motif that may constitute a DNA recognition component of the complex.[177] Footprinting studies indicate that SL1 remains bound to the core element through numerous rounds of initiation by Pol I.[178]

A stable SL1 interaction with DNA requires the presence of a second DNA-binding protein, UBF, which itself binds weakly to an upstream site (UCE) localized roughly between −90 and −150.[179,180] The synergism in binding of SL1 and UBF to the promoter appears to involve a direct protein–protein interaction between TBP and UBF.[181] Two kinds of evidence argue that the UCE should not be considered part of the promoter: (1) Pol I-specific promoters of protozoa, fungi, and plants lack a UCE (they contain only a core element); and (2) UBF is not essential for *in vitro* transcription.[182]

[173] S. M. Lobo and N. Hernandez, *Cell* **58,** 55 (1989).
[174] I. W. Mattaj, N. A. Dathan, H. D. Parry, P. Carbon, and A. Krol, *Cell* **55,** 435 (1988).
[175] F. Waibel and W. Filipowicz, *Nature* (*London*) **346,** 199 (1990).
[176] L. Comai, N. Tanese, and R. Tjian, *Cell* **68,** 965 (1992).
[177] L. Comai, J. C. Zomerdijk, H. Beckmann, S. Zhou, A. Admon, and R. Tjian, *Science* **266,** 1966 (1994).
[178] E. Bateman, C. T. Iida, P. Kownin, and M. R. Paule, *Proc. Natl. Acad. Sci. U.S.A.* **82,** 8004 (1985).
[179] S. P. Bell, R. M. Learned, H.-M. Jantzen, and R. Tjian, *Science* **241,** 1192 (1988).
[180] R. M. Learned, T. K. Learned, M. M. Haltiner, and R. T. Tjian, *Cell* **45,** 847 (1986).
[181] H. Kwon and M. Green, *J. Biol. Chem.* **269,** 30140 (1994).
[182] T. Moss and V. Y. Stefanovsky, *Progr. Nucl. Acid Res* **50,** 25 (1995).

It is not clear whether the initiation factors so far defined are sufficient for transcription initiation by Pol I. Because preinitiation complex formation by Pol II depends on TFIIB and an analogous factor (BRF) is required for promoter initiation by Pol III, a Pol I-specific TFIIB equivalent may yet be identified. At least two additional DNA-binding proteins, core protein-binding factor (CPBF)[183] and enhancer 1-binding factor (E1BF),[184] are candidate participants in initiation in that they interact with the core element and their removal severely reduces basal rRNA synthesis.

Activation

Activators. Determinants of eukaryotic promoter activity are typically short sequences located at some distance from the core promoter. Regulatory sequences that exhibit both orientation and position independence have been designated "enhancers"; the function of these sequences is to bind specific DNA recognition proteins (activators) that interact directly or indirectly with the components of the basal transcription machinery. However, not all activator-binding sites qualify as enhancers, and there are no apparent mechanistic differences between enhancers and other types of activator-binding sites.

Transcriptional activators are typically divisible into at least two functional units: (1) a DNA-binding domain and (2) an "activation" domain. A few activators such as the well-characterized VP16 protein of herpes simplex virus 1 lack DNA-binding activity, but associate with the DNA by interacting with a DNA-bound protein.[185] Activation domains are typically characterized by their most prominent amino acid(s); however, it is unclear that this method of categorization has any functional significance.[186–187] The function of the activation domain is to mediate interaction (direct or indirect; see below) with the basal transcription complex. In addition to DNA-binding and activation functions, activators are also sensors of physiological signals. Therefore, the presence of specific activator-binding sites in a regulatory region confers responsiveness to specific stimuli.

Adaptors. Adaptors (also known as coactivators or mediators) are not DNA-binding proteins, but they contribute to activation by physically coupling transcriptional activators to the basal transcription apparatus. Two lines of evidence suggested the existence of adaptors. (1) Basal transcription complexes formed with a partially purified HeLa cell TFIID fraction re-

[183] Z. Liu and S. T. Jacob, *J. Biol. Chem.* **269,** 16618 (1994).

[184] H. Niu, J. Zhang, and S. T. Jacob, *Gene Expression* **4,** 111 (1995).

[185] S. J. Triezenberg, K. L. La Marco, and S. L. McKnight, *Genes Dev.* **2,** 730 (1988).

[186] M. Tanaka and W. Herr, *Mol. Cell. Biol.* **14,** 6056 (1994).

[187] J. L. Regier, F. Shen, and S. J. Triezenberg, *Proc. Natl. Acad. Sci. U.S.A.* **90,** 883 (1993).

sponded to activators, but a complex formed with recombinant TBP did not,[188] indicating that some TAFs might be required for activation; (2) squelching (inhibition of activation of a test promoter by a strong activator) could be blocked by depletion of a factor that had no apparent role in basal transcription.[189–191] Several proteins with adaptor function specific for particular classes of activators have since been identified among the TBP-associated proteins of TFIID by protein purification[104,105,192] and by genetic selection in yeast.[193]

Potential Mechanisms of Activation. The mechanism of activation will have two components: (1) the sensing of the activator by the basal transcription apparatus, and (2) the response of the basal transcription apparatus to the activator.

RNA Polymerase Activation. Demonstration of direct physical interaction between upstream regulatory proteins and components of the Pol II basal transcription apparatus supports the idea that, at least in some cases, activation is mediated by protein–protein interactions between activators and/or adaptors and components of the basal transcription apparatus.[194–198] Physical interactions between distant proteins requires the looping of intervening DNA for which there is evidence in both prokaryotes (see above) and eukaryotes.[199,200] More complex models in which chromatin domains have been superimposed on the simple looping idea have also been postulated.[201,202]

Activators identified so far do not stimulate transcription initiation by

[188] B. F. Pugh and R. Tjian, *Cell* **61,** 1187 (1990).

[189] S. L. Berger, W. D. Cress, A. Cress, S. J. Triezenberg, and L. Guarente, *Cell* **61,** 1199 (1990).

[190] R. J. I. Kelleher, P. M. Flanagan, and R. D. Kornberg, *Cell* **61,** 1209 (1990).

[191] P. M. Flanagan, R. J. Kelleher III, W. J. Feaver, N. F. Lue, J. W. La Pointe, and R. D. Kornberg, *J. Biol. Chem.* **265,** 11105 (1990).

[192] T. Hoey, R. O. J. Weinzieri, G. Gill, J.-L. Chen, B. D. Dynlacht, and R. Tjian, *Cell* **72,** 247 (1993).

[193] S. L. Berger, B. Pina, N. Silverman, G. A. Marcus, J. Agapite, J. L. Regier, T. S. J. Triezenberg, and L. Guarente, *Cell* **70,** 251 (1992).

[194] K. F. Stringer, C. J. Ingles, and J. Greenblatt, *Nature (London)* **345,** 783 (1990).

[195] Y.-S. Lin, I. Ha, E. Maldonado, D. Reinberg, and M. R. Green, *Nature (London)* **353,** 569 (1991).

[196] C. J. Brandl and K. Struhl, *Proc. Natl. Acad. Sci. U.S.A.* **86,** 2652 (1989).

[197] W. S. Lee, C. C. Kao, G. O. Bryant, X. Liu, and A. J. Berk, *Cell* **67,** 365 (1991).

[198] C. J. Ingles, M. Shales, W. D. Cress, S. J. Triezenberg, and J. Greenblatt, *Nature (London)* **351,** 588 (1991).

[199] I. A. Mastrangelo, A. J. Courey, J. S. Wall, S. P. Jackson, and P. V. C. Hough, *Proc. Natl. Acad. Sci. U.S.A.* **88,** 5670 (1991).

[200] W. Su, S. Jackson, R. Tjian, and H. Echols, *Genes Dev.* **5,** 820 (1991).

[201] R. Kellum and P. Schedl, *Cell* **64,** 941 (1991).

[202] L. Phi-Van and W. H. Stratling, *EMBO J.* **7,** 655 (1988).

preassembled preinitiation complexes.[203] Rather, they appear to function by affecting the kinetics or equilibrium of one or more steps in preinitiation complex assembly. Activators are known to affect TBP binding[204] and TFIIB recruitment[195,205] by direct interaction between the activator and the transcription factor. Direct interaction of activators with TFIIH, TFIIF, and TFIIA have also been reported.[206–208] In general, the expectation is that the activator will reduce the rate of dissociation of a general factor or RNAP, thereby increasing the probability that a template molecule will complete preinitiation complex assembly.[209] If there is more than one rate-limiting step, later steps in preinitiation complex assembly can also be facilitated by contacts with the same or different activators.[210] Mechanisms in which activators interfere with the formation of repressing chromatin or increase the affinity of general transcription factors for chromatin templates have also been proposed.[211–214]

RNA POLYMERASE III ACTIVATION. Class 3 Pol III snRNA promoters contain upstream elements that fit the definition of an enhancer both by virtue of the flexibility of their position and because they consist of a variable number of modules for the binding of different factors, each of which contributes to activation. In that Pol III snRNA promoters are highly related to the mRNA promoters of Pol II,[170] mechanisms of activation of Pol III class 3 genes are expected to be similar to those discussed previously.[215]

RNA POLYMERASE I ACTIVATION. UBF-mediated activation of Pol I genes is mechanistically unlike that of Pol II and Pol III gene activation. UBF is a member of the HMG (high mobility group) box transcription factor family, a family of nonhistone chromosomal proteins that are highly charged. HMG factors consist of reiterated N-terminal HMG DNA-binding

[203] Y.-S. Lin and M. R. Green, *Cell* **64,** 971 (1991).
[204] P. M. Lieberman and A. J. Berk, *Genes Dev.* **5,** 2441 (1991).
[205] S. G. E. Roberts and M. R. Green, *Nature (London)* **371,** 717 (1994).
[206] J. Ozer, P. A. Moore, A. H. Bolden, A. Lee, C. A. Rosen, and P. M. Lieberman, *Genes Dev.* **8,** 2324 (1994).
[207] H. Xiao, A. Pearson, B. Coulombe, R. Truat, S. Zhang, J. L. Regier, S. J. Triezenberg, D. Reinberg, O. Flores, C. J. Ingles, and J. Greenblatt, *Mol. Cell. Biol.* **14,** 7013 (1994).
[208] H. Zhu, V. Joliet, and R. Prywes, *J. Biol. Chem.* **269,** 3489 (1994).
[209] R. E. Kingston and M. R. Green, *Curr. Biol.* **4,** 325 (1994).
[210] B. Choy and M. R. Green, *Nature (London)* **366,** 531 (1993).
[211] G. E. Croston, L. M. Kerrigan, D. R. Lira, D. R. Marshak, and J. T. Kadonaga, *Science* **251,** 643 (1991).
[212] G. E. Croston, P. J. Laybourn, S. M. Paranjape, and J. T. Kadonaga, *Genes Dev.* **6,** 2270 (1992).
[213] J. L. Workman, I. C. A. Taylor, and R. E. Kingston, *Cell* **64,** 533 (1991).
[214] J. L. Workman and R. E. Kingston, *Science* **258,** 1780 (1992).
[215] I. M. Willis, *Eur. J. Biochem.* **212,** 1 (1993).

domains and a C-terminal acidic domain.[216] Fusions of HMG factor domains to a heterologous DNA-binding domain establish that HMG-type transcription factors have no potential for *trans*-activation.[103] Activation by UBF may involve formation of an "enhancesome," a structure visualized by electron spectroscopic imaging, consisting of 180 bp of DNA wrapped completely around a UBF dimer, and thought to be generated by DNA bending due to the binding of each HMG motif to consecutive 20-bp stretches.[182] Two UBF complexes, one centered at +1 and the other at the UCE,[179,217] may generate a DNA superstructure that contributes to the binding of the TBP complex and subsequent Pol I recruitment.

Maximal levels of Pol I transcription initiation require the presence of one or more reiterated enhancer elements located upstream from the UCE.[218] The binding of UBF to the enhancer as well as to the UCE has led to speculation that the DNA structure itself may recruit TBP complexes that, in turn, stimulate initiation.[182] The effect of the enhancers is likely to be to promote stable preinitiation complex formation because (1) enhancer-binding sequences added in *trans* reduce transcription of enhancerless rDNA templates due to titration of UBF or, perhaps, UBF–SL1 complexes; (2) by the use of concatamers that could be rapidly resolved it was possible to show that the transient presence of enhancer sequences mediated the same degree of activation as templates containing permanent *cis*-linked enhancers.[219,220]

Repression

The general weakness of eukaryotic promoters in the absence of activators and the widespread use of distinct activators that respond to individual signals would seem to render superfluous mechanisms for specifically inactivating gene expression. Nevertheless, examples of negative control in eukaryotes abound. Although repression is classically defined as a protein–DNA interaction that prevents transcription initiation,[79] a broader definition would include inhibition mediated by protein–protein interactions or nucleosome positioning.

Inhibition of Preinitiation Complex Assembly. The existence of overlapping promoter and operator sites, which led to the promoter occlusion

[216] R. Grosschedl, K. Giese and J. Pagel, *Trends in Genet.* **10,** 94 (1994).
[217] B. Leblanc, C. Read, and T. Moss, *EMBO J.* **12,** 513 (1993).
[218] R. H. Reeder, *in* "Transcriptional Regulation" (S. L. McKnight and K. R. Yamamoto, eds.), p. 315. Cold Spring Harbor Laboratory Press, Cold Spring Harbor, New York, 1992.
[219] C. S. Pikaard, L. K. Pape, S. L. Henderson, K. Ryan, M. H. Paalman, M. A. Lopata, R. H. Reeder, and B. Sollner-Webb, *Mol. Cell. Biol.* **10,** 4816 (1990).
[220] M. Dunaway and P. Droge, *Nature* (*London*) **341,** 657 (1989).

repressor model in bacteria, is also seen in eukaryotes. Examples include the simian virus 40 (SV40) T antigen, which, after accumulating in the early stages of SV40 lytic infection, stimulates SV40 replication and represses transcription of the viral early genes by binding to the SV40 early promoter[221]; the *Drosophila* engrailed homeodomain protein, which competes with TBP for binding to the TATA box element[222]; and the yeast ARGR complex, which depending on its position upstream from, or overlapping, the TATA element, mediates induction or repression, respectively, by arginine.[223]

The promoter may also be occluded by specific nucleosome positioning. For example, in yeast, when the Mcm1-Matα2 heterodimer is bound to its recognition site upstream of **a**-specific genes, adjacent nucleosomes become positioned,[224] leading to the inclusion of the TATA element in a nucleosome. Nucleosome positioning (and repression) at **a**-specific promoters (and elsewhere) is dependent on the Tup1 protein[225,226] and on histone H4.[227] Consistent with the relative resistance of Pol III promoters to chromatin-mediated repression,[228] an actively transcribed tRNA gene placed downstream of an **a**-specific promoter is not assembled into positioned nucleosomes, nor is it repressed. In contrast, Pol I- and other Pol II-specific genes placed in the same position are repressed.[229,230]

General nucleosomal structure probably also acts as a repressor by preventing access of the general transcriptional machinery to the promoter. The degree of nucleosome depletion *in vivo* is correlated with the extent of derepression,[231,232] and order-of-addition experiments *in vitro* show that TFIID binding prior to nucleosome assembly prevents nucleosome-mediated inhibition of transcription.[233,234]

[221] R. Tjian, *Cell* **26,** 1 (1981).
[222] Y. Ohkuma, M. Horikoshi, R. G. Roeder, and C. Desplan, *Proc. Natl. Acad. Sci. U.S.A.* **87,** 2289 (1990).
[223] M. De Rijcke, S. Seneca, B. Punyammalee, N. Glasdorrf, and M. Crabell, *Mol. Cell. Biol.* **12,** 68 (1992).
[224] M. Shimizu, S. Y. Roth, C. Szent-Gyorgyi, and R. T. Simpson, *EMBO J.* **10,** 3033 (1991).
[225] J. P. Cooper, S. Y. Roth, and R. T. Simpson, *Genes Dev.* **8,** 1400 (1994).
[226] D. Tzamarias and K. Struhl, *Nature (London)* **369,** 758 (1994).
[227] S. Y. Roth, *Curr. Opin. Genet. Dev.* **5,** 168 (1995).
[228] A. Wolffe, *Embo. J.* **8,** 527 (1989).
[229] B. M. Herschbach and A. D. Johnson, *Mol. Cell. Biol.* **13,** 4029 (1993).
[230] R. H. Morse, S. Y. Roth, and R. T. Simpson, *Mol. Cell. Biol.* **12,** 4015 (1992).
[231] M. Han, U. J. Kim, P. Kayne, and M. Grunstein, *EMBO J.* **7,** 2221 (1988).
[232] M. Han and M. Grunstein, *Cell* **55,** 1137 (1988).
[233] J. L. Workman and R. G. Roeder, *Cell* **55,** 613 (1987).
[234] J. A. Knezetic, G. A. Jacob, and D. S. Luse, *Mol. Cell. Biol.* **8,** 3114 (1988).

TBP binding to DNA is also regulated by mechanisms other than occlusion. ATP-dependent inhibitor (ADI) was identified as an activity in yeast that inhibits TBP binding in the presence of ATP.[235] ADI-mediated TBP dissociation from the promoter is accomplished through interactions with the conserved basic domain of TBP as well as with sequences upstream of the TATA element and may involve ATP hydrolysis. Because TFIIA also interacts with the TBP basic domain, inhibition of ADI activity by TFIIA may be due to competition between the two factors. Alternatively, TFIIA may stabilize the TBP–DNA complex.[235]

Interference with preinitiation complex formation at steps following TBP binding has also been observed. Negative cofactors 1 and 2 (NC1 and NC2) were identified in HeLa cell extracts as activities that interact with TBP and inhibit basal transcription *in vitro* by preventing the association of TBP with other general transcription factors.[236,237] The NC2 activity (also known as Dr1) is encoded by a 19-kDa TBP-binding protein whose phosphorylation regulates its interaction with TBP.[238]

Inhibition of Activation. Regulatory proteins affect the ability of activators to stimulate transcription at several steps: (1) nuclear localization, (2) activator assembly, (3) activator binding to DNA, and (4) activator interaction with the general transcription machinery.[239] To a large extent, each of these effects is mediated by protein–protein interactions rather than interactions between a repressor (negative regulator) and the promoter itself. For example, IκB proteins interact with specific Rel activator proteins, thereby masking the Rel nuclear localization sequence.[240] Multimeric activators, such as those of the bHLH family, which depend for DNA binding on a basic domain in each of two subunits of the dimeric activator,[241] are prevented from DNA binding by dimerization with members of the family lacking a basic domain.[242] DNA binding of the activator can also be subject to interference by a competing protein with binding sites overlapping that of the activator.[243] Finally, interaction of the yeast Gal80 and Gal4 proteins is an example of inhibition of the activator at a step subsequent to DNA

[235] D. T. Auble and S. Hahn, *Genes Dev.* **7,** 844 (1993).
[236] M. Meisterernst and R. G. Roeder, *Cell* **67,** 557 (1991).
[237] M. Meisterernst, A. L. Roy, H. M. Leiu, and R. G. Roeder, *Cell* **66,** 981 (1991).
[238] J. A. Inostroza, F. H. Mermelstein, I. H. Ha, W. S. Lane, and D. Reinberg, *Cell* **70,** 477 (1992).
[239] B. M. Herschbach and A. D. Johnson, *Annu. Rev. Cell. Biol.* **9,** 479 (1993).
[240] A. A. Beg, S. M. Ruben, R. I. Scheinman, S. Haskill, C. A. Rosen, and A. S. Baldwin Jr., *Genes Dev.* **6,** 1899 (1992).
[241] A. Voronova and D. Baltimore, *Proc. Natl. Acad. Sci. U.S.A.* **87,** 4722 (1990).
[242] M. Van Doren, H. M. Ellis, and J. W. Posakony, *Development* **113,** 245 (1991).
[243] S. Small, A. Blair, and M. Levine, *EMBO J.* **11,** 4047 (1992).

binding. Presumably Gal80 masks a domain required for Gal4 interaction with the basal transcription machinery.[244]

Summary

In eubacteria, the interaction of each of several (5–10) σ factors with a single core polymerase determines the promoter specificity of the corresponding holoenzyme; in contrast, in eukaryotes the most obvious specificity determinant is the existence of three "core" polymerases, which differ in the specificity with which they interact with particular TBP-containing complexes. For Pol II, the interaction is mediated by TFIIB, which contacts both RNAP and specific TAFs; for Pol III, the same function may be served by BRF (B-related factor), a component of TFIIIB. Presumably, a similar factor is necessary for interaction of TBP-containing complexes with Pol I. In both eubacteria and prokaryotes, promoter specificity is further determined by the presence of regulators of initiation—activators or repressors—capable, usually, of responding to particular metabolic, developmental, or environmental cues.

We define a eubacterial promoter as the minimal set of sequence elements necessary for specific recognition and initiation by RNA polymerase holoenzyme. The problem in applying a similar definition to eukaryotic promoters is the difficulty in identifying a eukaryotic holoenzyme that is functionally analogous to $E\sigma^{70}$. However, the plethora of factors required for full expression of eukaryotic transcription units compels us to find a common way of conceptualizing the roles of each set of factors in promoter site selection (Fig. 1). We start from the following premises: (1) because the common task at all eukaryotic promoters seems to be getting TBP to the DNA—in either a specific, TATA-binding or a nonspecific TATA-less-binding form—the eukaryotic counterpart to the eubacterial holoenzyme ought to include what is minimally needed for TBP binding (specifically or nonspecifically) to DNA and its association with RNAP; (2) promoters that contain no apparent TATA sequence are functionally analogous to weak eubacterial promoters that require activators for maximal expression. According to this view, Pol I, II, and III are analogous to eubacterial core polymerase, while TBP and the protein(s) that mediate its association with the core are analogous to σ. Although elongation factors may bind to the eukaryotic polymerases, they are equivalent to proteins such as the termination factor ρ and Nus proteins, which bind at various times to $E\sigma^{70}$, but are not part of the holoenzyme. Thus, if fully assembled, functional counterparts to *E. coli* holoenzymes exist in eukaryotes, they should consist

[244] N. F. Lu, D. I. Chasman, A. R. Buchman, and R. G. Kornberg, *Mol. Cell. Biol.* **7,** 3446 (1987).

of (1) Pol I "core" enzyme, TBP, one or more additional components of the SL1 complex, and possibly a TFIIB-like factor, (2) Pol II "core" enzyme, TBP and one or more TAFs, TFIIB, Inr-recognition protein(s), and possibly TFIIA and TFIIF, and (3) Pol III "core" enzyme, TBP, BRF, and one or more additional components of TFIIIB.

The fit to the eubacterial holoenzyme model is least comfortable in the case of Pol III. It seems reasonable to exclude TFIIIA, which is necessary only for class 1 promoters, and is dispensible once TFIIIB is bound to the DNA; similar reasoning leads us to exclude TFIIIC, which is not necessary for TATA-containing class 3 promoters. In some cases, the distinction between proteins that function essentially as activators and those that are essential components of an initiation-competent holoenzyme may turn out to be semantic, but it seems useful to bear the distinction in mind to help anticipate what kinds of preassembled, fully functional complexes might be found in eukaryotic cells.

Conceptual analogies between eubacterial and eukaryotic enzymes may also be helpful in understanding eventually how the enzymes evolved, including the evolution of pathways for assembling different functional units. For example, the carboxy-terminal domain of the *E. coli* RNAP α subunit is functionally equivalent to eukaryotic coactivators, the main differences being that it is covalently attached to the holoenzyme and has evolved a sequence-specific DNA-recognition capability. Strong functional and amino acid sequence similarities between eubacterial and eukaryotic polymerases should not be surprising because the characteristic activities of the enzymes are the same. The basal transcription machinery should have evolved relatively early, but should be constrained in evolution by selection against even small changes in recognition specificity, in order to preserve fidelity in the quantity and quality of RNA molecules synthesized. Discoveries of strong similarities between transcription factors and polymerases of Archaea and those of eukaryotes[245] provide an argument in favor of the idea that once the basic transcription apparatus emerged, the ways in which it could evolve were limited.[248]

Acknowledgment

We thank Lucia Rothman-Denes for helpful discussions.

[245] P. Baumann, S. A. Qureshi, and S. P. Jackson, *Trends Genet.* **11,** 279 (1995).
[246] R. Waldschmidt, I. Wanandi, and K. H. Seifart, *EMBO J.* **10,** 2595 (1991).
[247] S. P. Bell, H.-M. Jantzen, and R. Tjian, *Genes Dev.* **4,** 943 (1990).
[248] We regret that we could not cite explicitly a vast amount of work that has contributed to the principles and facts outlined in this review. Readers are encouraged to explore the literature in detail to come to their own conclusions about the complex roles of various protein factors in eukaryotic transcription and their possible relation to eubacterial counterparts.

[2] *Escherichia coli* Promoter Sequences: Analysis and Prediction

By GERALD Z. HERTZ and GARY D. STORMO

Escherichia coli promoters are among the most well-studied regulatory signals in DNA. An important question concerns how the polymerase obtains its specificity for promoters; others concern how to describe and represent that specificity to aid in the prediction of promoters in genomic DNA sequences. *Escherichia coli* has several different sigma (σ) factors that confer specificity to different promoter classes.[1,2] σ^{70} is the sigma factor responsible for most mRNA, rRNA, and tRNA synthesis in *E. coli*. This chapter analyzes the σ^{70} promoters and describes some methods for predicting their occurrence in DNA sequences.

In 1975, Pribnow[3] first described the similarity of promoters just upstream of the transcriptional initiation site. This region was originally called the "Pribnow box" but is now generally referred to as the -10 region. The Pribnow consensus, TATRATG (R = A/G), was based on only six promoters. However, it was realized early on that the -10 region was only part of the sequence requirement for promoter activity. In 1977, Seeburg *et al.*[4] described a consensus sequence for a second conserved region starting approximately 35 bases upstream of the initiation site. This latter consensus, GTTGACACTTTA, was based on 15 promoters and has become known as the -35 region. Both of these consensus sequences are remarkably close to current consensuses based on about 300 promoters. In 1979, Rosenberg and Court[5] compiled 46 promoters along with many mutations that affected promoter activity. From this it was clear that most mutations that affected transcription fell into one of the two previously recognized partially conserved regions. They proposed the still accepted consensus sequences of TTGACA and TATAAT for the -35 and -10 regions, respectively. The spacing between these two hexamers varied from 16 to 19 bases. They also suggested consensus sequences for additional positions that were less conserved.

[1] R. R. Burgess, A. A. Travers, J. J. Dunn, and E. K. F. Bautz, *Nature* (*London*) **221,** 43 (1969).
[2] J. D. Helmann and M. J. Chamberlin, *Annu. Rev. Biochem.* **57,** 839 (1988).
[3] D. Pribnow, *Proc. Natl. Acad. Sci. U.S.A.* **72,** 784 (1975).
[4] P. H. Seeburg, C. Nüsslein, and H. Schaller, *Eur. J. Biochem.* **74,** 107 (1977).
[5] M. Rosenberg and D. Court, *Annu. Rev. Genet.* **13,** 319 (1979).

METHODS IN ENZYMOLOGY, VOL. 273

By 1983, Hawley and McClure[6] had compiled 112 promoters from both the *E. coli* genome and its phage, including some mutants. They aligned the promoters on the basis of the previously published[5] 6-base pair (bp) consensus sequences of the −35 and −10 regions. In their alignment, the −35 and −10 regions were spaced from 15 to 21 bases, although 100 of the 112 promoters had spacings of 17 ± 1 bases. They also observed that the −10 region preceded the start of transcription by 4 to 8 bases, with the preferred spacings being 6 and 7 bases, and that an A/T-rich sequence of approximately 10 bases preceded the −35 region. In 1984, Raibaud and Schwartz[7] observed that the sequences of promoters requiring activator proteins for full expression were less conserved than promoters not requiring activators.

By 1987, Harley and Reynolds[8] had compiled a list of 263 promoters, including phage and some mutants. The major difference from the Hawley and McClure alignment was that they allowed the −10 region and the start of transcription to be spaced by as much as 12 bases. In 1993 Lisser and Margalit[9] compiled 300 *E. coli* mRNA promoters, all wild type and having known transcriptional starts. With the entire sequence of *E. coli* to be completed shortly, we can imagine that before too long we will know the position and sequence of all of its promoters.

Representations of Specificity

The oldest and simplest method of representing the specificity in promoters is with a consensus sequence. These are easy to remember and search for by eye. A number of more sophisticated programs have been written that can identify more subtle consensus sequences with alignment variability.[10,11] However, unlike restriction enzymes for which single "words" (often ambiguous) are accurate descriptions of functional patterns, for regulatory sites it is usually (if not always) the case that sequences with mismatches to the consensus can still be functional sites. That is, the consensus sequence is only the most common base at each position but others are often observed as well. In fact, in all of the extensive compilations of *E. coli* promoters,[6,8,9] no positions are completely conserved. In fact, all bases are observed to occur at each position and there are no perfect matches to all 12 consensus positions. Most mRNA promoters match from

[6] D. K. Hawley and W. R. McClure, *Nucleic Acids Res.* **11,** 2237 (1983).
[7] O. Raibaud and M. Schwartz, *Annu. Rev. Genet.* **18,** 173 (1984).
[8] C. B. Harley and R. P. Reynolds, *Nucleic Acids Res.* **15,** 2343 (1987).
[9] S. Lisser and H. Margalit, *Nucleic Acids Res.* **21,** 1507 (1993).
[10] D. J. Galas, M. Eggert, and M. S. Waterman, *J. Mol. Biol.* **186,** 117 (1985).
[11] G. Mengeritsky and T. F. Smith, *Comput. Appl. Biosci.* **3,** 223 (1987).

7 to 9 of the 12 positions of the consensus sequence, but the observed range is 5 to 11.[9]

While one could use the consensus sequence to search for promoters, allowing for some number of mismatches, this is not a reliable strategy. The reason is that different positions have different degrees of conservation. For example, in the −10 region, the first, second, and sixth positions are much more conserved than the other three, so that a more accurate consensus might be TA*nnn*T. The consensus base at some positions occurs only about 50% of the time in some of the compilations. While this is clearly significantly different from a random occurrence of 25%, it also means that another base is approximately as likely to occur at that position as is the consensus base. For this and other reasons, consensus sequences have largely been replaced by other methods of representing specificity and for searching for new promoters.

The most common approach, after consensus sequences, is to represent specificity with a weight matrix. This is a matrix with four rows, one for each possible base, and as many columns as there are positions in the site. Thus a matrix for the −10 region would be 4×6, with the elements each providing a score for a particular base at each position. The score for a particular sequence, such as a potential −10 region, is the sum of the scores of the bases that occur (Fig. 1). A variety of methods for determining the matrix elements have been proposed. The weight-matrix concept was first introduced in the representation of *E. coli* ribosome-binding sites and the elements were determined by a simple neural network learning method called a *perceptron.*[12] Later, different statistical approaches, based on the base frequencies of known promoters, were proposed.[13–15] In Mulligan *et al.*,[15] it was even shown that the scores obtained by their approach were highly correlated ($r = 0.83$) with the quantitative activities of a set of promoters.

Since then, methods have been developed to find the matrix that gives the best possible correlation between scores and measured activities.[16,17] In the absence of sites with measured activities, some theoretical work has shown that the matrix most likely to give the best possible correlation

[12] G. D. Stormo, T. D. Schneider, L. Gold, and A. Ehrenfeucht, *Nucleic Acids Res.* **10,** 2997 (1982).

[13] R. Harr, M. Häggström, and P. Gustafsson, *Nucleic Acids Res.* **11,** 2943 (1983).

[14] R. Staden, *Nucleic Acids Res.* **12,** 505 (1984).

[15] M. E. Mulligan, D. K. Hawley, R. Entriken, and W. R. McClure, *Nucleic Acids Res.* **12,** 789 (1984).

[16] G. D. Stormo, T. D. Schneider, and L. Gold, *Nucleic Acids Res.* **14,** 6661 (1986).

[17] D. Barrick, K. Villanueba, J. Childs, R. Kalil, T. D. Schneider, C. E. Lawrence, L. Gold, and G. D. Stormo, *Nucleic Acids Res.* **22,** 1287 (1994).

-28	18	1	12	10	-29
-15	-31	-12	-10	-2	-22
-18	-50	-11	-7	-11	-36
17	-17	10	-10	-5	18

= **-60**

A C T A T A A T C G

-28	18	1	12	10	-29
-15	-31	-12	-10	-2	-22
-18	-50	-11	-7	-11	-36
17	-17	10	-10	-5	18

= **85**

A C T A T A A T C G

-28	18	1	12	10	-29
-15	-31	-12	-10	-2	-22
-18	-50	-11	-7	-11	-36
17	-17	10	-10	-5	18

= **-59**

A C T A T A A T C G

FIG. 1. Matrix evaluations of a sequence. The matrix contains an element for each possible base at six positions of a binding site. The matrix rows are in the order A, C, G, T from top to bottom. For each alignment of the matrix above the sequence, a score is calculated as the sum of the matrix elements, which are boxed, corresponding to the sequence. The sequence TATAAT scores the highest with this matrix. (Reprinted with permission from *Methods Enzymol.,* Vol. 183, p. 211.)

between scores and measured activities is the same as that obtained by a maximum likelihood statistical approach.[18,19] In this case, the observed frequencies of the bases at each position are divided by the frequencies expected by chance, about 0.25 for each base in *E. coli*, and the logarithms of those values used in the matrix (Fig. 2). The promoter prediction method described by Staden in 1984 is equivalent to this maximum likelihood approach.[14] We discuss and analyze Staden's method in greater detail below.

However, a number of other approaches for finding good weight matrices have been attempted. Several have used neural network approaches that separate the training set into promoter and nonpromoter sequences.[20–22] Cardon and Stormo[23] used an "expectation maximization" algorithm to determine alignments and weight matrices. The results of all of these approaches are comparable; in fact, the weight matrices themselves have

[18] O. G. Berg and P. H. von Hippel, *J. Mol. Biol.* **193,** 723 (1987).
[19] G. D. Stormo, *Methods Enzymol.* **183,** 211 (1990).
[20] B. Demeler and G. Zhou, *Nucleic Acids Res.* **19,** 1593 (1991).
[21] M. C. O'Neill, *Nucleic Acids Res.* **20,** 3471 (1992).
[22] P. B. Horton and M. Kanehisa, *Nucleic Acids Res.* **20,** 4331 (1992).
[23] L. R. Cardon and G. D. Stormo, *J. Mol. Biol.* **223,** 159 (1992).

similar characteristics. Differences may have more to do with the choice of training data than with the method of obtaining the weight matrix.[22]

A major difficulty is that, while the two conserved regions at −10 and −35 can be represented by ungapped matrices, the spacing between those regions is variable. O'Neill[21] treated the three most common spacings (16, 17, and 18) as separate cases and developed weight matrices for each one. Several have penalized nonoptimal spacings depending on their observed frequencies.[13–15] Others determined spacer penalities using the neural network[22] or expectation maximization[23] algorithm that was also determining the weight matrices.

Unfortunately, weight-matrix approaches are relatively mediocre in discriminating promoters from nonpromoters. In a comparison by Horton and Kanehisa[22] of several methods for identifying promoters, the best methods identified 90% of the test promoters only when the number of false positives was approximately 2%. Because nonpromoter sequences outnumber promoter sequences by presumably several thousand, such discrimination would identify many more nonpromoters than promoters in genomic DNA, i.e., the number of false positives exceeds the number of true positives. One possible reason the weight-matrix approaches are not better predictors is that many promoters require a protein activator for full activity. For example, in reviews of 114 promoters, Collado-Vides and co-workers[24,25] observed that 46% of the promoters required an activator for full activity. Because all current methods develop models from examples of both activated and nonactivated promoters, the features essential to promoter function are partially obscured. Unfortunately, the number of examples of well-characterized, nonactivated promoters is still fairly small.

Another possible reason weight-matrix approaches are not better is that promoter activity is not additive across the positions of the promoter. This could be due to interactions between the polymerase-binding activities of different promoter positions. In addition, promoter expression is not simply correlated with the strength of polymerase binding because, unlike simple DNA-binding proteins, polymerase must perform a number of kinetic steps at a promoter. For example, the polymerase needs to bind the DNA, open the DNA, initiate transcription, and release the promoter for elongation. Particular mutations may have different effects on these various processes so that combinations of mutations may have nonadditive results.[26] Thus, nonadditive interactions between the polymerase and the promoter may be expected, but are not modeled by standard weight-matrix methods.

[24] J. Collado-Vides, B. Magasanik, and J. D. Gralla, *Microbiol. Rev.* **55,** 371 (1991).
[25] J. Collado-Vides, *BioSystems* **29,** 87 (1993).
[26] H. Moyle, C. Waldburger, and M. M. Susskind, *J. Bacteriol,* **173,** 1944 (1991).

One way to notice nonadditivity is if positions in promoter sequences were correlated in some way. O'Neill concluded that the length of the spacing between the −35 and −10 regions was correlated with sequences in the conserved regions[27] and it was part of the justification for building separate weight matrices for different spacing classes.[21] However, these correlations may have been largely due to O'Neill not accounting for the relative frequencies of the different spacings between the −35 and −10 regions. Cardon and Stormo,[23] using a larger sample size and a more rigorous statistical analysis, found no significant correlations between any positions and spacing class. Still, Beutel and Record[28] observed that longer spacers contained more purine–pyrimidine and pyrimidine–purine heterodinucleotides than shorter spacers; thus, the structure of the spacer might correlate with its length in such a way that would not be apparent from sequence analysis unless adjacent correlations among the bases were also considered.

If we knew exactly what nonindependent interactions were important to the representation of promoter specificity, we could simply modify the weight-matrix method to account for them and that would improve the performance. However, at this time we do not have enough information of the appropriate kind to take advantage of it. Instead we may be able to capture higher order information through the use of a multilayer neural network. Unlike the single-layer perceptrons, these networks are capable of discriminating on the basis of nonadditive information in the training sets. Several attempts have been made. Unfortunately, at this time there is no evidence that these methods perform significantly better than simple weight matrices.[22] Thus, while we suspect that nonadditive effects may be important and taking them into account may improve our predictive ability, so far we have little to support those suppositions.

Another limitation of current approaches for identifying promoters is their focus on the −35 and −10 regions and sometimes the region around the transcriptional start. While these regions are the most conserved with current alignments, promoter activity has been shown to depend on regions up to 60 bases upstream of the transcriptional start in some promoters.[29] Different forms of the σ^{70}-containing RNA polymerase arise with different growth conditions and have different specificities[30] that presumably depend on the regions outside the generally recognized conserved regions. These

[27] M. C. O'Neill, *J. Biol. Chem.* **264,** 5522 (1989).

[28] B. A. Beutel and M. T. Record Jr., *Nucleic Acids Res.* **18,** 3597 (1990).

[29] L. Rao, W. Ross, J. A. Appleman, T. Gaal, S. Leirmo, P. J. Schlax, M. T. Record Jr., and R. L. Gourse, *J. Mol. Biol.* **235,** 1421 (1994).

[30] M. Ozaki, N. Fujita, A. Wada, and A. Ishihama, *Nucleic Acids Res.* **20,** 257 (1992).

additional regions are also responsible for the differential recognition of promoters by σ factors, such as σ^{38}, that are structurally related to σ^{70}.[31]

Making Alignments

Whether we generate a consensus sequence or a simple weight matrix or any other representation, we either explicitly or implicitly utilize an alignment. Some methods develop an alignment first and then secondarily construct a weight matrix from that alignment. In the early days, alignments were done by eye on the small data sets that existed and consensus sequences were developed from those alignments.[3,4] Later alignments were also done by finding best matches to consensus sequences, but then those alignments were turned into weight matrices.[6] Harley and Reynolds[8] aligned their promoter compilation using the Staden method (described in detail below), which uses weight matrices derived from the alignment of Hawley and McClure.[6] Harley and Reynolds then used their new alignment to derive modified weight matrices. They then cycled between updating their alignment and modifying their weight matrices until their alignment converged. This cycling between updating an alignment and modifying a weight matrix is the essence of the expectation maximization algorithm, which was later introduced into sequence alignment from the field of computer science.[32]

Only more recently have the two steps of creating an alignment and a weight matrix been completely automated. Newer alignment methods have the objective of identifying the alignment that maximizes a log-likelihood information content, which is equivalent to the sum of the weight-matrix scores of the aligned sequences. This is the basis of several methods of multiple alignment based on expectation maximization, such as the method of Cardon and Stormo[23] and methods based on hidden Markov models.[33,34] It is also the basis of other sequence alignment algorithms such as a greedy algorithm used by us[35] and of a Gibbs sampling method.[36]

[31] K. Tanaka, S. Kusano, N. Fujita, A. Ishihama, and H. Takahashi, *Nucleic Acids Res.* **23,** 827 (1995).

[32] C. E. Lawrence and A. A. Reilly, *Proteins* **7,** 41 (1990).

[33] A. Krogh, M. Brown, I. S. Mian, K. Sjölander, and D. Haussler, *J. Mol. Biol.* **235,** 1501 (1994).

[34] P. Baldi, Y. Chauvin, T. Hunkapiller, and M. A. McClure, *Proc. Natl. Acad. Sci. U.S.A.* **91,** 1059 (1994).

[35] G. Z. Hertz and G. D. Stormo, *in* "The Third International Conference on Bioinformatics and Genome Research" (H. A. Lim and C. R. Cantor, eds.). World Scientific Publishing, Singapore, p. 199 (1994).

[36] C. E. Lawrence, S. F. Altschul, M. S. Boguski, J. S. Liu, A. F. Neuwald, and J. C. Wootton, *Science* **262,** 208 (1993).

Analysis of Staden Method

Horton and Kanehisa[22] compared several computational methods for identifying *E. coli* promoters and found that the relatively simple weight-matrix methods of Staden[14] and Mulligan *et al.*[15] worked as well as more complicated neural network methods. The only significant differences appeared when the number of false positives was allowed to become large. We have chosen to examine the Staden method in greater detail to illustrate some of the considerations and limitations of current methods for identifying *E. coli* promoters. The Staden method determines a simple log-likelihood score in which the score is the sum of the logarithm of the frequency with which the indicated base is observed at the indicated position of the alignment.

The Staden method uses three weight matrices directly derived from the manually determined alignment of Hawley and McClure[6]: a 25-bp matrix encompassing the −35 region, a 19-bp matrix encompassing the −10 region, and a 12-bp matrix encompassing the +1 site—i.e., the start of transcription (Fig. 2). The alignment is converted to weight matrices by taking the logarithm of the frequency with which the corresponding base occurs at the corresponding position. One base at one position of the −10 region occurs with a zero frequency and, therefore, is estimated to occur once in the alignment to avoid taking the logarithm of zero. The spacing between the 6-bp highly conserved portions of the −35 and −10 regions is allowed to vary from 15 to 21 bases, with different spacings being penalized approximately equal to the logarithm of their relative frequencies. The spacing between the 6-bp highly conserved portion of the −10 region and the start of transcription is allowed to vary from 4 to 8 bases without any spacing penalties. The scoring method is justified by log-likelihood statistics and can be theoretically related to the expected activity of a promoter.[18,19]

For a computational method to have substantial predictive value, the frequency of false positives should be less than the occurrence of promoters. If *E. coli* promoters occur approximately once every 1000 bp, then the false-positive rate should be less than once every 2000 bases because the *E. coli* genome is double stranded. Such an estimate is probably an overestimate of the number of promoters, but will suffice for this discussion. As a negative control, we scored both strands of a 100,000-bp randomized sequence, having equal frequencies of each of the four bases, to approximate the frequency of false positives corresponding to various cutoff scores. As a positive control, we scored two different sets of promoter sequences. The first set contained 47 promoter sequences that are believed not to require activation by any activator proteins. These sequences were chosen on the basis of a review of the regulation of 114 promoters by Collado-Vides *et*

−35 REGION WEIGHT MATRIX

position	A	C	G	T
1	0.05	-0.20	-0.51	0.43
2	0.33	-0.01	-0.91	0.19
3	0.10	-0.41	0.06	0.16
4	0.30	0.05	-0.37	-0.10
5	0.24	-0.32	-0.27	0.21
6	0.71	-0.68	-0.42	-0.22
7	0.42	-0.32	-0.75	0.24
8	0.42	-0.83	-0.27	0.24
9	0.30	-0.22	-1.12	0.42
10	0.41	-0.19	-0.38	-0.03
11	0.34	-0.55	-0.15	0.14
12	-0.43	-0.10	-0.07	0.41
13	-0.10	-1.02	0.04	0.53
14	-0.07	0.43	0.04	-0.69
15 core	-2.64	-1.39	-0.93	1.19
16 core	-1.54	-1.54	-1.54	1.21
17 core	-2.64	-0.93	1.15	-0.93
18 core	0.94	-0.44	-2.23	-0.39
19 core	-0.07	0.76	-0.93	-0.62
20 core	0.58	-1.25	-0.50	0.28
21	-0.07	-0.11	-0.62	0.50
22	0.19	-0.20	-0.29	0.19
23	-0.11	-0.20	-0.07	0.31
24	-0.07	-0.50	-0.29	0.54
25	0.10	-0.34	-0.04	0.19

−10 REGION WEIGHT MATRIX

position	A	C	G	T
1	-0.34	0.19	-0.20	0.22
2	0.33	-0.29	-0.15	0.00
3	0.16	-0.15	-0.04	0.00
4	0.16	-0.04	-0.11	-0.04
5	0.33	-0.85	-0.24	0.33
6	-0.20	-0.11	-0.77	0.60
7	0.04	-0.34	0.04	0.19
8	-0.56	-0.11	0.00	0.43
9	-0.20	-0.34	0.43	-0.07
10	-0.39	-0.04	0.22	0.10
11 core	-2.64	-1.03	-0.93	1.16
12 core	1.33	-2.64	-3.33	-2.23
13 core	0.04	-0.56	-0.44	0.56
14 core	0.86	-0.69	-0.50	-0.62
15 core	0.71	-0.24	-0.69	-0.39
16 core	-3.33	-2.23	-3.33	1.35
17	0.22	-0.77	0.16	0.10
18	0.00	-0.36	0.04	0.23
19	0.10	0.07	0.07	-0.29

SPACER PENALTY BETWEEN THE −35 AND −10 CORE REGIONS

spacer	penalty
15	-3.91
16	-1.61
17	0
18	-1.61
19	-3.00
20	-3.91
21	-4.61

+1 REGION WEIGHT MATRIX

position	A	C	G	T
1	-0.07	0.30	-0.02	-0.30
2 core	-0.89	0.80	-1.01	0.00
3 start	0.72	-1.70	0.52	-2.40
4 core	-0.32	0.13	-1.48	0.65
5	0.09	0.13	-0.61	0.20
6	0.13	-0.53	0.20	0.04
7	0.24	-0.20	0.00	-0.10
8	0.09	0.00	-0.26	0.13
9	0.09	-0.26	-0.10	0.20
10	0.37	-0.26	0.09	-0.38
11	0.46	-0.32	-0.05	-0.32
12	0.17	-0.26	-0.32	0.28

FIG. 2. Weight matrices and spacer penalties for the Staden method.[14] Weight matrices were calculated by taking the natural logarithm of the observed frequency of each base (Hawley and W. R. McClure)[6] normalized by its expected frequency of 0.25. A base occurring with its expected frequency of 0.25 would be represented by a zero in the weight matrix. Staden did not normalize the frequencies by 0.25, but that difference simply shifts the scores by a constant so that they are not always negative. The spacer penalty was determined by taking the natural logarithm of an approximated spacer frequency normalized by the approximated frequency of the most frequently occurring spacer class. "core" indicates which positions occur within the more highly conserved core regions of the alignments. Transcription starts at the third position of the +1 region weight matrix.

al.[24,25] 61 (54%) of these 114 promoters are believed not to involve activators. However, 2 were eliminated because the transcriptional starts had not been sufficiently mapped, 1 was eliminated because promoter activity was not detected *in vivo*, and 11 were eliminated because they were also contained in the Hawley and McClure training set. The second test set contained 222 mRNA promoter sequences collected by Lisser and Margalit.[9] The Lisser and Margalit collection contains 300 promoter sequences, but 10 were eliminated because of insufficient sequence around the transcriptional start, 34 were eliminated because they appear in the Hawley and McClure training set, and an additional 34 were eliminated because they appear in the set of 47 nonactivated promoters. To account for inaccuracies in the mapping of transcriptional start sites, scores for the test sequences were taken to be the maximum score within ±5 bases of the supposed major transcriptional start.

In the larger test set, only 26% of the sequences scored equal to or greater than the score corresponding to a false-positive frequency of 1 in 2000 bases. However, the sequences in the smaller test set containing only nonactivated promoters scored above that cutoff with a frequency of 60% (Fig. 3 and Table I). In the Collado-Vides collection of 114 promoters,[24,25] 54% of the promoters are not associated with upstream activators. Thus, if only nonactivated promoters can be identified by computational methods, we would expect an unselected set of promoters to have $0.54 \times 0.6 = 32\%$ of the promoters scoring above the cutoff. Because the larger test has been depleted of promoters that were contained in the Hawley and McClure training set and the test set of nonactivated promoters, the 26% of promoters scoring above the cutoff is consistent with activated promoters being largely undetected by standard computational methods.

However, we scored 22 of the activatable promoters in the Collado-Vides collection and found that 45% of these promoters scored above the cutoff. This latter result suggests that there may be an overall bias for stronger promoters in the Collado-Vides collection, which was selected on the basis of how well the regulation of the promoters was understood. In the Collado-Vides promoter collection,[24,25] 40% of the activated promoters are exclusively activated by the cyclic AMP receptor protein (CRP); thus, it may be possible to improve the ability to recognize a significant portion of the activated promoters by specifically accounting for activation by this single protein.

We further analyzed the Staden method by examining whether the weight matrices could be simplified without affecting their ability to distinguish promoter from nonpromoter sequences. Figure 4 contains plots of the information content[37] at each position of the three aligned regions

[37] T. D. Schneider, G. D. Stormo, L. Gold, and A. Ehrenfeucht, *J. Mol. Biol.* **188,** 415 (1986).

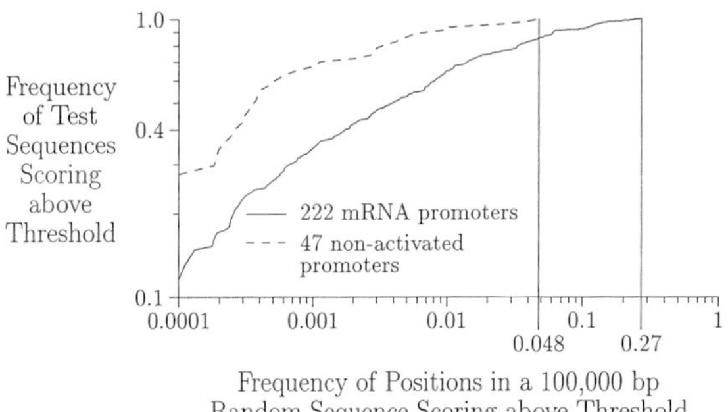

FIG. 3. The Staden scores of promoter sequences relative to the scores of random sequences. The horizontal axis is the frequency with which the positions of a random sequence score at or above a particular threshold value. The vertical axis is the frequency with which promoter sequences score at or above a particular threshold value.

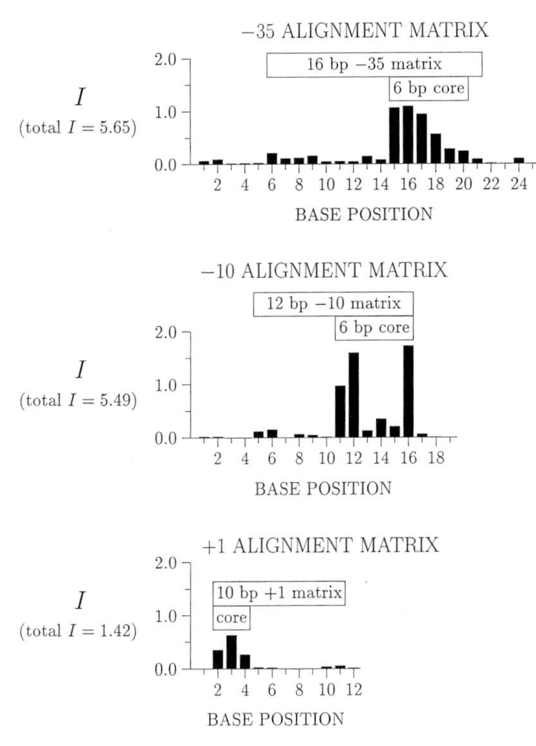

FIG. 4. Plots of the sample size-corrected information content at each position of the three aligned regions determining the Staden weight matrices. Also indicated are the truncated matrices referred to in Table I. The sample size-corrected information content is calculated as described in Schneider *et al.*[37]

TABLE I
COMPARISON OF DIFFERENT VARIATIONS OF STADEN METHOD FOR IDENTIFYING *Escherichia coli* PROMOTERS[a]

Width of weight matrix (bp)			222 mRNA promoters	47 Nonactivated promoters
−35 matrix	−10 matrix	+1 matrix		
25	19	12	0.26	0.60
16	19	12	0.24	0.55
6	19	12	0.23	0.40
25	12	12	0.25	0.53
25	6	12	0.26	0.53
25	19	3	0.24	0.52
25[b]	19	12	0.17	0.36
25	19[c]	12	0.27	0.60
25	19[d]	12	0.23	0.49
25	19[e]	12	0.27	0.57
16	6	3	0.18	0.43
16	12	3	0.20	0.51
16	6	10	0.24	0.53
16	12	10	0.25	0.60

[a] Numbers indicate the fraction of promoters scoring in the top 0.0005 of scores obtained from both strands of a randomized 100,000-bp sequence.
[b] Without penalizing suboptimal spacers between the −35 and −10 regions.
[c] Penalizing suboptimal spacers between the −10 region and the +1 site.
[d] Spacing of 4 to 12 bases rather than 4 to 8 bases between the −10 region and the +1 site.
[e] Penalizing suboptimal spacers of 4 to 12 bases between the −10 region and the +1 site.

determining the weight matrices. The information content is a measure of how well sequences are conserved within a sequence alignment. From these plots, it appears that the weight matrices used by Staden[14] contain extraneous information that should not contribute to the ability to recognize promoters. When the −35 weight matrix was truncated from 25 to 16 bp, there was a relatively slight decrease in the ability to distinguish promoters from the random sequences (Table I). However, when the −35 matrix was truncated to its 6-bp highly conserved core, there was a more significant decrease. Truncating the −10 matrix from 19 bp to either 12 bp or its 6-bp highly conserved core resulted in only a slight decrease in the ability to distinguish promoters from the random sequences. Truncating the 12-bp + 1 matrix to its 3-bp core also resulted in only a slight decrease in the ability to resolve promoters. When these simplifications were combined (16-bp −35 region, 6-bp −10 region, and a 3-bp +1 region), there was a substantial decrease in the ability to distinguish promoters from the random sequences. However, full resolution was restored when the −10 matrix was expanded to 12 bp and the +1 matrix was expanded to 10 bp.

We also tested the importance of penalizing suboptimal spacings between the −10 and −35 regions. When suboptimal spacings were not penalized, there was a significant decrease in the ability to resolve promoters. However, the addition of a spacer penalty between the −10 and +1 matrices did not affect promoter resolution. However, increasing the permitted spacer length between the −10 region and +1 site to 12 bases slightly decreased the ability to resolve promoters unless this increase was combined with a spacer penalty.

In conclusion, the most significant variable identified is the quality of the *E. coli* promoter sequences. For example, promoters that require activation are harder to resolve than promoters that do not require activation. Modifications to the weight matrices often had a relatively minor effect on the ability to discriminate promoters, although sequences outside the generally recognized consensus regions appear important. In particular, the A/T-rich region upstream of the −35 region has a significant effect on the ability to discriminate the nonactivated promoters. Surprisingly, expanding the +1 matrix from 3 to 10 bp also improved the ability to resolve promoters.

Acknowledgments

We thank Hanah Margalit for providing us with the *E. coli* promoter sequences in a computer-readable form. This work was supported by Public Health Service Grant HG-00249 from the National Institutes of Health.

Section II

Biochemical Assays of Transcription Initiation

[3] Kinetic Analysis of RNA Polymerase–Promoter Interactions

By GARY N. GUSSIN

Two-Step Model for Formation of Open Complexes

Unlike most sequence-specific DNA recognition proteins, RNA polymerase (RNAP) participates in a series of molecular isomerizations subsequent to DNA binding. This process can be described as a pathway consisting of stepwise chemical reactions characterized by appropriate equilibrium and rate constants. In the minimal two-step mechanism [Eq. (1)], the major *Escherichia coli* RNAP (containing σ^{70}) and promoter DNA (P) form a specific, closed complex (RP_c), which isomerizes to the transcriptionally competent open complex (RP_o); during isomerization, the DNA strands separate to permit base pairing between substrate NTPs and the template strand.[1–4]

$$R + P \underset{k_{-1}}{\overset{k_1}{\rightleftharpoons}} RP_c \underset{k_{-2}}{\overset{k_2}{\rightleftharpoons}} RP_o \tag{1}$$

Nonspecific DNA binding by RNAP prior to formation of the closed complex[2,5] is omitted, as are steps subsequent to open complex formation, including initiation of RNA chain synthesis and movement of RNAP away from the promoter (promoter clearance).[6,7]

Determination of K_B *and* k_f

K_B ($\cong k_1/k_{-1}$) and k_f ($\cong k_2$) can be determined, as outlined below, according to Eq. (2):

$$\tau_{obs} = 1/k_f + 1/(K_B k_f[RNAP]) \tag{2}$$

[1] W. R. McClure, *Proc. Natl. Acad. Sci. U.S.A.* **77,** 5634 (1980).
[2] M. J. Chamberlin, *in* "RNA Polymerase" (R. Losick and M. Chamberlin, eds.), pp. 17–67. Cold Spring Harbor Laboratory Press, Cold Spring Harbor, NY, 1976.
[3] W. R. McClure, *Annu. Rev. Biochem.* **54,** 171 (1985).
[4] S. Leirmo and M. T. Record, *in* "Nucleic Acids and Molecular Biology" (F. M. Eckstein and D. M. J. Lilley, eds.), Vol. 4, pp. 123–151. Springer Verlag, Berlin. 1990.
[5] P. H. Von Hippel and O. Berg, *J. Biol. Chem.* **264,** 675 (1989).
[6] R. Knaus and H. Bujard, *EMBO J.* **7,** 2919 (1988).
[7] B. Krummel and M. J. Chamberlin, *Biochemistry* **28,** 7829 (1989).

where τ_{obs} is the average time required for open complex formation at a particular RNA polymerase concentration ([RNAP]).[1] Each step in Eq. (1) may represent a series of reactions; thus, regardless of the details of the mechanism, K_B and k_f are phenomenological constants that characterize, respectively, a rapidly reversible, RNAP concentration-dependent step (or series of reversible steps) followed by an irreversible, RNAP concentration-independent step (and all subsequent steps).

Rationale for Assay

Determination of kinetic parameters by assaying products of abortive initiation takes advantage of the fact that RNAP does not translocate during the formation of the first several phosphodiester bonds in a nascent RNA chain.[8] If one or more substrates are omitted, oligonucleotides are produced continuously at a rate proportional to the number of open complexes present, without RNAP dissociation from the template or movement away from the promoter. Substrate NTPs corresponding to nucleotides at positions +1 and +2 (and in some cases subsequent positions) in the growing RNA chain will lead to abortive initiation products of the type $ppp(N_1)p(N_2)$. Added specificity can be attained[9] by replacing the initiating NTP with the dinucleotide XpN_1, where X corresponds to the base at position −1; the resulting product is at least the trinucleotide Xp(N_1)p(N_2).

When RNAP is added to DNA at time zero, the time course of oligonucleotide synthesis follows Eq. (3)[1]:

$$N = k_c P_T[t - \tau_{obs}(1 - e^{-t/\tau_{obs}})] \quad (3)$$

where N is the number of moles of product synthesized, P_T is the concentration of promoter-containing DNA at zero time, t is the time of incubation, and k_c is the intrinsic rate of catalysis (number of moles of product synthesized per minute per mole of DNA). In such an experiment, called a *lagtime* assay, two important parameters are measured: τ_{obs}, which is in fact the lagtime (see Fig. 1) and is used to determine kinetic constants defined in Eqs. (1) and (2); and $k_c P_T$, the steady state rate of oligonucleotide synthesis. Within experimental error, the steady state rate is closely approximated by the rate achieved at 3–5 times τ_{obs}. Inspection of the steady state rate often can indicate whether experimental conditions are appropriate (see below). The basic experiment consists of a series of lagtime assays (see Fig. 1) performed at several RNAP concentrations, with the values of τ_{obs} obtained in each assay plotted according to Eq. (2) to yield a typical τ plot

[8] D. E. Johnston and W. R. McClure, *in* "RNA Polymerase" (R. Losick and M. Chamberlin, eds.), pp. 413–428. Cold Spring Harbor Laboratory Press, Cold Spring Harbor, NY, 1976.
[9] D. K. Hawley and W. R. McClure, *Proc. Natl. Acad. Sci. U.S.A.* **77,** 6381 (1980).

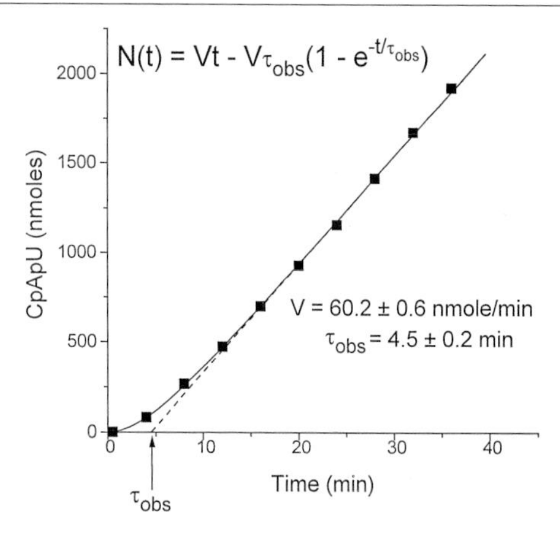

FIG. 1. Data from a lagtime assay. The data were obtained for a mutant derivative of the λP_R promoter with the substitution of an A:T bp for the consensus T:A bp at position −10. For this experiment, RNAP and DNA concentrations were 12.5 and 0.25 nM, respectively. Values of the steady state rate of abortive synthesis, V ($= k_c P_T$), and τ_{obs} are indicated. τ_{obs} can be calculated by fitting all the data points to Eq. (3) or by extrapolating the linear portion of the graph to the x axis (dashed line).

(Fig. 2). When τ_{obs} is plotted as a function of 1/[RNAP], the intercept ($1/k_f$) and the slope ($1/K_B k_f$) can be used to determine kinetic parameters for both steps in Eq. (1).

Assumptions

Derivation of Eqs. (2) and (3) is based on several simplifying assumptions[1,9]: (1) RNAP must be in excess so that its concentration is effectively

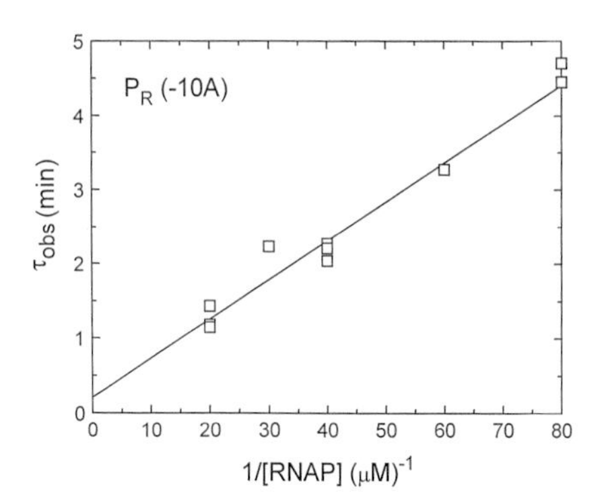

FIG. 2. τ plot analysis. Data similar to those illustrated in Fig. 1 were plotted according to Eq. (2) to yield a typical τ plot. Calculated values of k_f and K_B are 8.9 ± 6.9 × 10^{-2} sec^{-1} and 3.5 ± 2.9 × 10^6 M^{-1}, respectively.

constant during the course of the reaction. In practice, this condition is met for [active RNAP] > 1–5 nM and five times P_T. The fraction of RNAP molecules that are active can be determined according to methods of Cech and McClure[10] or Chamberlin and co-workers[11]; (2) early in the course of accumulation of open complexes, RP_c must reach steady state; in fact, whenever the rapid equilibrium condition [see point (3), below] is met, the steady state assumption is valid because the concentration of closed complexes, $[RP_c]$, is negligible, and $d[RP_c]/dt$ is also effectively zero; (3) determination of K_B requires that the closed complex be in rapid equlibrium with free RNAP, with a much higher probability of dissociation than isomerization to the open complex. {Otherwise, the slope of the τ plot [Eq. (2)] is $(k_2 + k_{-1})/k_2k_1$.} A way to test whether the assumption is justified is outlined below; (4) Eq. (2) is valid only if formation of open complexes is essentially irreversible ($k_2 \gg k_{-2}$), a condition that holds for most promoters, even when formation of open complexes is slow.[12,13] However, open complexes at certain *Bacillus subtilis* promoters[14] and at the *rrnB* P1 promoter[15] are unstable in the absence of abortive initiation substrates, which requires modification of the two-step model (see Ref. 15).

Experimental Methods

Abortive Initiation Lagtime Assays

Transcription initiation is assayed using a DNA fragment 200–1000 bp in length to avoid interference from binding to the ends of the fragment (if the fragment is too short)[16] and to reduce the relative number of nonspecific binding sites if the fragment is too long.[17] DNA should be isolated in the absence of ethidium bromide or ultraviolet (UV) light to avoid nicking. For analysis of open complex formation at the λP_R promoter, DNA fragment (0.05–1.0 nM) and RNAP (5–100 nM active enzyme) are combined in transcription buffer [0.1 M KCl, 0.01 M $MgCl_2$, 0.04 M Tris-HCl (pH 8.0), 1 mM dithiothreitol, and 100 μg of bovine serum albumin per milliliter] plus 0.5 mM CpA and 50 μM [α-^{32}P]UTP (specific activity, 0.4–0.8 Ci/

[10] C. L. Cech and W. R. McClure, *Biochemistry* **19,** 2440 (1980).
[11] M. J. Chamberlin, W. C. Nierman, J. Wiggs, and N. Neff, *J. Biol. Chem.* **254,** 10061 (1979).
[12] M.-C. Shih and G. N. Gussin, *J. Mol. Biol.* **172,** 489 (1984).
[13] J.-J. Hwang, Ph.D. Dissertation, University of Iowa, Iowa City, IA, 1987.
[14] F. W. Whipple and A. L. Sonenshein, *J. Mol. Biol.* **223,** 399 (1992).
[15] S. Leirmo and R. L. Gourse, *J. Mol. Biol.* **220,** 555 (1991).
[16] P. Melancon, R. R. Burgess, and M. T. Record, Jr., *Biochemistry* **22,** 5169 (1983).
[17] S. H. Shanblatt and A. Revzin, *Nucleic Acids Res.* **12,** 5287 (1984).

mmol) at 37°. [RNAP] and P_T are concentrations in the final reaction mixture.

At various times, 20-μl aliquots are removed and added to 5 μl of transcription stop solution [7 *M* urea, 0.1 *M* EDTA, 0.4% (w/v) sodium dodecyl sulfate (SDS), 40 m*M* Tris-HCl (pH 8.0), 0.5% (w/v) bromphenol blue, and 0.5% (w/v) xylene cyanol] to terminate the reaction. The product is separated from the substrates on a 20% (w/v) polyacrylamide gel[18]; samples of uniform size are excised from the gel and counted in a liquid scintillation counter. Data (counts per minute) for each time point are corrected by subtracting background (B) to obtain N. Background can be determined by cutting out samples from several regions of the gel, counting them, and taking the average value. The process can be made less tedious through the use of computerized detectors to quantify the amount of product in each lane of the gel. Other methods of separating the reaction product from the substrates include paper chromatography,[1] retention of DNA–RNAP–product complexes on nitrocellulose filters,[19] and production of a fluorescent side chain on hydrolysis of a derivatized substrate (γ-[1-aminonaphthalene-5-sulfonyl]-UTP).[20]

Data Analysis

SigmaPlot (Jandel Scientific, Corte Madera, CA) or any of several other similar computer programs can be used to perform a nonlinear least-squares fit of lagtime data to Eq. (3). Values of N can be weighted by multiplication by $1/(N + B/n)$, where n is the number of samples taken to yield an average value for the background (B). This assumes a Poisson sampling error in the determination of N, with an additional contribution due to variation in the background, which will be especially significant for early time points (low values of N). Alternatively, if sampling error is assumed to be proportional to N, the appropriate weighting factor would be $1/N^2$. In this case, without knowing the proportionality constant, it is difficult to incorporate the error in background into the weighting factor. In practice, the choice of weighting factor does not significantly affect the calculated value of τ_{obs} unless there is a great deal of scatter in the data points.

K_B and k_f are determined by plotting values of τ_{obs} as a function of 1/[RNAP], based on a least-squares fit to Eq. (2). By analogy with Michaelis–Menten kinetics, data can also be fit to the reciprocal or logarithmic forms of this equation.[21] If one assumes that experimental error is propor-

[18] A. J. Carpousis and J. D. Gralla, *Biochemistry* **19,** 3245 (1980).
[19] J.-H. Roe, R. R. Burgess, and M. T. Record, Jr., *J. Mol. Biol.* **176,** 495 (1984).
[20] W. C. Suh, S. Leirmo, and M. T. Record, Jr., *Biochemistry* **31,** 7815 (1992).
[21] W. W. Cleland, *Methods Enzymol.* **63,** 103 (1979).

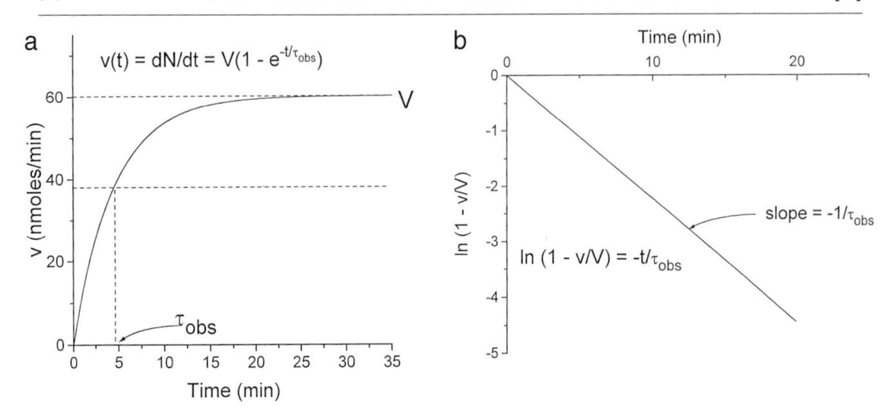

FIG. 3. Plots of hypothetical fixed-time assays. (a) This theoretical plot of a fixed-time assay corresponds to the lagtime experiment shown in Fig. 1. Values of V and τ_{obs} obtained from the lagtime assay are substituted into Eq. (4). In principle, the data from any fixed-time assay can be fit to Eq. (4). In this case, τ_{obs} is the time at which $v(t) = V(1 - 1/e)$. (b) The theoretical data from (a) are plotted to fit the logarithmic form of Eq. (4) in which $\ln(1 - v/V) = -t/\tau_{obs}$.

tional to τ_{obs}, the appropriate weighting factor for Eq. (2) as written should be the reciprocal of the sum of squares of the residuals for each value of τ_{obs}. If the data are fit to the conventional Michaelis–Menten equation (with $v = 1/\tau_{obs}$), the appropriate weighting factor is in fact the sum of squares of the residuals, while for the logarithmic form of the equation no weighting factor is necessary (see Ref. 21 for a discussion of the appropriateness of particular forms of the equation).

Fixed-Time Assays

For promoters with long lagtimes (>30–45 min), it may be useful to use the fixed-time assay[22] rather than the lagtime assay to determine τ_{obs}. In these experiments, RNAP and DNA are mixed at time zero in the absence of substrates. At various times thereafter, aliquots are removed and incubated with substrates for a short period of time (~5 min). The reaction mixtures are then assayed as described above and the data are plotted (Fig. 3a) to fit Eq. (4):

$$dN/dt = k_c P_T[1 - e^{-t/\tau_{obs}}] \tag{4}$$

which is the differentiated form of Eq. (3) (the amount of product synthesized during each short time interval is proportional to the average rate of

[22] D. K. Hawley and W. R. McClure, *J. Mol. Biol.* **157,** 493 (1982).

synthesis during that interval). The logarithmic form of Eq. (4) can be rearranged to yield $\ln(1 - v/V) = -t/\tau_{obs}$, where $V(= k_c P_T)$ is the maximal (steady state) rate of di- or trinucleotide synthesis, and $v(\equiv dN/dt)$ is the average rate of synthesis during the interval centered at time t. Thus a plot of $\ln(1 - v/V)$ as a function of t is linear with slope $-1/\tau_{obs}$ (Fig. 3b).

Measuring Dissociation Rate Constants

The use of Eq. (2) to determine k_f and K_B requires that open complexes be stable ($k_{-2} \ll k_2$). Stability can be assayed by challenging preformed open complexes with a reagent such as heparin or poly[d(AT)] that sequesters free RNAP. In a typical challenge experiment, open complexes are allowed to form for an extended period (3–5 times τ_{obs}), then diluted in the presence of heparin or poly[d(AT)]. At various times thereafter, aliquots are removed from the reaction mixture and incubated with substrates for a short fixed-time period. The decay in the ability to form di- or trinucleotides should be exponential; because heparin at high concentrations can inhibit open complexes, it is best in practice to determine a dissociation rate constant (k_{obs}) at each of several heparin concentrations ([H]), and plot values of k_{obs} as a function of [H].[10] Extrapolation to [H] = 0 yields k_{-2}. Extrapolation is based on the assumption that the primary mechanism of inactivation is the binding of heparin to free RNAP. That is, this method measures k_{-2} [as defined by Eq. (1)] only when closed complexes rapidly dissociate ($k_{-1} \gg k_{-2}$). Although poly[d(AT)] is the preferred challenge regent because it does not attack open complexes,[10] it cannot be used in any reaction containing UTP and ATP (or UpA) as substrates for abortive initiation.

Testing for Rapid Equilibrium

The rapid equilibrium assumption ($k_2 \ll k_{-1}$) is tested by challenging mixtures of RNAP and template DNA with heparin or poly[d(AT)] at various times during a lagtime assay. The question is whether at early times (Fig. 1), when only a small fraction of the templates are in open complexes, there is a significant fraction of closed complexes present in the reaction. Heparin is added at concentrations (10–50 μg/ml) too low to inhibit binary complexes significantly during the time course of the reaction; therefore, further formation of closed complexes should be inhibited, but isomerization of already-formed closed complexes to open complexes should still occur. Consequently, if the rapid equilibrium assumption is valid, the slope of abortive product synthesis in a lagtime assay should remain constant after heparin addition; if closed complexes are stable (and relatively resistant to heparin), then the slope should continue to increase.[9] It is also possible to use a repressor known to inhibit closed complex formation specifically to interrupt a lagtime or fixed-time assay to demonstrate rapid equilibrium.[22]

Complications and Cautionary Notes

Possible Errors in Determining τ_{obs}. Accurate determination of τ_{obs} in lagtime assays requires that reactions reach steady state (limiting slope); when lagtimes are long (30–45 min), an apparent limiting slope may be reached, but the total time of the assay may nevertheless be insufficient. This will lead to an underestimate of τ_{obs}. Achievement of steady state will be indicated by (1) agreement (within 10–20%) of slopes of lagtime assays performed at several different RNAP concentrations (taking into account day-to-day variations in [RNAP] introduced by dilution errors), (2) no systematic correlation between the slope and [RNAP], and (3) agreement of all slopes with the slope obtained following preincubation of enzyme (at high concentration) and DNA for a period approximating 5 times τ_{obs} prior to addition of NTPs.[1]

Determination of Background. Background activity for a DNA fragment known to contain only one promoter capable of utilizing the substrates provided in the reaction mixture can be determined by counting radioactivity of several sections of the gel close to the section containing the product of the reaction. As stated previously, sections from several regions of the gel are excised and their average is used to yield corrected values of N. Background determined in this way appears to be independent of [RNAP] or the extent of the reaction, and is not significantly different from that found when reactions are stopped immediately on addition of substrates. Errors in determining background may have small effects on the calculation of τ_{obs} by altering the point of extrapolation of the steady state portion of the lagtime plot to the x axis (see Fig. 1).

Overutilization of Substrate. Because substrate concentrations are the same order of magnitude as the K_{m} for the catalytic reaction,[9,22] it is important to assure that no more than 5–10% of the substrate be consumed during the course of the assay. Overutilization of substrate can lead to a reduction in slope and, as outlined above, can cause an underestimation of τ_{obs}. Remedies for overutilization of substrate include use of the fixed-time assay (substrates are added only during the time interval assayed),[22] reduction of P_{T} (the concentration of template DNA), and reduction in the concentration of the nonradioactive substrate.[22]

τ Plots with Negative or Decreasing Slopes. τ plots with negative slopes are difficult to interpret with any confidence because they are indicative

of a technical or conceptual problem. In some situations, a negative slope may be due to failure of lagtime reactions to reach steady state at low enzyme concentrations. Alternatively, there could be an inhibitor in the RNAP preparation, or the enzyme may be aggregating at high concentrations.

Another aberration of the τ plot is a decreasing slope at low [RNAP]. This will result if RP_o is unstable (k_{-2} is not $\ll k_2$).[23] In this case the intercept of the τ plot will equal $1/(k_2 + k_{-2})$.

Problems with Slow or Fast Promoters. Promoters that form open complexes slowly cause difficulties with the lagtime assay because of overutilization of substrate and inaccuracies in determining τ_{obs}. Both problems can be solved by the fixed-time assay, first because substrates are not present throughout the entire time course of the reaction, and second, because the approach to a plateau level of product is more accurately determined than an approach to a limiting slope.[22] However, promoters that form open complexes rapidly reach steady state very quickly, making determination of the limiting slope fairly accurate. If they also have a high rate of catalysis, k_c, overutilization of substrate must be avoided by one of the remedies discussed previously (but not by the fixed-time assay). In addition, for fast promoters, determination of the product, $K_B k_f$, may be fairly accurate, but the individual parameters cannot be determined accurately when τ ($=1/k_f$) is on the order of a few seconds, as it is for the λP_R promoter.[19,24]

Multiple Start Sites. Many promoters initiate abortive synthesis at two or more adjacent nucleotides, leading to the formation of more than one product. Frequently, the choice of substrates, including possible dinucleotides, can be used to limit initiation to only one of site. In the case of P_{lac} and the λP_{RE} promoter, it has been shown that the ratio of the products emanating from two different start sites is not affected by changes in the rate of open complex formation or presence of an activator, suggesting that a single open complex can initiate at more than one site.[12,25] Determination that two different products come from the same promoter (and not two overlapping promoters) is an obvious precondition to the determination of K_B and k_f.

Overlapping Promoters. A single DNA fragment may contain more than one promoter. Usually, if the two transcription start sites are separated by more than 90 base pairs, the kinetics of open complex formation at the two promoters can be determined by choosing substrates that restrict abortive initiation to one promoter or the other. However, if the promoters

[23] S. Strickland, G. Palmer, and V. Massey, *J. Biol. Chem.* **250,** 4048 (1975).
[24] R. S.-C. Fong, S. Woody, and G. N. Gussin, *J. Mol. Biol.* **240,** 119 (1994).
[25] A. J. Carpousis, J. E. Stefano, and J. D. Gralla, *J. Mol. Biol.* **157,** 619 (1982).

actually overlap (and therefore compete for RNAP), special problems arise that make determination of kinetic parameters for each promoter individually difficult, if not impossible.[26,27] In such cases, it is essential that one promoter or the other be completely inactivated by mutation, but the mutations must be known to affect only one promoter. The proper choice of mutations is not always straightforward.[24]

Three-Step Model

Nature of Model

The complex dependence of open complex formation on temperature and salt implicates additional intermediates in the formation of open complexes. Equation (5) is based on studies of λP_R[19] and P_{lacUV5}[28]:

$$R + P \underset{}{\overset{K_1}{\rightleftharpoons}} RP_c \underset{k_{-2}}{\overset{k_2}{\rightleftharpoons}} RP_i \underset{}{\overset{K_3}{\rightleftharpoons}} RP_o \tag{5}$$

Step 2 in Eq. (5) is thought to involve an isomerization of RNAP that "nucleates" strand separation. This formulation is supported by footprinting studies that distinguish three types of complexes depending on temperature.[29–31] RP_i has properties common to both open and closed complexes. The complex is not yet unwound, but because the second step is irreversible, k_2 is a component of k_f in Eq. (1); also, RP_i is resistant to polyanions, a property that frequently has been used as a distinguishing characteristic of open complexes. The three-step model represented by Eq. (3) has been deduced from two types of experiments: (1) studies of dissociation of open complexes, in which the rate of dissociation was an inverse function of temperature, and (2) temperature-shift experiments in which stable open complexes reequilibrated at low temperature between two polyanion-stable forms (RP_i and RP_o).[19,28]

Kinetic Parameters

The relationships between the parameters of the three-step model and those of the minimal two-step model are[28] as follows:

[26] T. P. Malan, A. Kolb, H. Buc, and W. R. McClure, *J. Mol. Biol.* **180,** 881 (1984).
[27] J. A. Goodrich and W. R. McClure, *J. Mol. Biol.* **224,** 15 (1992).
[28] H. Buc and W. R. McClure, *Biochemistry* **24,** 2712 (1985).
[29] A. Spassky, K. Kirkegaard, and H. Buc, *Biochemistry* **24,** 2723 (1985).
[30] G. Duval-Velentin and R. Ehrlich, *Nucleic Acids Res.* **15,** 575 (1987).
[31] P. Schickor, W. Metzger, W. Werel, H. Lederer, and H. Heumann, *EMBO J.* **9,** 2215 (1990).

$$K_B = K_1(k_2 + k_3 + k_{-3})/(k_3 + k_{-3}) \quad (6a)$$
$$1/k_f = 1/k_2 + 1/(k_3 + k_{-3}) \quad (6b)$$
$$k_d = k_{-2}/(1 + K_3) \quad (6c)$$

where k_d is the phenomenological dissociation rate constant measured as outlined above. Note that $K_B k_f = K_1 k_2$, and when $k_2 \ll k_3$ or k_{-3}, $K_B = K_1$ and $k_f = k_2$. Two methods of measuring k_3 and k_{-3} were described by Buc and McClure[28]: (1) temperature downshift experiments that assay these parameters at the lower temperature and (2) comparison of steady state rates on linear and supercoiled templates. The latter experiments, which can be performed at any temperature, yield K_3, but do not permit calculation of separate values of k_3 and k_{-3}.

Because K_3 for most promoters is far to the right at 30–37°, measurement of separate parameters for the three steps in Eq. (5) can ordinarily be accomplished only at lower temperatures, at which k_3 is small enough to affect the overall rate of open complex formation[4]; such measurements have been reported in detail only for the *lacUV5* promoter.[28] However, it is conceivable that effects of mutations or regulatory proteins on promoter function would be manifest at the unwinding step in open complex formation. Two indications that this was the case would be (1) a change in the steady state rate of lagtime assays (or plateau levels in fixed time assays) due to a change in the ratio of RP_i to RP_o, and (2) reciprocal effects on measured values of K_B and k_f [see Eqs. (6a) and (6b)].

Activators and Repressors

Activators

Effects of activators and repressors on kinetic parameters are ordinarily assayed by preincubating the regulatory protein with the DNA template for 10–20 min prior to addition of RNAP. In the case of the activator, one first determines the amount of protein necessary to achieve maximal stimulation of the extent of open complex formation in a single-round assay of transcription initiation (heparin is added along with substrates after a fixed time of preincubation of RNAP with the promoter-containing fragment); this assay requires the use of low [RNAP] and short preincubation times to ensure that a peak of activity is obtained. Alternatively, a minimum value of τ_{obs} can be determined in lagtime assays by varying the concentration of activator at a single (low) [RNAP]. An important reason for low [RNAP] in either assay is that, because of synergy in binding of activators and RNA polymerase,[32] the saturation concentration of the activator may vary with [RNAP].

An additional complication arises if, as in the case of the λP_{RM} promoter, the activator functions as a repressor at high concentrations.[33] This may preclude the use of saturating levels of activator and result in an underestimate of its effects on open complex formation.[34]

Repressors

Few detailed kinetic studies of the effects of repressors on open complex formation have been reported.[33,35,36] Such assays are difficult because they must be performed at several intermediate repressor concentrations, sufficient merely to inhibit, rather than abolish, open complex formation. When λ repressor acts at P_R, it affects K_B; because open complexes are much more stable than repressor–operator complexes, repressor reduces the rate, but not the extent, of open complex formation. Therefore, steady state rates of lagtime assays are the same in the presence and absence of repressor.[33] If a repressor acted at a step subsequent to RNAP binding to the promoter, or formed stable complexes with the operator under the conditions assayed, interpretation of kinetic data would be much more difficult.

Other Derivations and Additional Steps

Alternative Kinetic Analyses

In principle, any direct assay for the formation of open complexes can be used to analyze the kinetics of the association of RNAP with promoter-containing DNA.[37,38] With respect to the kinetic analysis, a single-round transcription assay is the same as a fixed-time assay. Usually, RNAP and the promoter DNA are incubated for varying amounts of time, after which substrate NTPs are added along with heparin to limit transcription to a single round. The amount of transcript is thus proportional to $[RP_o]$ at the time of substrate addition. The data essentially take the form of Fig. 3a. An interesting variant of this approach is the template competition

[32] J. Fassler and G. N. Gussin, *Methods Enzymol.* **273,** Chap. 1, 1996 (this volume).
[33] D. K. Hawley, A. D. Johnson, and W. R. McClure, *J. Biol. Chem.* **260,** 8618 (1985).
[34] R. S.-C. Fong, S. Woody, and G. N. Gussin, *J. Mol. Biol.* **232,** 792 (1993).
[35] E. Bertrand-Burggraf, S. Hurstel, M. Daune, and M. Schnarr, *J. Mol. Biol.* **193,** 293 (1987).
[36] P. J. Schlax, M. W. Capp, and M. T. Record, Jr., *J. Mol. Biol.* **245,** 331 (1995).
[37] T. R. Kadesch, S. Rosenberg, and M. J. Chamberlin, *J. Mol. Biol.* **155,** 1 (1982).
[38] J. E. Stefano and J. D. Gralla, *J. Biol. Chem.* **257,** 13924 (1982).

assay,[39] which was used to analyze the formation of stable intermediates in the pathway to RP_o at the T7 A1 promoter.

Additional Steps

Kinetic analyses of wild-type and mutant promoters have led several authors to suggest additional intermediates in open complex formation. Many of these intermediates have been inferred from differences between expected and observed rates of formation of kinetically detectable complexes,[39] but there is no simple way to assay additional steps beyond those schematized in Eqs. (1) and (5). Two kinds of open complex (RP_{o1} and RP_{o2}) have been distinguished by $KMnO_4$ footprinting.[40] The formation of RP_{o2}, which results in somewhat more extensive strand separation than is found in RP_{o1}, depends on Mg^{2+} (or certain other divalent cations).

Discussion

The kinetic analysis based on Eq. (1) is not applicable to all promoters, but most promoters tested do satisfy the conditions necessary for the determination of K_B and k_f. These assays provide two measures of relative promoter strength, $K_B k_f$[41] and τ_{obs} at [RNAP] = 30–50 nM,[3] both of which correlate well with promoter activity *in vivo*[42] except in the case of consensus promoters, for which promoter clearance is a problem.[6]

The validity of Eq. (2) and the determination of K_B and k_f require a rapid equilibrium between RP_c and free enzyme and DNA. Verification of rapid equilibrium has been reported for several promoters, including P_{lac}, P_R, P_{RM}, and P_{RE}.[9,12,22] In the case of P_R, both the wild-type and *x3* mutant promoter were deduced to satisfy the rapid equilibrium assumption. (For wild-type P_R, rapid equilibrium was inferred from the sensitivity of the formation of open complex formation to salt[19]; for P_R*x3*, the rapid equilibrium assumption was validated by polyanion challenge.[9]) However, this assumption has rarely been tested for mutant and wild-type versions of the same promoter or for the same promoter in the presence or absence of an activator. This may not be a serious problem because, if $k_{-1} \gg k_2$ for a wild-type promoter, it seems unlikely that a down mutation would decrease k_{-1} or increase k_2; similarly, if closed complexes at an activated

[39] S. Rosenberg, T. R. Kadesch, and M. J. Chamberlin, *J. Mol. Biol.* **155,** 31 (1982).
[40] W. C. Suh, W. Ross, and M. T. Record, Jr., *Science* **259,** 358 (1993).
[41] M. E. Mulligan, D. K. Hawley, R. Entriken, and W. R. McClure, *Nucleic Acids Res.* **12,** 789 (1984).
[42] H. Moyle, C. Waldburger, and M. M. Susskind, *J. Bacteriol.* **173,** 1944 (1991).

promoter are in rapid equilibrium, then it is likely that the same will be true for closed complexes that form at the unactivated promoter.

If one assumes that steps in open complex formation are the same at all promoters, then differences in nucleotide sequence may change both the overall rate of open complex formation (the inverse of the slope of a τ plot), and the relative magnitudes of the rate constants associated with each step. A substantial change in the relative magnitudes of k_{-1} and k_2 would make it impossible to determine K_B precisely. For $k_{-1} = k_2$, the error in determining K_B would be a factor of 2. A much larger change in the kinetic parameters could conceivably shift the slope of the τ plot from $1/K_B k_f$ (rapid equilibrium; $k_{-1} \gg k_2$) to $1/k_1$ (bind-and-stick; $k_{-1} \ll k_2, k_1$).[1,4] This would not affect the measurement of k_f according to Eq. (2), but would require reassessment of the validity of the assumptions on which derivation of the equation is based. This is an argument that the rapid equilibrium assumption should be retested whenever an up mutation or activator causes a substantial increase in the measured value of K_B.

Similarly, in the three-step model, k_2 is limiting at high temperature, but at low temperatures k_3 is limiting. The cross-over point is about 23° for P_{lacUV5} and about 13° for the λP_R promoter.[4,19,28] For mutant promoters, the temperature-shift experiments described earlier can be used to test whether there is a change in the cross-over point, especially in the event the mutation affects the steady state rate of abortive synthesis (see above).

In spite of these complications, the abortive initiation assay has been an extremely useful indicator of promoter strength for both strong and weak promoters under a variety of conditions. Although the focus of this chapter has been on determination of kinetic parameters that characterize open complex formation, the abortive initiation assay provides a simple way to quantitate the activity of a promoter even when a complete kinetic analysis is not necessary. As noted above, both $K_B k_f$ and τ_{obs} are reliable measures of relative promoter strength. Moreover, unlike a single-round transcription assay, the abortive initiation assay detects products from every open complex, independent of the frequency of natural abortive initiation (see Ref. 7).

Finally, differences between abortive and productive initiation kinetics can be used to examine changes in the frequency of natural abortive initiation as a result of mutation or the presence of an activator.[43] Similarly, abortive initiation was used elegantly to dissect the role of purified protein factors in initiation of basal transcription in eukaryotes.[44] In particular, it

[43] M. Menendez, A. Kolb, and H. Buc, *EMBO J.* **6,** 4227 (1987).
[44] J. A. Goodrich and R. Tjian, *Cell* **77,** 145 (1994).

was possible to distinguish between factors necessary for formation of the first phosphodiester bond and those factors necessary for chain elongation.

Acknowledgments

Data presented in Figs. 1 and 2 were obtained by Raymond Shiaw-Ching Fong (deceased) when he was a graduate student in the author's laboratory; Igor Sidorenkov provided significant help in preparing the figures. Support for the research came from NIH Grant GM50577. I am grateful to Lucia Rothman-Denes for comments on the manuscript and to Will McClure for numerous helpful discussions I have had with him over the past several years. His extensive contributions to our understanding of the kinetics of open complex formation are evident in the references.

[4] Quantitative Parameters for Promoter Clearance

By LILIAN M. HSU

Transcription initiation by *Escherichia coli* RNA polymerase entails the sequential processes of open complex formation and RNA chain initiation, both of which can influence the rate of promoter expression. Open complex formation has been widely investigated and its rate is proportional to the product $K_B k_2$.[1] Knowledge of open complex formation, however, is insufficient to account for the activity of many promoters.[2] One reason is that, during the subsequent RNA chain initiation reaction, repetitive cycling occurs on some promoters to delay the clearance of the polymerase from the promoter.[3-5]

Promoter clearance is accomplished when RNA polymerase enters the elongation phase. The process leading up to this transition involves an extended series of RNA chain initiation reactions during which RNA polymerase repetitively cycles through the open complex to produce abortive transcripts in high molar abundance from the initially transcribed region of the template. During this phase, RNA polymerase undergoes several structural changes that include translocation downstream, relinquishing the open promoter contacts, and release of σ factor.[6,7] The release of σ factor

[1] W. R. McClure, *Annu. Rev. Biochem.* **54,** 171 (1985).
[2] U. Deuschle, W. Kammerer, R. Gentz, and H. Bujard, *EMBO J.* **5,** 2987 (1986).
[3] J. E. Stefano and J. Gralla, *Biochemistry* **18,** 1063 (1979).
[4] A. J. Carpousis and J. D. Gralla, *Biochemistry* **19,** 3245 (1980).
[5] L. M. Munson and W. S. Reznikoff, *Biochemistry* **20,** 2081 (1981).
[6] A. J. Carpousis and J. D. Gralla, *J. Mol. Biol.* **183,** 165 (1985).
[7] B. Krummel and M. J. Chamberlin, *Biochemistry* **28,** 7829 (1989).

appears to coincide with the cessation of abortive initiation and marks the transition of the polymerase from the initiation to the elongation phase.[7,8] Biochemically, one can gain a quantitative handle on promoter clearance by analyzing the level and pattern of abortive and productive transcripts.

This chapter describes a method for quantitating the recycling probabilities of the polymerase at a promoter. The quantitation yields a parameter termed *abortive probability*, P_i, which is the probability that a polymerase that has synthesized a nucleotide chain to the ith template position in the initial transcribed sequence (ITS) will abort at this position and not synthesize the nucleotide $i + 1$ in length. The collective profile of P_i describes the pattern of repetitive cycling for each promoter and likely will aid in understanding the clearance transition.

Rationale of Method

To obtain the P_i values, an assay protocol was developed to facilitate the quantitative estimation of all transcripts—abortive and productive—from a single promoter. By performing *in vitro* transcription from short single-promoter templates prepared by polymerase chain reaction (PCR) amplification, one can specify the size of the runoff RNA such that the full-length product (around 50 nucleotides) can be fractionated in the same high-percentage denaturing polyacrylamide gel as the abortive RNAs.

To obtain the full range of abortive products from dinucleoside tetraphosphates to longer oligonucleotides, transcription was initiated using NTPs rather than di- or trinucleotide primers. (The use of primers gives rise to altered abortive patterns and is generally avoided.[7,9–11]) By performing the electrophoretic separation in an electrolyte gradient buffer,[12] abortive transcripts are collected toward the bottom of the gel while the full-length RNA becomes well included. This makes possible the simultaneous display, exposure, and quantitation of all transcripts from the same gel.

To facilitate quantitation, transcripts are labeled once per chain at the 5′ end through the use of γ-^{32}P-labeled initiating nucleoside triphosphate. Because 5′ end labeling limits the amount of signal incorporated per transcript, [γ-^{32}P]ATP was used at a high specific activity, about 10–20 cpm/fmol. This leads to high background radioactivity in the gel lanes, which

[8] U. M. Hansen and W. R. McClure, *J. Biol. Chem.* **255,** 9564 (1980).

[9] J. R. Levin and M. J. Chamberlin, *J. Mol. Biol.* **196,** 61 (1987).

[10] J. R. Levin, B. Krummel, and M. J. Chamberlin, *J. Mol. Biol.* **196,** 85 (1987).

[11] W. Metzger, P. Schickor, T. Meier, W. Werel, and H. Heumann, *J. Mol. Biol.* **232,** 35 (1993).

[12] J.-Y. Sheen and B. Seed, *Biotechniques* **6,** 942 (1988).

can be reduced by a prior ethanol precipitation step to remove partially the unincorporated labeled NTP from the transcription products. However, to ensure quantitative recovery of all abortive products, especially the dinucleoside tetraphosphate, a precipitation condition with glycogen carrier and an appropriate salt is necessary.

Stepwise Procedure

In Vitro Transcription

Transcription reactions are routinely performed in a 10 to 20-μl volume under extensive synthesis conditions[13] with a 20 nM concentration of a single-promoter template DNA in 50 mM Tris-HCl (pH 8.0), 10 mM $MgCl_2$, 10 mM 2-mercaptoethanol, acetylated bovine serum albumin (BSA, 10 μg/ml) at an optimal KCl concentration previously determined for each promoter. NTP is supplemented to the desired final concentration (e.g., 20 μM) with the starting nucleoside triphosphate carrying the γ-^{32}P label at a specific activity of 10–20 cpm/fmol. Reactions are commenced with the addition of RNA polymerase to 50 nM and incubated at 37° for 10 min. Reactions are subsequently terminated by the addition of 100 μl of a 10 mM EDTA solution containing glycogen (1 mg/ml), supplemented to 0.3 M sodium acetate, and precipitated with 3 vol of ethanol. The ethanol pellet is recovered and treated as described below.

Reagents and Ingredients

All stock solutions and reagents used for transcription are made with diethyl pyrocarbonate (DEPC)-treated autoclaved water.[14]

- Promoter template (10×): 200 nM in TE buffer [10 mM Tris-HCl (pH 8.0), 1 mM Na_2EDTA]
- Transcription buffer (10×): 0.5 M Tris-HCl (pH 8.0), 0.1 M $MgCl_2$, 0.1 M 2-mercaptoethanol (2-ME), acetylated BSA (0.1 mg/ml)
- KCl solution (10×; concentration varies according to promoter)
- [γ-^{32}P]ATP (10×): 200 μM at 10–20 cpm/fmol for A-starting promoters
- GTP/CTP/UTP mixture (10×): 200 μM each
- RNA polymerase (~500 nM): RNA polymerase is diluted by adding diluent to concentrated enzyme as described.[15]

[13] K. M. Arndt and M. J. Chamberlin, *J. Mol. Biol.* **202,** 271 (1988).

[14] J. Sambrook, E. F. Fritsch, and T. Maniatis, "Molecular Cloning," 2nd Ed. Cold Spring Harbor Laboratory Press, Cold Spring Harbor, NY, 1989.

[15] N. Gonzalez, J. Wiggs, and M. J. Chamberlin, *Arch. Biochem. Biophys.* **182,** 404 (1977).

RNA polymerase diluent[16]: 10 m*M* Tris-HCl (pH 8.0), 10 m*M* 2-ME, 10 m*M* KCl, 5% (v/v) glycerol, 0.1 m*M* Na_2EDTA, acetylated BSA (0.4 mg/ml), 0.1% (v/v) Triton X-100

Glycogen–EDTA solution: Glycogen (1 mg/ml), 10 m*M* Na_2EDTA

Sodium acetate (3 *M*): *Note:* Do not adjust pH. Low pH (~pH 5) precipitates EDTA in the mixture to form an insoluble pellet

Formamide loading buffer (FLB): 80% (v/v) deionized formamide, 1× TBE, 10 m*M* Na_2EDTA, 0.04% (w/v) xylene cyanol, 0.04% (w/v) bromphenol blue; 1× TBE is 89 m*M* Trizma base, 89 m*M* boric acid, 2.5 m*M* Na_2EDTA, pH 8.3

Electrophoresis buffers[12]: Top reservoir, 0.5× TBE; bottom reservoir, 0.67× TBE containing 1 *M* sodium acetate

Reaction Protocol

1. Reaction components are kept on ice and drawn and mixed at room temperature. For a 20-μl reaction, add 2 μl of 10× transcription buffer, 2 μl of 10× KCl, 2 μl of 200 n*M* template DNA, 2 μl of 10× GTP/CTP/UTP, and 2 μl of 10× [γ-^{32}P]ATP to 8 μl of DEPC–H_2O. Mix the contents and spin briefly in a microcentrifuge.
2. Add 2 μl of 500 n*M* RNA polymerase; mix immediately with a pipette tip. Incubate in a 37° water bath for 10 min.
3. Add 100 μl of EDTA–glycogen solution to terminate the reaction, then 10 μl of 3 *M* sodium acetate and 400 μl of ethanol; mix well after each addition by pipetting.
4. Allow precipitation overnight (12 hr) at −20°. Spin for 15 min in a 4° microcentrifuge to pellet the RNA. (Glycogen carrier yields a visible firm pellet, enabling a thorough removal of the ethanol supernatant.)
5. Carefully pipette off and discard the ethanol supernatant. Do not wash the pellet.
6. Dry the pellet (15 min) in a SpeedVac rotary (Savant, Hicksville, NY) vacuum desiccator.
7. Redissolve the pellet with FLB in the same original volume as the reaction aliquot.

Gel Electrophoresis

1. Prepare a 20 cm × 40 cm × 4 mm 17% (19:1, w/w) 7 *M* urea polyacrylamide gel in 1× TBE buffer.

[16] M. Chamberlin, R. Kingston, M. Gilman, J. Wiggs, and A. DeVera, *Methods Enzymol.* **101,** 540 (1983).

2. Preelectrophorese the gel in electrolyte gradient buffer for 30 min at 30–35 W. Use an aluminum plate to dissipate the heat evenly. (The aluminum plate is clamped to the front gel plate without contacting the bottom buffer.)
3. Heat the samples for 4 min at 90° and keep on ice.
4. Rinse out the wells immediately before loading by squirting with the upper reservoir buffer. Load 8 μl per sample.
5. Electrophorese at 35 W until xylene cyanol has migrated 17 cm from the well.
6. Carefully separate the plates. Cover the gel with Saran Wrap. To autoradiograph, expose the gel without drying to an X-ray film [Kodak (Rochester, NY) X-Omat AR] with an intensifying screen [Du Pont (Wilmington, DE) Cronex Lightning Plus] at −80° overnight. For quantitation, the wrapped gel is exposed directly to a phosphor screen at room temperature for varying lengths of time (30 min to overnight). Quantitation is performed using ImageQuant (Molecular Dynamics) or equivalent software. (With the high-percentage gels, overnight exposure at room temperature results in negligible diffusion of the small oligonucleotides.[17])

Detailed Comments

RNA Polymerase

Escherichia coli RNA polymerase holoenzyme is purified according to published procedures[15,18] or purchased commercially (e.g., Epicentre Technologies). Preparations of RNA polymerase obtained by this conventional method were shown to be contaminated variably with residual Gre factors.[19,20] Although the contaminating Gre factors can lead to the formation of cleavage products and influence the choice of ^{32}P label used (see discussion below), their submolar amounts do not appear to alter the pattern of abortive initiation and promoter clearance.[21]

Preparation of Single-Promoter Templates by Polymerase Chain Reaction Amplification

Promoter DNA was prepared by PCR amplification and pooled from 6–10 reactions of 100-μl volume. The PCR-amplified DNA can be recov-

[17] N. V. Vo, personal communication (1995).

[18] R. R. Burgess and J. J. Jendrisak, *Biochemistry* **14,** 4634 (1975).

[19] S. Borukhov, A. Polyakov, V. Nikiforov, and A. Goldfarb, *Proc. Natl. Acad. Sci. U.S.A.* **89,** 8899 (1992).

[20] S. Borukhov, V. Sagitov, and A. Goldfarb, *Cell* **72,** 459 (1993).

[21] L. M. Hsu, N. V. Vo, and M. J. Chamberlin, *Proc. Natl. Acad. Sci. U.S.A.* **92,** 11588 (1995).

ered free of unincorporated dNTP and primers by ethanol precipitation from 2.5 *M* ammonium acetate at room temperature, followed by one extraction each with phenol–chloroform and chloroform, and a final ethanol precipitation from 0.3 *M* sodium acetate.[14] The final rinsed pellet was dried and redissolved in 200 μl of TE buffer. DNA concentration was determined by absorbance at 260 nm. The conversion factors used to calculate DNA concentration are DNA (50 μg/ml per A_{260} unit) and 650 Da per base pair.

The procedure as described above gave >99% removal of unincorporated dNTPs and primers; thus, concentration determination of the promoter DNA is accurate. The removal of these residual components can be more quickly achieved with the use of commercial kits and spin columns [e.g., Millipore (Bedford, MA)-MC, FMC (Philadelphia, PA) spinbind kit, Promega (Madison, WI) Wizard PCR cleanup kit]. However, such purified DNA is recovered in solutions of unspecified composition, all of which interfere with A_{260} reading.[17] Despite the time-saving feature of the commercial kits, further manipulations (phenol extraction, ethanol precipitation) are necessary to recover the DNA in a quantifiable form.

Variable Reaction Parameters

To achieve accurate quantitation of abortive and productive yields from a given promoter, the concentration of several reaction components can be varied with the aim of increasing the total number of initiations obtained. The components include the RNA polymerase-to-DNA ratio, the NTP concentration, and the time of incubation. A higher concentration of NTP (than the 20 μM specified) was found to increase the frequency of initiation without altering significantly the relative abortive versus productive yields.[22] Elevated levels of RNA polymerase, DNA template, and NTP were necessary to achieve accurate quantitation of several weak promoters.[17]

KCl Titration to Maximize the Specific Transcripts

When transcribing from short promoter fragments, titration to optimize the concentration of KCl is necessary to minimize the nonspecific transcripts, abundant at low KCl concentrations, and maximize the promoter-derived transcript, which emerges as the major product as KCl concentration rises.[3] The specific promoter-derived transcript is usually referenced by size and can be confirmed by a chain-terminating ladder.[23] The optimal

[22] L. M. Hsu and M. J. Chamberlin, in preparation (1996).

[23] K. M. Arndt and M. J. Chamberlin, *J. Mol. Biol.* **213,** 79 (1990).

KCl concentration can vary significantly according to promoter; for example, it was found to be 190 m*M* for T7 A1, 250 m*M* for T5 N25, and 150 m*M* for T5 $N25_{antiDSR}$.[22]

Titration experiments showed that the relative extent of abortive versus productive synthesis from a number of promoters does not change with KCl concentration.[22] Rather, KCl concentration appears to influence the absolute activity of RNA polymerase. This was true also with $MgCl_2$, potassium acetate, and potassium glutamate titrations. For T7 A1 promoter, Cl^- was the preferred counteranion for K^+ *in vitro,* yielding more transcripts per unit time than acetate or glutamate.[24] However, weak P_{RM} mutant promoters were more active with potassium glutamate.[17]

Quantitative Precipitation Using Glycogen Carrier

Glycogen was used previously as the carrier molecule in ethanol precipitation to recover oligonucleotide transcripts, but not in a quantitative manner.[25] We tested the effectiveness of glycogen as the precipitation carrier for abortive products and found that at a final concentration of 0.77–0.83 mg of glycogen per milliliter and 0.3 *M* sodium acetate, ~80% of dinucleoside tetraphosphate (the smallest abortive product) could be reproducibly recovered in the ethanol pellet. [The final glycogen concentration of 0.77–0.83 mg/ml results from mixing 100 μl of 10 m*M* Na_2EDTA solution containing glycogen (1 mg/ml) with 10 or 20 μl of the transcription reaction, plus 10 μl of 3 *M* sodium acetate, prior to the addition of ethanol.] The recovery of longer oligonucleotides (3-mer or larger) was between 90 and 100%. The effectiveness of recovering the di- or trinucleotide abortive products is sharply reduced with lower glycogen concentrations, and the loss is further compounded by the use of 1 *M* ammonium acetate as the precipitating salt. A typical gel result is shown in Fig. 1. For quantitation purposes, glycogen–ethanol precipitation was carried out routinely at $-20°$ overnight.

Molecular biology-grade glycogen can be purchased from Boehringer Mannheim Biochemicals (Indianapolis, IN). Alternatively, nuclease-free glycogen can be prepared from a solution of crude glycogen [Sigma (St. Louis, MO), 10% in water] by five or six extractions with equal volumes of H_2O-saturated phenol (no chloroform or pH adjustment) until the milky white interface is completely removed. Solid glycogen can be precipitated by ethanol and redissolved in DEPC–H_2O for use in transcription reactions.

[24] S. Leirmo, C. Harrison, D. S. Cayley, R. R. Burgess, and M. T. Record, Jr., *Biochemistry* **26,** 2095 (1987).

[25] S. Borukhov, V. Sagitov, C. A. Josaitis, R. L. Gourse, and A. Goldfarb, *J. Biol. Chem.* **268,** 23477 (1993).

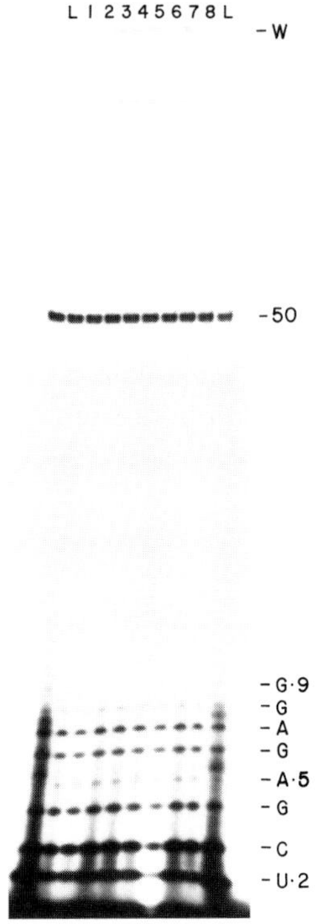

FIG. 1. Quantitative recovery of abortive products by ethanol precipitation with glycogen carrier. To test the efficiency of precipitation, two similar reaction aliquots were ethanol precipitated under a given condition either once or twice. The once-precipitated reactions (odd-numbered lanes) serve as the 100% controls; by comparison, the twice-precipitated reactions (even-numbered lanes) reveal the efficiency of recovery obtained with each precipitation condition. To perform this test, a 100-μl transcription reaction was carried out with T7 A1 promoter DNA at 190 m*M* KCl in transcription buffer and 20 μM of NTP as described in text; [γ-^{32}P]ATP was labeled at ~20 cpm/fmol. After a 10-min incubation at 37°, 10-μl aliquots were mixed with 100 μl of a 10 m*M* EDTA solution containing either 0.1 or 1 mg of glycogen per milliliter and 10 μl of either 3 *M* sodium acetate or 10 *M* ammonium acetate. Thus, four sets of ethanol precipitation conditions were set up: lanes 1 and 2, 10 μg of glycogen with 0.3 *M* sodium acetate; lanes 3 and 4, 100 μg of glycogen with 0.3 *M* sodium acetate; lanes 5 and 6, 10 μg of glycogen with 1 *M* ammonium acetate; and lanes 7 and 8, 100 μg of glycogen with 1 *M* ammonium acetate. After mixing, 400 μl of ethanol was added to each tube and the precipitation was allowed overnight at −20°. After spinning, the even-numbered tubes were set up for a second ethanol precipitation under the same condition by redissolving the pellet in 100 μl of DEPC-treated H_2O, 10 μl of the appropriate salt solution (3 *M* sodium acetate or 10 *M* ammonium acetate), and 400 μl of ethanol. The second precipitation was left for 2 hr at −20° and the pellet collected by microfuge spinning. L, The unprecipitated control, in which a reaction aliquot (10 μl) was mixed directly with an equal volume of the formamide loading dye. PAGE analysis was carried out in a 15% (19:1) 7 *M* urea gel.

γ-^{32}P versus α-^{32}P Labeling

The use of γ-^{32}P labeling of transcripts proves to be the crucial parameter in obtaining an unambiguous profile of abortive products and giving an accurate quantitative estimation of the relative proportion of abortive versus productive synthesis from a given promoter. A comparison of the transcript pattern from three promoters obtained with γ-^{32}P or α-^{32}P labeling is shown in Fig. 2. Qualitatively, the proportion of abortive and productive RNA is drastically altered depending on the method of labeling. Many minor contaminant bands, whose presence is undetectable with γ-^{32}P labeling, are revealed by body labeling of the transcripts. Although in theory the multiple incorporations into a body-labeled transcript can be corrected during quantitation, the disproportionately high molar representation of paused RNA and other nonspecific transcripts is problematic. For example, paused RNA species about 10–15 nucleotides in length can complicate the assignment of abortive product bands. Longer nonspecific transcripts, if closely migrating to the full-length RNA, can lead to an overestimation of the productive RNA. In the short-oligomer range, the products from transcript cleavage reactions also emerge (refer to Fig. 2). These extraneous bands can confuse the assignment of the true abortive products.[22] From the above comparison, the triphosphate-labeled RNA ladder generated from γ-^{32}P labeling of transcripts emerges as the best reference for abortive RNA bands.

Criteria of Abortive RNA

To study promoter clearance, it is necessary to identify for each promoter its set of abortive products. For purposes of quantitation and mechanistic consideration, abortive RNAs must be distinguished from paused RNAs[9] or RNAs associated with transcriptionally arrested complexes[23] present variably among transcripts obtained from different promoters.

Abortive RNAs are prematurely released from initial transcription complexes and differ from paused RNAs in that they cannot be chased to the full length in the presence of high concentration of NTP.[10] Because their release does not lead to dissociation of the enzyme–template complex, abortive RNAs accumulate steadily even in the presence of heparin due to polymerase recycling.[4] This linear kinetics of accumulation distinguishes abortive transcripts from RNAs associated with transcriptionally arrested complexes.[23] Transcriptional arrest at a given template position leads to a plateaued level of synthesis of an associated RNA and impedes the formation of RNA from template positions beyond the arrest site.

The most definitive proof of a released RNA is obtained by demon-

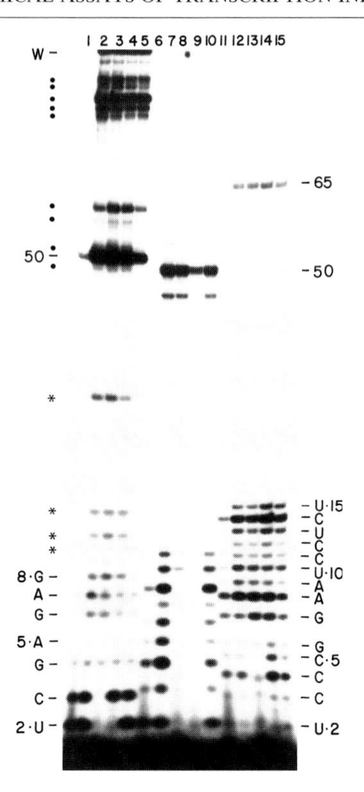

FIG. 2. Comparison of transcription product profiles obtained with γ-^{32}P versus α-^{32}P labeling. Three sets of *in vitro* transcription were performed with T7 A1 (lanes 1–5), T5 N25 (lanes 6–10), and T5 N25anti (lanes 11–15) promoters under their individual optimal KCl concentration of 190, 250, and 150 m*M*, respectively. In each set, the reactions differed in the labeling nucleoside triphosphate used: lanes 1, 6, and 11, [γ-^{32}P]ATP; lanes 2, 7, and 12, [α-^{32}P]ATP; lanes 3, 8, and 13, [α-^{32}P]GTP; lanes 4, 9, and 14, [α-^{32}P]CTP; and lanes 5, 10, and 15, [α-^{32}P]UTP. W, Position of the sample wells. The abortive products from the T7 A1 promoter are indicated on the left; from the T5 $N25_{antiDSR}$ promoter, on the right. (In designating the abortive RNA, the number indicates the length and the letter indicates the 3′-most base of the transcript.) The abortive products for the T5 N25 promoter (not shown) are U2, A3, A4, A5, U6, U7, U8, G9, and A10 (see lane 6). The full-length RNAs for T7 A1, T5 N25, and T5 $N25_{antiDSR}$ promoters are 50, 50, and 65 nucleotides, respectively. Note the sharp difference in intensity of the full-length RNA depending on γ-^{32}P or α-^{32}P labeling; the amount of full-length RNA is barely visible with γ-^{32}P labeling, particularly for the two T5 promoters. In T7 A1 transcription, nonspecific RNAs (indicated by solid dots) are disproportionately illuminated by α-^{32}P labeling, as are paused RNAs (indicated by asterisks). In lanes 8, 9, and 13, no abortive RNAs are expected at the short lengths (between 2 and 4 nucleotides). The bands detected in these positions are transcript cleavage products due to contaminating Gre factors in the RNA polymerase preparation; they are detected only with α-^{32}P labeling.[22]

strating the physical separation of the RNA from the enzyme–template complex. Molecular sieve chromatography (i.e., Sephacryl S-300) and filter binding were used to prove unequivocally that the unprecedented long abortive products from $N25_{antiDSR}$ promoter (up to 16 nucleotides) are indeed released RNAs.[22]

By the above criteria, the abortive products from T7 A1 promoter range from 2 to 9 nucleotides, from T5 N25, 2–10 nucleotides, and from T5 $N25_{antiDSR}$, 2–16 nucleotides. The identity of each abortive product was further ascertained by nearest-neighbor analysis.[22]

Quantitative Description of Promoter Clearance

By displaying the triphosphate-labeled transcripts—abortive and productive—on the same gel, one can quantitate the amount of each RNA and obtain the sum. The sum corresponds to the total number of initiations in a given unit of time. By expressing the amount of each RNA as a percentage of total, the *abortive yield* of individual abortive RNAs and the *productive yield* of full-length RNA can be determined. In calculating these parameters, one assumes little or no synthesis of paused or arrested RNAs in the elongation region. Using short promoter templates, the presence of RNA species other than abortive and full-length RNAs was made negligible.[22]

When such a quantitation was performed with T7 A1, T5 N25, and T5 $N25_{antiDSR}$ promoters, all three promoters were found to undergo abortive initiation extensively. The summary abortive yield was 92, 98, and 99.7%, respectively, for the three promoters at 20 μM NTP. The productive yield corresponding to each promoter was 8, 2, and 0.3%, respectively. These results indicate that RNA polymerase is able to clear the promoter once every 10–15 initiations from T7 A1, once every 50–100 initiations from T5 N25, and once every 300–500 initiations from T5 $N25_{antiDSR}$. Thus, there can be enormous variation in the clearance parameters from promoter to promoter.

With the gel display and quantitation, one can obtain for each promoter the following information: (1) the maximum size of abortive RNA, (2) abortive and productive yields, and (3) abortive probability associated with each template position and the overall abortive profile of a promoter. The maximum abortive RNA size indicates the span of the clearance transition and the probable position of σ release. The productive yield reflects the extent of clearance. Abortive probability, as calculated below, further identifies the barrier to clearance embedded in the ITS region.

Calculation of Abortive Probabilities

From abortive yields, the *abortive probability* (P_i) can be calculated for each abortive RNA by correcting for the fraction of RNA polymerase reaching that template position during repetitive cycling. For dinucleoside tetraphosphate, abortive probability is identical to abortive yield because it is the first abortive product. For all other abortive RNAs longer than 2 nucleotides, the abortive probability can be calculated using the following equation:

$$P_i = \frac{X_i}{100\% - \sum_{2}^{i-1} X_i} \times 100\%$$

where X_i is the abortive yield, in percent, of the ith RNA species. The denominator gives the fraction of RNA polymerase–promoter complexes that survives to the ith position. P_i, in percent, is the probability that the transcript will be aborted from the initial transcription complex at the ith position. A position of high abortive probability corresponds to a high barrier to clearance.

Concluding Remarks

Presently, we know that abortive probability is dependent on two types of factors: intrinsic and extrinsic.[22] The intrinsic, sequence-dependent factors include signals embedded in the promoter recognition region (PRR)[7] and the initial transcribed sequence (ITS).[2,26] The exact contribution of each of these elements to promoter clearance is not clearly known and may vary from promoter to promoter.[7,9,10,27] The extrinsic factors that might affect the clearance transition include NTP concentration, salt effect, ionic composition, temperature, and presence or absence of ancillary factors, etc. The contribution of a number of extrinsic factors to promoter clearance has been analyzed.[22] While the salt and triphosphate concentrations do not alter the efficiency of clearance, several regulatory proteins can exert their control at the abortive cycling stage; these include the cAMP–CRP complex,[28] LacI,[29,30] and the transcript cleavage factors GreA and GreB.[21] The determinants of abortive probability require further elucidation.

[26] W. Kammerer, U. Deuschle, R. Gentz, and H. Bujard, *EMBO J.* **5,** 2995 (1986).
[27] A. J. Carpousis, J. E. Stefano, and J. D. Gralla, *J. Mol. Biol.* **157,** 619 (1982).
[28] M. Menendez, A. Kolb, and H. Buc, *EMBO J.* **6,** 4227 (1987).
[29] H. E. Choy and S. Adhya, *Proc. Natl. Acad. Sci. U.S.A.* **89,** 11264 (1992).
[30] J. Lee and A. Goldfarb, *Cell* **66,** 793 (1991).

Acknowledgments

I thank Caroline Kane for suggesting to the editor that I write this article, Mike Chamberlin for encouraging me, Nam Vo for sharing preliminary data, and Susan Uptain for critical reading of the manuscript. I am grateful to the National Science Foundation (Grant No. HRD 9252956) and National Institutes of Health (Grant No. R15 GM 49478) for research support that made possible the findings reported here.

[5] *In Vitro* Assay for Reiterative Transcription during Transcriptional Initiation by *Escherichia coli* RNA Polymerase

By FENGXIA QI, CHONGGUANG LIU, LUCIE S. HEATH, and CHARLES L. TURNBOUGH, JR.

Reiterative transcription (also referred to as pseudotemplated transcription, transcriptional slippage, and RNA polymerase stuttering) is a reaction catalyzed by *Escherichia coli* and viral RNA polymerases[1] and presumably by RNA polymerases of many other organisms. In this reaction, a homopolymeric stretch of RNA (from 4 to >100 nucleotides) is introduced into the nascent transcript by repetitive transcription within a shorter stretch of (≥3) complementary bases in the DNA or RNA template. Apparently, the mechanism involves one or more rounds of upstream slippage of the nascent transcript relative to the template so that the same position in the template can specify multiple residues in the transcript. Reiterative transcription can occur during transcriptional initiation or elongation. In both cases, the resulting transcripts can either be released from the transcription complex or extended into productive transcripts after a switch to nonreiterative nucleotide addition.[1] Although reiterative transcription involving A, G, and U addition can occur, and probably C addition as well, reactions in which A and particularly U residues are added to the nascent transcript are strongly preferred and more efficient.[1,2] These characteristics presumably reflect the requirement for disruption of a hybrid between the nascent transcript and DNA or RNA template (which should be relatively easier with U:A and A:T base pairing) during reiterative transcription.

Studies have indicated that reiterative transcription plays an important

[1] J. P. Jacques and D. Kolakofsky, *Genes Dev.* **5,** 707 (1991).

[2] L. A. Wagner, R. B. Weiss, R. Driscoll, D. S. Dunn, and R. F. Gesteland, *Nucleic Acids Res.* **18,** 3529 (1990).

role in the expression and regulation of a large number of genes through a variety of mechanisms ranging from transcriptional frameshifting to enhancement of translational initiation.[1,3,4] One example is regulation of promoter clearance by UTP-sensitive reiterative transcription within the initially transcribed region of the *pyrBI* operon of *E. coli.*[3] The *pyrBI* operon encodes the two nonidentical subunits of the pyrimidine biosynthetic enzyme aspartate transcarbamoylase. Expression of this operon is regulated over a roughly 50-fold range by transcriptional attenution and independently, over a 7-fold range, by UTP-sensitive reiterative transcription at the *pyrBI* promoter. The *pyrBI* promoter region contains the sequence 5′ TATAATGCCGGACAATTTGCCG (nontemplate strand); the −10 region and predominant transcriptional start site are underlined. According to the current model for regulation by reiterative transcription, after the synthesis of the nascent transcript 5′ AAUUU, rapid and reversible slippage occurs between the three U residues in the transcript and three A residues in the template strand of DNA. A high UTP concentration strongly favors the addition of one or more U residues and the synthesis of $AAUUUU_n$ (n = 1 to >30) transcripts, which are not extended downstream to include structural gene sequences. In contrast, a low UTP concentration suppresses U addition(s) and allows the addition of a G residue to the AAUUU transcript, which leads to the synthesis of full-length *pyrBI* transcripts. The net effect is production of high levels of aspartate transcarbamoylase only under conditions of pyrimidine limitation.

In this chapter, we use the *pyrBI* promoter as a model to describe methods for detecting and characterizing reiterative transcription during initiation by *E. coli* RNA polymerase. Although this example is UTP dependent, the same procedures and principles apply to reiterative transcription involving any nucleoside triphosphate. In addition, many of the basic methods and concepts described here should be useful for the study of reiterative transcription during elongation and the study of this reaction catalyzed by other types of RNA polymerase.

Materials

Chemicals and Reagents

^{32}P-Labeled nucleoside 5′-triphosphates (~3000 Ci/mmol) in aqueous solution containing 5 m*M* 2-mercaptoethanol (Amersham, Arlington Heights, IL)

[3] C. Liu, L. S. Heath, and C. L. Turnbough, Jr., *Genes Dev.* **8,** 2904 (1994).
[4] F. Qi and C. L. Turnbough, Jr., *J. Mol. Biol.* **254,** 552 (1995).

Unlabeled nucleoside 5′-triphosphates supplied as 100 m*M* ultrapure rNTP solutions or as sodium salts of each rNTP (Pharmacia Biotech, Piscataway, NJ)
L-Glutamic acid, monopotassium salt (Sigma, St. Louis, MO)
Heparin, sodium salt from porcine intestinal mucosa (Sigma)
Diethyl pyrocarbonate (DEPC; Sigma)
Agarose (Sigma)
Gelatin (Difco, Detroit, MI)
Polyacrylamide gel electrophoresis reagents (Bio-Rad, Hercules, CA)
Buffer-saturated phenol, pH 7.9 (Amresco, Solon, OH)
Standard salts and other chemicals: Purchase from Sigma or Fisher (Fair Lawn, NJ); they should be of the highest purity available

Materials

Qiagen plasmid purification kits (Qiagen, Chatsworth, CA)
Spectra/Por dialysis tubing, molecular weight cutoff of 3500 (Spectrum, Houston, TX)
X-Ray film, X-Omat AR, 35 × 43 cm (Eastman Kodak, Rochester, NY)

Buffers and Solutions

Transcription buffer I (10×): 200 m*M* Tris–acetate (pH 7.9 at 25°), 100 m*M* magnesium acetate, 1 m*M* Na_2EDTA, 1 m*M* dithiothreitol, 1 *M* potassium L-glutamate. Filter sterilize and store at −20°
Transcription buffer II (10×): 200 m*M* Tris–HCl (pH 7.9 at 25°), 100 m*M* $MgCl_2$, 1 m*M* Na_2EDTA, 1 m*M* dithiothreitol, 0.5 *M* KCl. Filter sterilize and store at −20°
TAE buffer (20×): 0.8 *M* Tris–acetate (pH 7.6), 20 m*M* Na_2EDTA, 0.1 *M* sodium acetate. Store at room temperature
TBE buffer (20×): 1 *M* Tris–borate (pH 8.3), 20 m*M* Na_2EDTA. Filter and store at room temperature
TE buffer: 10 m*M* Tris–HCl (pH 8.0) and 1 m*M* Na_2EDTA (EDTA added from stock solution at pH 8.0). Autoclave and store at room temperature
Stop solution: 7 *M* urea, 2 m*M* Na_2EDTA, bromphenol blue (0.025%, w/v), xylene cyanol (0.025%, w/v). Store at −20°
Heparin stock solution: 1 mg/ml. Filter sterilize and store at −20°
Gelatin solution: 1% (w/v) in TAE buffer. Autoclave and store at room temperature
DEPC-treated water: Add 2 ml of DEPC per liter of water; shake solution vigorously. Allow it to stand for 12 hr, and autoclave the solution for 30 min. Store at room temperature

NTP stock solutions: Prepare individual (0.2 to 10 m*M*) solutions of ATP, GTP, CTP, and UTP by diluting 100 m*M* ultrapure solutions or by dissolving sodium salts of the NTPs in DEPC-treated water. Measure concentrations spectrophotometrically using the molar absorbancy values provided by the supplier (Pharmacia Biotech). Store at −20°

RNA polymerase dilution buffer: 10 m*M* Tris–HCl (pH 7.9 at 25°), 0.1 m*M* Na_2EDTA, and 0.1 m*M* dithiothreitol. Filter sterilize and store at −20°

Enzymes

Restriction endonucleases (New England Biolabs, Beverly, MA)

Pfu DNA polymerase (Stratagene, La Jolla, CA)

RNA polymerase (σ^{70}) holoenzyme: purified from *E. coli* K12 strain MRE600 as previously described.[5,6] RNA polymerase holoenzyme purified from strain MG1655 using Mono Q (Pharmacia) high-resolution ion-exchange chromatography[7] works equally well. Concentrated solutions of RNA polymerase in storage buffer,[5] which contains 50% (v/v) glycerol, can be stored without significant loss of activity at −70° for several years and at −20° for at least 6 months

Methods

Creation of RNase-Free Environment

To avoid contaminating *in vitro* transcription reactions with RNases, the following precautions should be taken: Wear disposable gloves during the preparation of materials and solutions used in the reactions. Set aside chemicals to be used only for this work and handle them with baked spatulas. Prepare solutions with DEPC-treated water. Whenever possible, use sterile, disposable plasticware, which is essentially RNase free. Treat reusable glassware and plasticware with 0.2% (v/v) DEPC for 12 hr at room temperature and autoclave for 30 min. Alternatively, bake glassware at 225° for at least 5 hr. *Caution*: DEPC is a suspected carcinogen and should be used in a chemical fume hood and handled with gloves.

[5] R. R. Burgess and J. J. Jendrisak, *Biochemistry* **14,** 4634 (1975).
[6] N. Gonzalez, J. Wiggs, and M. J. Chamberlin, *Arch. Biochem. Biophys.* **182,** 404 (1977).
[7] D. A. Hager, D. J. Jin, and R. R. Burgess, *Biochemistry* **29,** 7890 (1990).

Preparation of DNA Templates

Linear, blunt-ended DNA fragments containing the promoter of interest are convenient templates for the *in vitro* transcription reaction. Blunt-ended fragments minimize nonspecific transcription initiated at the ends of the template. Typically, we prepare template DNA by restriction enzyme digestion of a high copy number plasmid propagated in *E. coli* K12 strain DH5α. As an example of the procedure, we describe here the preparation of the *pyrBI* promoter-containing template used in all experiments described in detail in this chapter. This template is a 758-bp *Pvu*II restriction fragment (originally derived from the *E. coli* K12 chromosome) that contains approximately 300 bp of sequence downstream from the *pyrBI* promoter.[8] This downstream sequence includes the *pyrBI* attenuator, which efficiently (98%) terminates *pyrBI* transcripts after approximately 135 nucleotides *in vitro*. Plasmid is purified by using either a Qiagen maxi kit or an alkaline lysis procedure[9] and digested to completion with *Pvu*II. The 758-bp fragment is isolated by 1% (w/v) agarose gel electrophoresis and electroelution into a dialysis bag that has been rinsed on the inside with 1% (w/v) gelatin and then with TAE buffer. Electrophoresis and eletroelution are done using TAE buffer. During these procedures, potentially damaging ethidium bromide staining and ultraviolet (UV) irradiation of the template are avoided by locating the appropriate DNA band in the agarose gel by staining flanking marker lanes. After electroelution, gel debris is removed by centrifugation. DNA is precipitated with 2.5 vol of ethanol in the presence of 0.3 *M* sodium acetate for 30 min at $-70°$ and collected by centrifugation (17,000 *g* for 15 min at 4°). The pellet is washed with 1 ml of ice-cold 70% (v/v) ethanol and the sample is spun as described above for 5 min. The pellet is dissolved in 200 μl of TE buffer, and this solution is extracted with an equal volume of buffer-saturated phenol and then with an equal volume of chloroform–isoamyl alcohol (24 : 1, v/v). DNA is precipitated and washed as described above. The pellet is dried *in vacuo*, dissolved in TE buffer or DEPC-treated water, and stored at $-20°$.

Templates can also be prepared from DNA synthesized *in vitro* by polymerase chain reaction (PCR) using *Pfu* DNA polymerase and standard reaction conditions for 30 cycles. For example, we have used two slightly different PCR-generated templates. One is the same 758-bp *Pvu*II fragment described above but excised from a longer DNA fragment synthesized by PCR. The second is a 777-bp primary PCR product containing the sequence

[8] C. L. Turnbough, Jr., K. L. Hicks, and J. P. Donahue, *Proc. Natl. Acad. Sci. U.S.A.* **80,** 368 (1983).

[9] J. Sambrook, E. F. Fritsch, and T. Maniatis, *in* "Molecular Cloning: A Laboratory Manual," 2nd Ed. Cold Spring Harbor Laboratory Press, Cold Spring Harbor, NY, 1989.

of the 758-bp *Pvu*II fragment plus 9 or 10 additional bp at the ends of the DNA fragment. These PCR-generated templates are gel purified as described above to remove salts and other materials that can inhibit the *in vitro* transcription reaction or interfere with transcript analysis. The PCR templates yield identical results compared to those with the plasmid-derived template.

Finally, a supercoiled plasmid can be used as template to study reiterative transcription, at least at the *pyrBI* promoter. We examined a plasmid that was constructed by replacing the 323-bp *Pvu*II fragment of plasmid pUC19 with the 758-bp *Pvu*II fragment carrying the *pyrBI* promoter region. The resulting plasmid, designated pCLT332, was propagated in strain DH5α and purified by using a Qiagen kit. Using the same molar concentration of template DNA, reiterative transcription with the plasmid template was qualitatively and nearly quantitatively (only a 20% reduction) the same as that observed with the 758-bp *Pvu*II fragment. To demonstrate that the putative AAUUUU$_n$ transcripts synthesized with plasmid pCLT332 originate at the *pyrBI* promoter, we constructed a nearly identical plasmid in which the *pyrBI* promoter is inactivated by a point mutation in the -10 region. When the mutant plasmid is used as template, synthesis of the putative AAUUUU$_n$ transcripts is eliminated. In these experiments, we demonstrated that supercoiled plasmid template was intact after *in vitro* transcription.

For all template preparations, the DNA concentration is determined spectrophotometrically assuming an A_{260} of 1.0 equals 50 μg/ml. The concentration is routinely confirmed by running a sample alongside similarly sized, linear DNA or plasmid standards in an agarose gel and comparing the intensity of ethidium bromide-stained bands. The purity of each DNA preparation is estimated from the A_{260}/A_{280} ratio, which should be 1.8.

Transcription Reaction Conditions

Reiterative transcription at the *pyrBI* promoter can be conveniently detected and measured by using a multiple-round transcription assay. For this assay, we have used several different reaction mixtures that yield similar results.[3,10] We recommend the use of the following reaction mixture, which should permit efficient transcriptional initiation at most *E. coli* promoters. This mixture (10 to 50 μl in a 0.6- or 1.5-ml microcentrifuge tube) contains 20 m*M* Tris–acetate (pH 7.9); 10 m*M* magnesium acetate; 0.1 m*M* Na_2EDTA; 0.1 m*M* dithiothreitol; 100 m*M* potassium glutamate; ATP, GTP, and CTP (200 μM each); 1000 μM UTP; 10 n*M* DNA template; and

[10] D. J. Jin and C. L. Turnbough, Jr., *J. Mol. Biol.* **236,** 72 (1994).

100 n*M* RNA polymerase. Individual NTP concentrations can be varied as required in the experiment. In the reaction mixture, one of the nucleoside triphosphates is ^{32}P labeled. This nucleotide is either the appropriate [γ-^{32}P]NTP to radiolabel the 5′ ends of transcripts or an appropriate [α-^{32}P]NTP to label transcripts at one or more positions. The specific activity of the ^{32}P-labeled NTP is typically 2.5 Ci/mmol except in the case of [α-^{32}P]UTP, for which the specific activity is usually 0.5 Ci/mmol.

In more detail, we prepare individual reaction mixtures by first combining stock solutions (i.e., 10× transcription buffer I, radiolabel, and DNA template and NTP stock solutions) to make a master mix containing a 1.25× concentration of each component except RNA polymerase. The master mix is kept on ice. Aliquots of the master mix are added to microcentrifuge tubes, which are then capped and placed in a 37° water bath for 5 min. The transcription reaction is initiated by the addition of RNA polymerase in a volume equal to one-quarter that in the assay tube (e.g., 2 μl added to 8 μl of 1.25× master mix). The RNA polymerase solution used here is prepared immediately before use by diluting a sample of concentrated stock solution (in storage buffer) with the required amount of dilution buffer. All RNA polymerase solutions are kept on ice.

After the addition of RNA polymerase to each aliquot of master mix, the sample is mixed gently and incubated at 37° for 15 min with the assay tube capped. Under certain circumstances (e.g., when it is necessary to quantitate the synthesis of long transcripts produced by nonreiterative transcription at low NTP concentrations that significantly slow the rate of transcript elongation), it is necessary to add 0.1 vol of heparin (1 mg/ml) after the initial 15-min incubation and then to continue incubation for 10 min. With or without the addition of heparin and extra incubation, the reaction is stopped by adding an equal volume of stop solution, mixing thoroughly, and placing the sample on ice.

Transcripts are separated by polyacrylamide gel electrophoresis. Immediately before this step, the capped assay tubes containing the reaction mixtures are placed in a boiling water bath for 3 min and then chilled on ice. The tubes are spun in a microcentrifuge for a few seconds to collect liquid from the sides of tubes. A sample (6 to 8 μl) of each mixture is loaded onto a 0.4-mm thick 25% (w/v) polyacrylamide [3.3% (w/v) bis acrylamide]–TBE gel containing 7 *M* urea. The dimensions of the gel are 31 × 38.5 cm (width by length) and the sample wells are 5 mm wide. Electrophoresis is performed at 2000 V for approximately 5 hr until the bromphenol blue marker dye migrates to 12 cm from the bottom of the gel. In some instances when it is necessary to analyze long transcripts, it may be useful to run a second, lower percentage (i.e., 6 to 10%) polyacrylamide gel. In this case, the duration of electrophoresis depends on the length of the transcript(s) of interest. After electrophoresis, the gel is transferred to a used, thoroughly scrubbed piece of X-ray film and covered with plastic wrap. Transcripts in the gel can be visualized by autoradiography using Kodak X-Omat AR film (placed next to the plastic wrap-covered side of the gel). The film is exposed in a cassette containing an intensifying screen (which is positioned next to the film) for 8 to 16 hr at −70°. Transcript levels can be quantitated by analysis of the gel with a phosphorimager (we use a Molecular Dynamics PhosphorImager, Sunnyvale, CA). This procedure can be done before autoradiography or after autoradiography with the wrapped gel warmed to room temperature and wiped dry. The plastic-wrapped gel can be stored in an autoradiography cassette at −70° for several weeks without distorting transcript bands.

An Example

As an example of the multiple-round transcription assay, an experiment demonstrating UTP-sensitive reiterative transcription at the *pyrBI* promoter is shown in Fig. 1. The UTP concentration was varied in the reaction mixture as indicated in Fig. 1. The lowest (20 μM) and the highest (1000 μM) concentrations roughly mimic those found in cells grown under conditions of severe pyrimidine limitation and pyrimidine excess, respectively.[11,12] The results show the induction by UTP of reiterative transcription (i.e., production of AAU_n transcripts, where $n > 3$). This induction is coincident with a reduction in the synthesis of the predominant aborted transcript AAUUUG and of normally elongated *pyrBI* transcripts, which are terminated at the *pyrBI* attenuator (see Ref. 3 for additional details). Changes in reaction conditions from those described above, which do not significantly alter the results, are described in Fig. 1.

Altering and Optimizing Reaction Conditions

Selection of Counterions and Salt Concentrations. As indicated above, the recommended reaction mixture can be altered to some extent without significantly affecting reiterative (as well as abortive and productive) transcription at the *pyrBI* promoter. For example, under certain conditions chloride or acetate can be used as the sole counterion for the buffer and salts (Fig. 2A). However, we recommend the use of glutamate as the princi-

[11] J. T. Andersen, K. F. Jensen, and P. Poulsen, *Mol. Microbiol.* **5,** 327 (1991).

[12] J. Neuhard and P. Nygaard, *in* "*Escherichia coli* and *Salmonella typhimurium*: Cellular and Molecular Biology" (F. C. Neidhardt, J. L. Ingraham, K. B. Low, B. Magasanik, M. Schaechter, and H. E. Umbarger, eds.), pp. 445–473. American Society for Microbiology, Washington, DC, 1987.

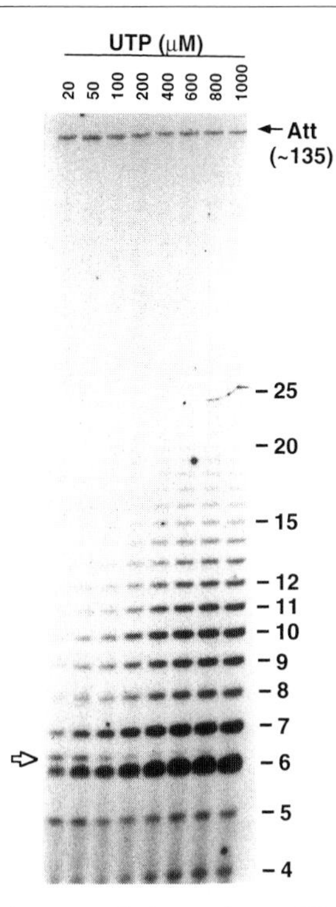

FIG. 1. UTP-induced reiterative transcription at the *pyrBI* promoter. The *pyrBI* promoter–leader region was transcribed in reaction mixtures containing the indicated concentration of UTP. Reaction conditions differed somewhat from those described for the recommended assay, in that the reaction time was increased to 30 min and transcription buffer II was used (thus chloride rather than acetate and glutamate was used as counterion for the buffer and salts, and the concentration of potassium was 50 m*M* rather than 100 m*M*). Transcripts were radiolabeled at their 5′ ends with [γ-^{32}P]ATP. Shown here is an autoradiograph of the 25% (w/v) polyacrylamide gel used to separate the transcripts. Lengths (in nucleotides) of transcripts with the sequence AAU_n are shown on the right. Bands corresponding to the aborted *pyrBI* transcript AAUUUG (open arrow) and the approximately 135-nucleotide long *pyrBI* transcripts terminated at the *pyrBI* attenuator (Att) are marked. (Adapted from Ref. 3 with permission of the publisher.)

pal anion in the reaction mixture (i.e., as the counterion for potassium) because it more closely mimics the *in vivo* ionic environment.[13] In addition, the use of glutamate instead of acetate or chloride as the counterion for potassium provides another advantage when assaying transcription from stringently controlled (i.e., negatively regulated by ppGpp) promoters like

[13] S. Leirmo, C. Harrison, D. S. Cayley, R. R. Burgess, and M. T. Record, Jr., *Biochemistry* **26,** 2095 (1987).

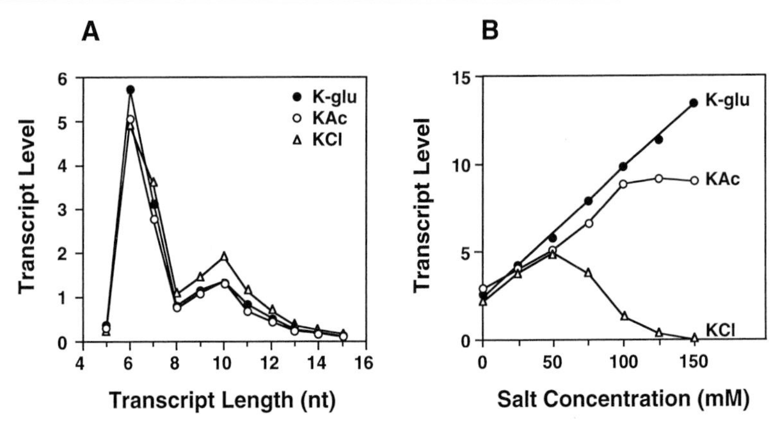

FIG. 2. Effects of different counterions and salt concentrations on reiterative transcription at the *pyrBI* promoter. (A) To examine counterion effects at noninhibitory concentrations, transcription reactions were performed using either transcription buffer II (KCl) or modified transcription buffer I containing either 50 m*M* potassium glutamate (K-glu) or 50 m*M* potassium acetate (KAc) as the potassium salt. Thus, all reaction mixtures contain 50 m*M* potassium. Reaction times were 30 min and transcripts were radiolabeled at their 5′ ends with [γ-^{32}P]ATP. Transcripts were separated by electrophoresis in a 25% (w/v) polyacrylamide gel and quantitated with a phosphorimager. Levels of AAU_n transcripts (in arbitrary units) were plotted against transcript length. (B) To demonstrate the effects of different salts and different salt concentrations, transcription reactions were performed as described above, except that the concentration of potassium salt was varied as indicated. Other details were also as described above. Levels of the AAUUUU transcript synthesized in each reaction mixture were plotted against salt concentration. An identical pattern is obtained by plotting the levels of any other single (or combination of) AAU_n transcript(s).

the *pyrBI* promoter.[14,15] Transcription from these promoters is inhibited by acetate and more strongly by chloride at concentrations at which glutamate is not inhibitory and at which potassium is at or near optimum levels (Fig. 2B). Thus, the use of glutamate as the counterion allows the inclusion in the reaction mixture of potassium at a concentration necessary for most efficient transcription. Because promoters can differ with respect to salt sensitivity, we recommend optimizing the potassium glutamate concentration for all promoters examined.

It should be noted that the presence of glutamate in samples subjected to polyacrylamide gel electrophoresis may cause distortion of the outermost sample lanes. In fact, this problem occurs with the recommended assay conditions and becomes more severe in assays using higher concentrations of potassium glutamate. The problem can usually be eliminated by running

[14] C. L. Turnbough, Jr., *J. Bacteriol.* **153,** 998 (1983).

[15] J. P. Donahue and C. L. Turnbough, Jr., *J. Biol. Chem.* **265,** 19091 (1990).

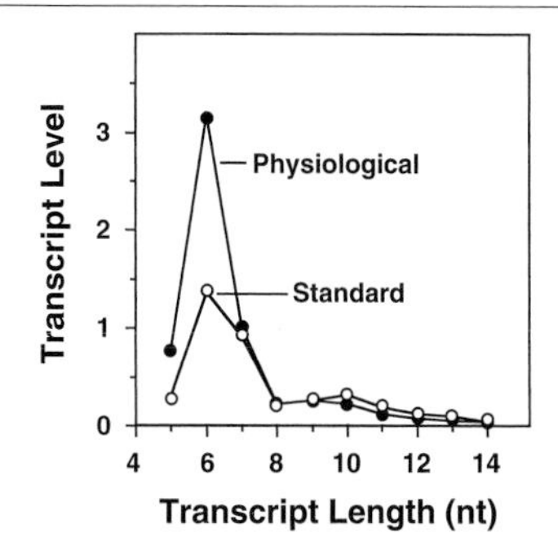

FIG. 3. Comparison of reiterative transcription at the *pyrBI* promoter at either standard assay or physiological NTP concentrations. Two assays were performed as described in the caption to Fig. 1, except that the NTP concentrations used in each reaction mixture were either as recommended for the standard assay or equal to physiological levels (i.e., 2.7 m*M* ATP, 1.4 m*M* UTP, 1.1 m*M* GTP, and 0.7 m*M* CTP). Levels of AAU_n transcripts synthesized in each reaction were measured (in arbitrary units) and plotted against transcript length.

samples of a mixture containing equal amounts of transcription buffer and stop solution in two or more of the lanes on each side of the sample lanes. In extreme cases, it may be necessary to dilute samples with a mixture containing equal amounts of water and stop solution.

Selection of NTP Concentrations and Reaction Times. The concentrations of ATP, GTP, and CTP (i.e., 200 μM each) in the recommended reaction mixture are much less than those present in rapidly growing cells of *E. coli.* Physiological NTP concentrations are roughly 2.7 m*M* ATP, 1.4 m*M* UTP, 1.1 m*M* GTP, and 0.7 m*M* CTP.[16] However, the pattern of reiterative transcription using the recommended NTP levels is not grossly different than that observed with the physiological concentrations, at least in the case of the *pyrBI* promoter. The primary difference is a shift toward the synthesis of shorter AAU_n transcripts with physiological NTP concentrations (Fig. 3). In addition, the synthesis of aborted AAUUUG transcripts and productive transcripts terminated at the *pyrBI* attenuator is not substantially different under the two reaction conditions (data not shown). We recommend the use of the lower NTP concentrations because it minimizes the dilution of the radiolabel, while permitting efficient transcriptional initiation and elongation. (*Caution*: The NTP concentration can have a major effect on the selection of transcriptional start sites at some promoters and this should be checked.)

In certain experiments it is necessary to vary the NTP concentration(s). For example, to demonstrate UTP-sensitive reiterative transcription at the

[16] C. K. Mathews, *J. Biol. Chem.* **247,** 7430 (1972).

pyrBI promoter, we varied the UTP concentration from 20 to 1000 μM. One potential problem in such experiments is the consumption of all or a significant fraction of UTP in assays containing the lowest concentrations of this nucleotide, which would result in an underestimation of transcription. This problem indeed occurs with strong promoters like the *pyrBI* promoter (Fig. 4A). However, the problem can be minimized by reducing the reaction time and/or using suboptimal concentrations of potassium to slow the rate of transcription.

As just indicated, the reaction time is an important variable in the assay. A reaction time that permits a valid assay should be empirically determined for all promoters and conditions examined. We recommend the use of

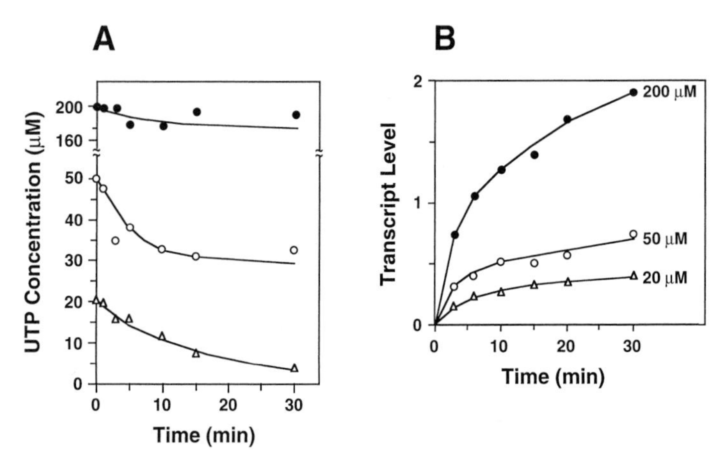

FIG. 4. Time course of UTP consumption and production of AAU_n transcripts in the assay for reiterative transcription at the *pyrBI* promoter. (A) Scaled-up (80-μl) assays were performed using transcription buffer II; ATP, GTP, and CTP (200 μM each); and either 20, 50, or 200 μM [α-^{32}P]UTP. Each of the three reaction mixtures contained 40 μCi of [α-^{32}P]UTP. Samples (10 μl) were removed at the indicated times, mixed with an equal volume of stop solution, and placed on ice. Each sample was diluted 20-fold with a mixture containing equal amounts of water and stop solution, and 6 μl of each diluted sample was loaded (every other lane) onto a 25% (w/v) polyacryamide gel. A sample of [α-^{32}P]UTP (0.075 μCi) was also loaded on the gel as a marker. The samples were subjected to electrophoresis until the bromphenol blue dye reached 20 cm from the bottom of the gel. The UTP bands were quantitated by phosphorimaging and plotted against reaction time. (B) Three scaled-up (80-μl) assays were performed as described in (A), except that the specific activity of [α-^{32}P]UTP was 0.5 Ci/mmol in each reaction. Samples (10 μl) were removed at the indicated times, mixed with an equal volume of stop solution, placed on ice, and analyzed by gel electrophoresis as described for the standard assay. Transcript levels were measured with a phosphorimager, and the levels of the major ladder transcript AAUUUU (in arbitrary units) were plotted against reaction time. Levels of all AAU_n, aborted, and full-length *pyrBI* transcripts change in parallel (data not shown).

reaction times of 15 min or less because of the apparent loss of RNA polymerase activity during the course of the reaction. In fact, the rate of transcript synthesis is not strictly linear even at early time points (Fig. 4B).

Initiation of Reactions by NTP Addition. In some instances, it may be necessary or more convenient to allow the formation of open transcription complexes before starting the reactions. This condition can be met by preincubating all reaction components minus NTPs at 37° (typically for 1 to 15 min) and initiating the reaction by the addition of an NTP mix. We have performed such experiments with the *pyrBI* promoter. In this case, we observe no differences in the results obtained when reactions are initiated by the addition of NTPs or by RNA polymerase addition as prescribed in the recommended assay (data not shown).

Verification of Reiterative Transcription

The hallmark of reiterative transcription like that occurring at the *pyrBI* promoter is the synthesis of a large set of transcripts that form a long, regularly spaced ladder of bands upon polyacrylamide gel electrophoresis. Adjacent bands in this ladder contain transcripts that differ in length by a single nucleotide; the shortest member of the ladder is only several nucleotides long, and the longest member can contain in excess of a hundred nucleotides. In general, and apparently always for transcripts containing $\geq$10 nucleotides, the abundance of individual transcripts decreases as a function of increasing transcript length. These characteristics usually provide a clear indication of reiterative transcription.

With many promoters, a short ladder of transcripts can also be produced by abortive initiation (i.e., synthesis and release of short transcripts that are entirely complementary to the DNA template). These transcripts can usually be distinguished from those produced by reiterative transcription because they rarely exceed nine nucleotides in length and their abundance does not change in a regular manner with transcript length. In addition, when abortive initiation produces a ladder of transcripts as observed after gel electrophoresis, the spacing between transcript bands is irregular. This banding pattern is due to the addition of nonidentical nucleotides to the growing transcript and the fact that these nucleotides have different effects on the electrophoretic mobility of the transcript.[4]

Another characteristic of at least some examples of reiterative transcription is that the reaction is induced by high concentrations of the repetitively added nucleotide. This effect can be observed in cases where the DNA template contains the minimum number of residues in the homopolymeric region necessary to permit transcript slippage (e.g., 3 A residues in the template strand of the *pyrBI* initially transcribed region). In this instance, a high concentration of the repetitively added nucleotide favors its addition to the nascent transcript immediately after transcription of the homopolymeric region, which commits the transcript to a pathway of repetitive nucleotide addition. In contrast, reiterative transcription is not significantly affected (at least directly) by the concentration of the repetitively added nucleotide at promoters containing a homopolymeric region that is much longer than necessary to allow transcript slippage (e.g., six A residues in the template strand of the *codBA* initially transcribed region).[4] In this case, a high concentration of the repetitively added nucleotide is unnecessary to direct the nascent transcript into the reiterative transcription pathway.

More direct evidence for reiterative transcription can be provided by demonstrating the nucleotide requirements for ladder transcript synthesis *in vitro*. This synthesis typically will require only a subset of the four NTPs. Specifically, ladder synthesis will require only the NTPs needed to transcribe into the homopolymeric region of the DNA template. For example, the ladder of *pyrBI* transcripts (minus the 6-mer AAUUUG) shown in Fig. 1 can be synthesized in a reaction mixture containing only ATP and UTP (and both nucleotides are required).[3] Addition of GTP, along with ATP and UTP, to the reaction mixture results in the synthesis of the same ladder plus the 6-mer AAUUUG, which is produced by abortive initiation. Ladder synthesis with ATP, UTP, and CTP is the same as with only ATP and UTP.

The most convincing way to prove reiterative transcription is to determine the sequences of the ladder transcripts (and compare with the sequence of the initially transcribed region of the template). We routinely determine these sequences by a combination of differential labeling of transcripts with [^{32}P]NTPs and quantitation of radiolabel incorporation.[3,4,17] In these experiments, transcripts are synthesized in a standard reaction mixture containing all four NTPs with one NTP labeled as indicated. We first establish the identity of the 5′ nucleotide of the ladder transcripts by labeling with a [γ-^{32}P]NTP. Most *E. coli* transcripts (i.e., ~75%) start with either ATP or GTP.[18] As demonstrated above, all *pyrBI* ladder transcripts are 5′ end labeled with [γ-^{32}P]ATP. Next, the nucleotide content of the ladder transcripts is shown by the incorporation of radiolabel from [α-^{32}P]NTPs. For example, the *pyrBI* ladder transcripts (excluding AAUUUG) are labeled with [α-^{32}P]ATP and [α-^{32}P]UTP but not with [α-^{32}P]GTP and [α-^{32}P]CTP. These experiments are complementary to those described above that identify the nucleotides required for ladder synthesis. Finally, the molar ratios of labeled nucleotides incorporated into ladder transcripts are determined. In the case of the *pyrBI* ladder tran-

[17] D. J. Jin, *J. Biol. Chem.* **269,** 17221 (1994).
[18] S. Lisser and H. Margalit, *Nucleic Acids Res.* **21,** 1507 (1993).

scripts, the molar ratio of [α-^{32}P]ATP to [γ-^{32}P]ATP is a constant 2 to 1 and the molar ratio of [α-^{32}P]UTP to [γ-^{32}P]ATP is $n - 2$ to 1, where n is the transcript length in nucleotides. These results provide an unambiguous sequence for the *pyrBI* ladder transcripts and a similar approach can be used to establish the identities of ladder transcripts synthesized at other promoters. One note of caution: some preparations of radiolabeled nucleotides contain substances that inhibit transcription or alter the pattern of reiterative transcription at certain promoters. Such preparations cannot be used to determine molar ratios of incorporated nucleotides without correcting for these effects.

Concluding Remarks

As the sequence requirements for the reiterative transcription reaction become more precisely defined and it becomes easier to identify promoters and other sites at which this reaction occurs, we anticipate that the number of examples of gene regulation involving reiterative transcription will grow rapidly. In this chapter, we have attempted to provide a starting point for the study of new examples of reiterative transcription in *E. coli* and closely related bacteria. We describe assay conditions and requirements, acceptable assay modifications, and methods to identify transcripts produced by reiterative transcription. The results presented focus entirely on reiterative transcription at the *pyrBI* promoter to provide a clear example of the methodology, but essentially the same procedures have been applied to the characterization of UTP-dependent reiterative transcription at the *codBA*,[4] *upp*, *pyrF*, *carAB(P1)*, *cya*, and *galETK(P2)*[17] promoters of *E. coli*. It is likely that some modification of the assays outlined here will be necessary to study other examples of reiterative transcription, perhaps occurring in nonenteric bacteria or even in eukaryotes; however, the general principles of the procedures described above should be applicable in all cases.

Acknowledgment

This work was supported by Grant GM29466 from the National Institutes of Health.

[6] Rigorous and Quantitative Assay of Transcription *in Vitro*

By HAILAN ZHANG, NANCY ILER, and CORY ABATE-SHEN

In vitro transcription assays permit investigations into the precise mechanisms of transcriptional regulation through the use of soluble cellular components in a clearly defined system. A basic eukaryotic *in vitro* transcription system for mRNA-encoding genes (class II genes) includes a purified DNA template driven by a promoter; the basal transcriptional machinery (the general transcription factors and RNA polymerase II), supplied by either crude nuclear extracts or purified proteins; ribonucleotide triphosphates; and cations (i.e., Mg^{2+}, K^{+}) optimized for RNA polymerase II activity. The first transcription system, developed in the late 1970s by Roeder and colleagues, used crude nuclear extracts to support transcription initiation from both viral and mammalian promoters.[1,2] In the early 1980s, transcription systems reconstituted with fractionated nuclear extract were established.[3,4] Since their inception, *in vitro* transcription systems have been continually refined, primarily due to technical advancements in the preparation of nuclear extracts, the purification and molecular cloning of transcription factors, and the analysis of the transcribed products. *In vitro* transcription systems have been used to define the precise functions of the general transcription factors (GTFs)[5–9] and specific transcription regulatory proteins,[10,11] and to map the functional domains of these proteins.[8,12,13] For example, the domains required for transcriptional activation and repression

[1] R. A. Weil, D. S. Luse, J. Segall, and R. G. Roeder, *Cell* **18,** 469 (1979).
[2] D. S. Luse and R. G. Roeder, *Cell* **20,** 691 (1980).
[3] T. Matsui, J. Segall, P. A. Weil, and R. Roeder, *J. Biol. Chem.* **255,** 11992 (1980).
[4] W. S. Dynan and R. Tjian, *Cell* **32,** 669 (1983).
[5] M. Sawadogo and R. G. Roeder, *Cell* **43,** 165 (1985).
[6] S. Buratowski, S. Hahn, P. A. Sharp, and L. Guarente, *Nature (London)* **334,** 37 (1988).
[7] O. Flores, H. Lu, and D. Reinberg, *J. Biol. Chem.* **267,** 2786 (1992).
[8] I. Ha, S. Roberts, E. Maldonado, X. Sun, L.-U. Kim, M. Green, and D. Reinberg, *Genes Dev.* **7,** 1021 (1993).
[9] D. Ma, H. Watanabe, F. Mermelstein, A. Admon, K. Orguri, X. Sun, T. Wada, T. Imai, T. Shiroya, D. Reinberg, and H. Handa, *Genes Dev.* **7,** 2246 (1993).
[10] B. F. Pugh and R. Tjian, *Cell* **61,** 1187 (1990).
[11] M. Meisterernst and R. G. Roeder, *Cell* **67,** 557 (1991).
[12] K. C. Yeung, J. A. Inostroza, F. H. Mermelstein, C. Kannabiran, and D. Reinberg, *Genes Dev.* **8,** 2097 (1994).
[13] C. Abate, D. Luk, and T. Curran, *Mol. Cell. Biol.* **11,** 3624 (1991).

by the Fos and Jun heterodimeric complex were defined by *in vitro* transcription assays using a HeLa nuclear extract system.[13] By using both HeLa nuclear extracts and a reconstituted system in conjunction with transient transfection assays, we have successfully characterized the transcriptional properties of the homeodomain protein Msx1, and identified its interactions with the GTFs.[14]

The two most commonly used *in vitro* transcription assays are the nuclear extract system and the reconstituted transcription system. The nuclear extract system allows one to assay the direct effects of an individual transcription factor and/or DNA element on transcription in a broadly defined system. The reconstituted system contains relatively purified and chemically defined protein factors, and has an added advantage in that this system facilitates examination of individual interactions among general transcription factors and putative interactions with specific transcriptional regulatory proteins. There are clear advantages to the use of the reconstituted system; however, it should be noted that the purification scheme for individual GTFs is extremely complex, time consuming, and costly. For most purposes, therefore, it is desirable to perform initial *in vitro* transcription studies employing the nuclear extract-based assays. Depending on these results, it can then be assessed whether a reconstituted system may or may not be required.

In this chapter, we discuss general considerations pertaining to *in vitro* transcription systems. We then describe the establishment and application of two common *in vitro* transcription systems: a HeLa nuclear extract system, and a reconstituted system composed of purified GTFs and RNA polymerase II (Fig. 1). Finally, we discuss preferred methods for data interpretation and quantitative analysis.

General Considerations

To ensure accurate and reproducible transcription reactions, it is essential to eliminate nuclease contamination. Accordingly, all appropriate reagents should be (1) prepared in autoclaved, deionized water, (2) filter sterilized, (3) aliquoted, and (4) stored at $-20°$ or lower. For the same purpose, work areas and pipette instruments should be thoroughly cleansed with 95% (v/v) ethanol before each use. Pipette tips and Eppendorf tubes must also be autoclaved. If possible, a set of pipette instruments should be designated for the sole purpose of transcription assays and should be calibrated regularly for accurate measurement. Do not treat the reagents

[14] K. M. Catron, H. Zhang, S. C. Marshall, J. A. Inostroza, J. M. Wilson, and C. Abate, *Mol. Cell. Biol.* **15,** 861 (1995).

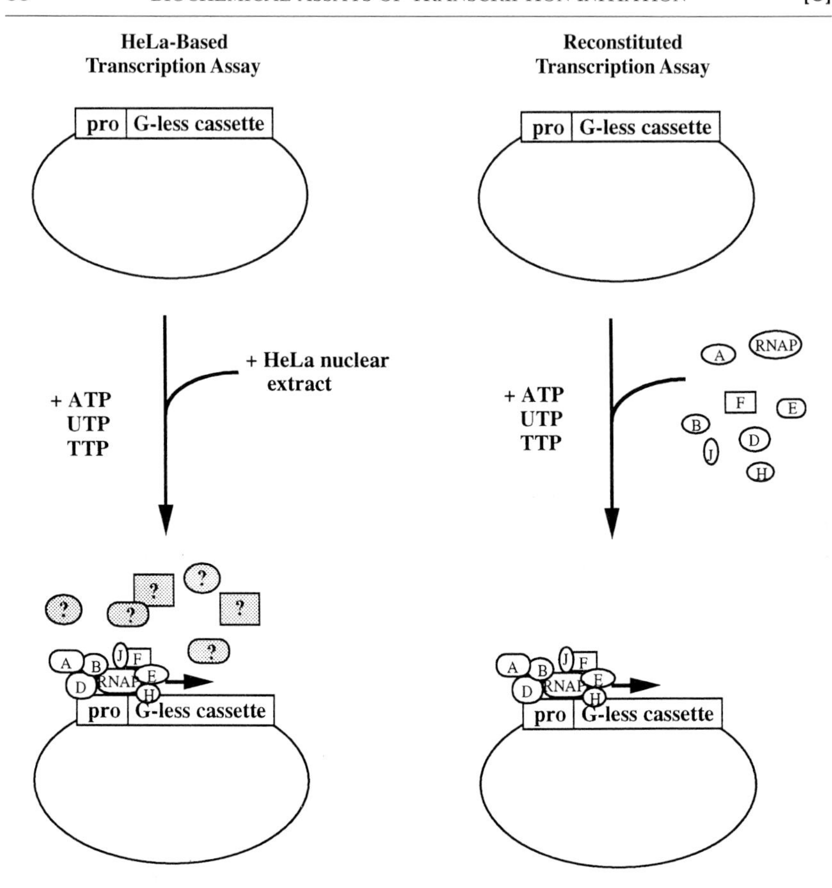

FIG. 1. Diagram of two *in vitro* transcription systems: HeLa nuclear extracts (left) and a system reconstituted from purified GTFs and RNA polymerase II (right). Pro, promoter; A, B, D, E, F, H, and J, TFIIA, TFIIB, TFIID, TFIIE, TFIIF, TFIIH, and TFIIJ; RNAP, RNA polymerase II; ?, unidentified nuclear proteins.

with diethyl pyrocarbonate (DEPC), because this may interfere with the transcription reaction.

Nuclear Extracts and Purified Proteins

Nuclear extracts and purified protein factors (i.e., GTFs, other regulatory proteins) are sensitive to temperature fluctuations. It is therefore critical to prepare these reagents at a constant 4°. Once prepared, nuclear extracts and/or purified factors should be thawed on ice immediately before addition to the reaction system. All manipulations should be performed expediently to prevent degradation of nuclear proteins. Because different batches of nuclear extracts or purified factors usually have different degrees of activity, their optimal concentrations in the transcription reaction should be determined experimentally. High molecular weight compounds such as polyethylene glycol 8000 (PEG 8000) or polyvinyl alcohol (PVA) are usually included in the system to increase viscosity. We generally pipette PEG 8000 or PVA before addition of other reagents to ensure even mixing.

DNA Templates

We recommend that DNA templates for *in vitro* transcription be in supercoiled form (e.g., plasmids) because, in our experience, the transcription reaction is generally less efficient with linear templates. The reporter plasmid contains a transcription template driven by the promoter. The choice of promoter depends on what is to be studied. Strong promoters (e.g., adenovirus-2 major late promoter, MLP) are preferred for studies to examine basal transcription and/or transcriptional repression, while weak promoters (e.g., adenovirus E4 promoter) may be favored for transcriptional activation studies. Transcription products can be detected indirectly, i.e., by primer extension, or directly, i.e., by visualizing the specific transcripts. The use of a G-less, or G-free, cassette has greatly facilitated transcript analysis in these assays.[15] As its name implies, a G-less cassette encodes a transcript lacking guanine nucleotides and therefore can be transcribed in a system containing only ATP, CTP, and UTP. To eliminate background interference of endogenous GTP contained in HeLa nuclear extracts and to enrich for the desired G-free transcripts, it is advisable to include RNase T_1, which specifically cleaves G-containing transcripts at GpN. The template plasmid may also contain binding sites for a specific DNA-binding transcription regulatory protein (activator or repressor), such that addition of the protein to the system permits investigations into its regulatory functions. The amount of DNA required for transcription should be determined experimentally for each template. Generally, an amount of 0.5–1.5 μg of MLP–G-less constructs is used in a 25-μl transcription system.

Regulated Transcription

To investigate the transcriptional regulatory properties of a specific protein, it must be purified before addition to the reaction system. Several methods have been developed to purify such proteins and these

[15] M. Sawadogo and R. G. Roeder, *Proc. Natl. Acad. Sci. U.S.A.* **82,** 4394 (1985).

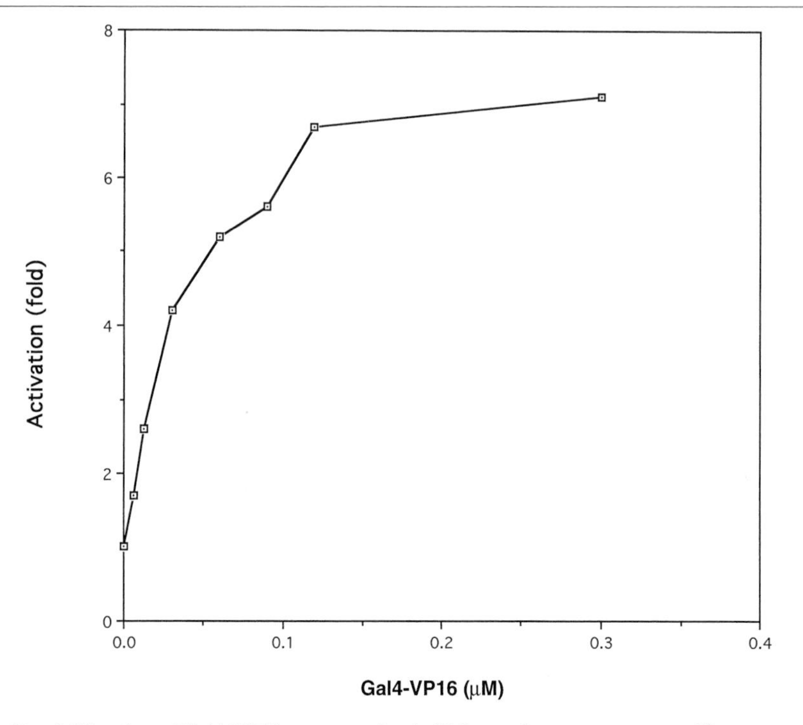

FIG. 2. Titration of Gal4-VP16 concentration in HeLa nuclear extract system. The reporter plasmid (0.5 μg) contained five tandem copies of the Gal4 DNA-binding sites upstream from the MLP and G-less cassette. Gal4-VP16 was purified as described.[14] After electrophoresis, transcripts on the gel were scanned and quantitated by using a PhosphorImager (Molecular Dynamics).

include expression in *Escherichia coli*[16] or baculovirus systems.[17] The optimal concentration of a transcriptional regulatory protein should be determined experimentally and is usually in a range of 0.1–1.0 μM. For example, 0.03–0.12 μM Gal4-VP16,[18] a transcriptional activator, can efficiently stimulate transcription from an MLP with upstream Gal4-binding sites (Fig. 2).

HeLa Nuclear Extract Transcription System

HeLa nuclear extracts have been used extensively for *in vitro* transcription studies owing to their ease of maintenance and the availability of the

[16] C. Abate, D. Luk, R. Gentz, F. J. Rauscher III, and T. Curran, *Proc. Natl. Acad. Sci. U.S.A.* **87,** 1032 (1990).

[17] G. E. Smith, M. D. Summers, and M. J. Fraser, *Mol. Cell. Biol.* **3,** 2156 (1983).

[18] I. Sadowski, J. Ma, S. Triezenberg, and M. Ptashne, *Nature* (*London*) **335,** 563 (1988).

HeLa cells, and because such extracts generally contain minimal RNase and DNase activities.[19]

Preparation of HeLa Nuclear Extracts

The following protocol is based on the procedure of Dignam *et al.*[20] with modifications.[13,21]

1. Harvest 20 liters of HeLa cells by centrifugation at 13,700 *g* (9000 rpm) in a Sorvall (Newtown, CT) GS-3 rotor at 4° for 15 min.
2. Resuspend the cell pellet in 120 ml of buffer H [10 m*M* Tris–HCl, (pH 7.9), 10 m*M* KCl, 1.5 m*M* $MgCl_2$, 1 m*M* dithiothreitol (DTT), 1 m*M* phenylmethanesulfonyl fluoride (PMSF), and 1 m*M* sodium metabisulfite] and leave on ice for 15–20 min.
3. Lyse the cells with 12 strokes of a Wheaton (Millville, NJ) Dounce homogenizer (#B). Work quickly to avoid leakage of nuclear proteins.
4. Centrifuge at 1500 *g* (3500 rpm) at 4° in a Sorvall SS34 rotor for 10 min to pellet the nuclei.
5. Resuspend the nuclei in 140 ml of buffer D [50 m*M* Tris–HCl (pH 7.4), 20% (v/v) glycerol, 20% (w/v) sucrose, 420 m*M* KCl, 5 m*M* $MgCl_2$, 0.1 *M* EDTA, 2 m*M* dithiothreitol, 1 m*M* PMSF, and 1 m*M* sodium metabisulfite] and stir in a beaker at 4° for 30 min.
6. Centrifuge at 95,900 *g* (24,000 rpm) in a Beckman (Fullerton, CA) SW28 rotor at 4° for 1 hr to remove cell debris.
7. Add ammonium sulfate slowly to the supernatant (0.33 g/ml) while stirring at 4°. Stir the mix for 30 min after the addition of ammonium sulfate.
8. Centrifuge at 27,000 *g* (15,000 rpm) in a Sorvall SS34 rotor at 4° for 15 min. Resuspend the pellet containing nuclear proteins in dialysis buffer [50 m*M* Tris–HCl (pH 7.9), 100 m*M* KCl, 12.5 m*M* $MgCl_2$, 1 m*M* EDTA, 1 m*M* DTT, 0.1 m*M* PMSF, and 1 m*M* sodium metabisulfite] and dialyze in the same buffer at 4° until the conductivity of nuclear extracts is the same as that of dialysis buffer.
9. Aliquot the dialyzed nuclear extracts and store in liquid nitrogen (long-term storage) or at −70° (short-term storage) until use.

Properly stored HeLa nuclear extracts prepared in this manner should be stable for at least 1 year without obvious loss of activity. It is recommended that the extracts be aliquoted prior to storage and that freshly thawed nuclear extracts be used in transcription assays for best results.

[19] J. L. Manley, *in* "Genetic Engineering: Principles and Methods" (J. K. Satlow and A. Hollaender, eds.), Vol. 4, p. 37. Plenum, New York, 1983.

[20] J. D. Dignam, R. M. Lebowitz, and R. G. Roeder, *Nucleic Acids Res.* **11,** 1475 (1983).

[21] J. Kadonaga, personal communication (1996).

Extracts can be refrozen and reused; however, this may result in reduced transcriptional efficiency.

Transcription Procedure

1. The reaction system (25 μl) contains 0.5–1.5 μg of supercoiled DNA template, 40 m*M* *N*-2-hydroxyethylpiperazine-*N'*-2-ethanesulfonic acid (HEPES, pH 7.6), 5 m*M* $MgCl_2$, 70 m*M* KCl, 1 m*M* dithiothreitol, 12% (v/v) glycerol, 2% (w/v) polyethylene glycol 8000 (PEG 8000), 20 units of RNasin (Promega, Madison, WI), 20 units of RNase T_1, 5 μl of HeLa nuclear extracts (approximately 50 μg), 0.5 m*M* ATP and CTP, 0.01 m*M* UTP, 0.01 m*M* 3'-*O*-methyl-GTP (Pharmacia, Piscataway, NJ), and 10 μCi of [α-^{32}P]UTP (>800 Ci/mmol). The system components should be added in the following order: (1) 2.5 μl of 20% (w/v) PEG 8000; (2) reaction mix [1 μl of 1 *M* HEPES (pH 7.6), 1.5 μl of 75 m*M* $MgCl_2$, 0.75 μl of 1 *M* KCl, 2 μl of 50% (v/v) glycerol, 1 μl of 25 m*M* dithiothreitol, 20 units of RNasin, and H_2O to 10–14 μl depending on the amount of DNA and/or proteins to be included in the system]; (3) 0.5–1.5 μg of DNA template; (4) regulatory protein or equal volumes of the buffer for dissolving the protein; (5) 5 μl of HeLa nuclear extracts; and (6) 20 units of RNase T_1, 0.5 μl of NTP mix (25 m*M* ATP, 25 m*M* CTP, 0.5 m*M* UTP, and 0.25 m*M* 3'-*O*-methyl-GTP), 1 μl of [α-^{32}P]UTP (10 μCi), and 1 μl of H_2O.
2. The reagents are mixed by gentle tapping followed by brief centrifugation (5 sec). The reaction is allowed to proceed at 30° for 40–60 min.
3. The reaction is terminated by the addition of 80 μl of stop solution [6.7 m*M* EDTA, 66 m*M* sodium acetate, 0.33% (w/v) sodium dodecyl sulfate (SDS), and 62.5 μg of yeast tRNA], followed by the addition of 100 μl of 5 *M* ammonium acetate.
4. An equal volume of phenol–chloroform (1:1, v/v) is added to the reaction mix, and vortexed for 1 min to extract RNA products.
5. Samples are centrifuged at 16,000 *g* (14,000 rpm) at room temperature in an Eppendorf microcentrifuge for 6 min to ensure complete separation of phenol and aqueous phases. The upper (aqueous) phase is transferred to a fresh tube. Extra care should be taken to avoid the interface, which contains proteins and may result in radioactive aggregates in gel wells.
6. An equal volume of chloroform is added, mixed by vortexing for 1 min, and centrifuged at 16,000 *g* (14,000 rpm) at room temperature for 4 min.
7. The upper phase is transferred to a fresh tube. Prechilled ethanol (2.5 vol) is added, mixed well, and incubated on dry ice for 20 min to precipitate RNA.
8. Samples are centrifuged at 16,000 *g* (14,000 rpm) at room temperature in an Eppendorf microcentrifuge for 10 min. RNA pellets are washed

with 80% (v/v) ethanol, and dried in a SpeedVac (Savant, Hicksville, NY) for 5–10 min. Prolonged drying may make the RNA pellet difficult to dissolve.

9. The RNA pellet is resuspended by vortexing in 10 μl of 10× formamide dye. The sample is then heated for 2 min at 90°, and 5 μl is electrophoresed on a 4.5–6% (w/v) polyacrylamide–6 *M* urea gel.

10. After electrophoresis, the gel is dried, and transcripts are visualized by autoradiography. Relative amounts of transcripts can be quantitated using a phosphorimager or densitometer (described below).

Reconstituted Transcription System

Several general transcription factors (GTFs) and RNA polymerase II have been isolated and purified from HeLa cells and other sources.[7,22–27] Of the human GTFs, TFIIA, TFIIB, TATA-binding protein (TBP) and several TBP-associated factors (TAFs), TFIIE, TFIIF, and three subunits of TFIIH have been cloned into bacterial expression vectors and are thus available in reasonable amounts.[9,28–36] These accomplishments have greatly facilitated the development of the reconstituted transcription systems. In this section, we discuss the *in vitro* transcription assays in the reconstituted system.

Purification of General Transcription Factors and RNA Polymerase II

In brief, the general scheme for purification of GTFs begins by fractionation of HeLa nuclear extracts by phosphocellulose chromatography.[3,7] The

[22] L. Zawell and D. Reinberg, *Prog. Nucleic Acid Res. Mol. Biol.* **44,** 67 (1993).
[23] A. G. Saltzman and R. Weinmann, *FASEB J.* **3,** 1723 (1989).
[24] O. Flores, I. Ha, and D. Reinberg, *J. Biol. Chem.* **265,** 5629 (1990).
[25] E. Maldonado, I. Ha, P. Cortes, L. Weis, and D. Reinberg, *Mol. Cell. Biol.* **10,** 6335 (1990).
[26] J. A. Inostroza, O. Flores, and D. Reinberg, *J. Biol. Chem.* **266,** 9304 (1991).
[27] P. Cortes, O. Flores, and D. Reinberg, *Mol. Cell. Biol.* **12,** 413 (1992).
[28] X. Sun, D. Ma, M. Sheldon, K. Yeung, and D. Reinberg, *Genes Dev.* **8,** 2336 (1994).
[29] I. Ha, W. S. Lane, and D. Reinberg, *Nature (London)* **352,** 689 (1991).
[30] A. Hoffmann, E. Sinn, T. Yamamoto, J. Wang, A. Roy, M. Horikoshi, and R. G. Roeder, *Nature (London)* **346,** 387 (1990).
[31] C. C. Kao, P. M. Lieberman, M. C. Schmidt, Q. Zhou, R. Pei, and A. J. Berk, *Science* **248,** 1646 (1990).
[32] M. G. Peterson, N. Tanese, B. F. Pugh, and R. Tjian, *Science* **248,** 1625 (1990).
[33] S. Ruppert, E. H. Wang, and R. Tjian, *Nature (London)* **362,** 175 (1993).
[34] M. G. Peterson, J. Inostroza, M. E. Maxon, O. Flores, A. Admon, and D. Reinberg, *Nature (London)* **354,** 369 (1991).
[35] L. Fischer, M. Gerard, C. Chalut, Y. Lutz, S. Humbert, M. Kanno, P. Chambon, and J.-M. Egly, *Science* **257,** 1392 (1992).
[36] S. Humbert, H. van Vuuren, Y. Lutz, J. H. J. Hoeijmakers, J.-M. Egly, and V. Moncollin, *EMBO J.* **13,** 2393 (1994).

GTFs are further purified from these fractions by subsequent chromatographic steps. For example, TFIIA and TFIIJ are purified from the 0.1 *M* (KCl) fraction from the phosphocellulose column, whereas TFIIH is purified from the 0.5 *M* fraction.[7] For detailed purification procedures, please see Refs. 24–27. RNA polymerase II can be purified as described.[37]

Transcription Procedure

Care should be taken in handling purified GTFs and RNA polymerase II because these proteins are usually available in limiting amounts after extensive purification procedures. These proteins should be aliquoted in small volumes and stored at −70°. They should be thawed on ice only before the transcription assay. Because protein concentrations of GTF preparations are usually low and are difficult to measure accurately, it is important to determine experimentally the optimal concentration for each batch of GTF in the transcription system. Some of the GTFs are interchangeable among various species whereas others are not. For example, yeast TATA-binding protein (TBP) can function with human GTFs in basal transcription.[6,38] However, yeast TBP cannot be used to reconstitute a system for activated transcription where a holo-TFIID composed of TBP and TBP-associated factors (TAFs) is required.

The procedure for transcription in the reconstituted system described below is based on a protocol from D. Reinberg's laboratory (University of Medicine and Dentistry of New Jersey) with minor modifications.

1. The reaction system (40 μl) consists of 0.15 μg of supercoiled plasmid template, 20 m*M* HEPES–KOH (pH 7.9), 8 m*M* $MgCl_2$, 50–60 m*M* KCl, 10 m*M* ammonium sulfate, 12% (v/v) glycerol, 10 m*M* 2-mercaptoethanol, 2% (w/v) PEG 8000, 20 units of RNase T_1 (Boehringer Mannheim, Indianapolis, IN), 0.6 μg of TFIIA/J, 0.03 μg of recombinant TFIIB, 0.18 μg of yeast TBP (for basal transcription) or 1.15 μg of TFIID (for activated transcription), 0.05 μg of recombinant TFIIE, 0.63 μg of TFIIF/H, 0.25 μg of RNA polymerase II, 0.6 m*M* ATP, 0.6 m*M* CTP, and 10 μCi of [α-^{32}P]UTP (800 Ci/mmol; New England Nuclear, Boston, MA). The reagents should be added sequentially as follows: (1) 3 μl of preinitiation buffer [200 m*M* HEPES (pH 7.9) and 80 m*M* $MgCl_2$], 4 μl of 20% (w/v) PEG 8000, 2 μl of 1% (w/v) ammonium sulfate, and 0.15 μg of DNA template;

[37] H. Lu, O. Flores, R. Weinmann, and D. Reinberg, *Proc. Natl. Acad. Sci. U.S.A.* **88,** 10004 (1991).

[38] B. Cavallini, J. Huet, J. L. Plassat, A. Sentenac, J. M. Egly, and P. Chambon, *Nature (London)* **334,** 77 (1988).

(2) GTFs and RNA polymerase II in BC100 [20 m*M* Tris–HCl (pH 7.9) at 4°, 20% (v/v) glycerol, 0.1 *M* KCl, 0.2 m*M* EDTA, 0.2 m*M* PMSF, and 0.5 m*M* DTT]; and (3) 4 μl of elongation solution (6 m*M* ATP, 6 m*M* CTP, and 0.15 m*M* UTP), 1 μl of preinitiation buffer, 20 units of RNase T_1, and 1 μl of [α-^{32}P]UTP (10 μCi).

2. The reagents are mixed by mild vortexing at low speed, spun down, and the reaction is allowed to proceed at 30° for 45 min.

3. Steps 3–10 of the HeLa system procedure are followed to terminate the transcription reaction, and to extract and resolve transcription products.

It should be emphasized that this procedure is for transcription reactions using GTFs and RNA polymerase II purified as described. For assays using GTFs and RNA polymerase II from other sources or purified by different methods, it may be necessary to optimize concentration of each factor experimentally. Theoretically, if a regulatory protein affects transcription levels in the reconstituted system, then this protein may regulate transcription by direct interactions with the GTFs and/or RNA polymerase II included in the system. It should be noted, however, that factors purified from nuclear extracts may still contain other nuclear proteins, and this possibility must be considered during data interpretation.

Analysis of *in Vitro* Transcription Products

Gel Electrophoresis and Autoradiography

In general, transcription products (i.e., transcripts) are visualized by denaturing polyacrylamide gel electrophoresis (PAGE). For resolving transcripts from a 380-bp G-less cassette, a 4.5–6% (w/v) polyacrylamide–6 *M* urea gel should be used. Samples are denatured by heating at 90° for 2 min. Radioactive molecular weight markers (RNA or DNA) should be treated the same way as the samples. If DNA markers are to be used, it should be noted that DNA migrates about 10% faster than RNA. For better resolution, individual wells should be rinsed immediately before loading samples to remove accumulated urea. It is usually appropriate to run a 5.5 × 5 in. gel (thickness, 0.8 mm) at constant voltage (200–300 V) in 1× Tris–borate–EDTA (TBE) at room temperature until the bromphenol blue reaches the bottom of the gel. After electrophoresis, the gel should be dried with a gel dryer, and the dried gel then exposed to autoradiography film [i.e., Kodak (Rochester, NY) X-Omat film]. With the use of two intensifying screens, autoradiography at −70° for 12–16 hr is usually sufficient to visualize the transcripts from MLP-driven templates.

Quantitative Analysis

There are several methods to analyze transcription products quantitatively. For example, after autoradiography, the density of the transcript bands on a developed film can be determined by a densitometer. Transcriptional levels can then be inferred by comparisons of individual densities. It should be noted, however, that the use of overexposed films will result in inaccurate density measurements and, therefore, inaccurate transcription results. For more accurate quantitation, one can align the developed film with the gel, cut gel slices corresponding to the transcript bands, and dissolve them in scintillation fluid for scintillation counting.

The most convenient and efficient method for the quantitation of transcription products is by use of a phosphorimager (e.g., PhosphorImager by Molecular Dynamics, Sunnyvale, CA). After electrophoresis, the gel should be dried as usual, wrapped with Sealwrap film (to prevent contamination to the facility), and covered by a phosphor screen. After a 2- to 4-hr exposure, the phosphors (e.g., crystals of $BaFBr:Eu^{2+}$) on the screen store energy released from the ^{32}P radiation. When the exposed screen is scanned by a laser beam in the phosphorimager, the phosphors release stored energy as blue lights, which can be measured by computer to form a digitized image of the gel. Because the pixel values of the image are proportional to the strength of the radioactive signal, relative amounts of transcripts in a selected area (i.e., a band) can be determined.[39] Therefore, a phosphorimager can be used for quantitative measurements of transcriptional levels. For example, we have shown that the murine homeodomain protein Msx1 is a transcriptional repressor using both HeLa and the reconstituted *in vitro* transcription systems (Fig. 3A and Fig. 4A). On the basis of a quantitative analysis by phosphorimager, we are able to show the levels of both transcriptional activation by Gal4-VP16 and transcriptional repression by Msx1, and to compare transcriptional activities of HeLa nuclear extract system and the reconstituted system (Fig. 3B and Fig. 4B).

Conclusion

In vitro transcription assays are powerful tools for characterizing the functions of general transcription factors and specific transcriptional regulators, and for studying protein–protein interactions among these transcription factors. Here, we have discussed two of the most commonly used *in vitro* transcription assays: the HeLa nuclear extract system and the reconstituted system. In addition to these described *in vitro* transcription

[39] Molecular Dynamics, Sunnyvale, CA, *PhosphorImager User's Guide (1991).*

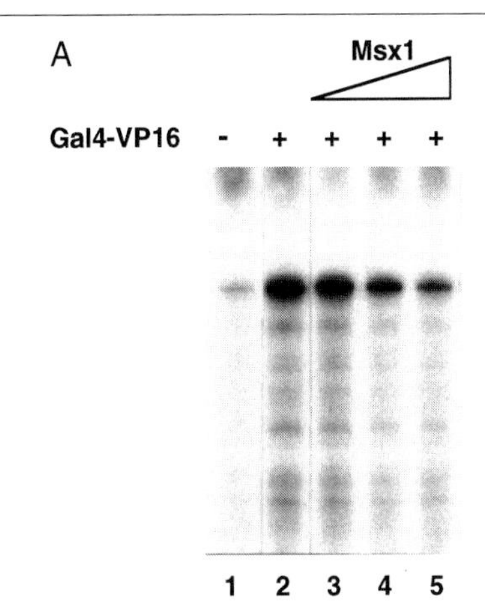

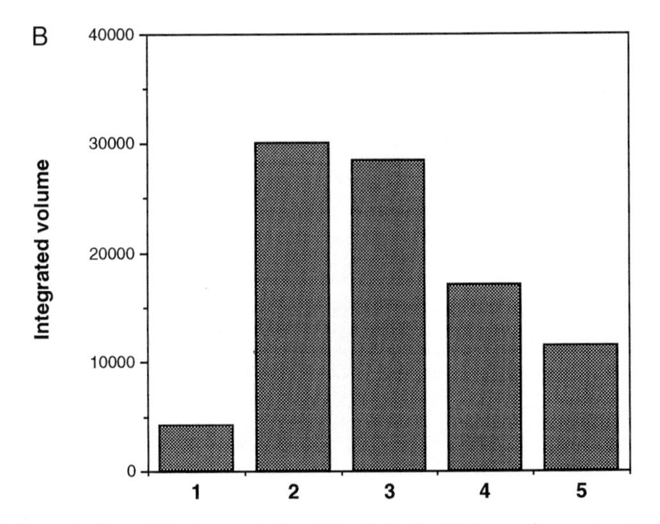

FIG. 3. Quantitation of the transcriptional levels in the HeLa nuclear extract system. (A) The reporter plasmid (0.5 μg) contained five tandem copies of the Gal4 DNA-binding sites upstream from the MLP and G-less cassette. Lane 1 shows the basal transcription. Addition of Gal4-VP16 resulted in activated transcription (lane 2), which was repressed by increasing concentrations of Msx1, a homeodomain protein as a transcriptional repressor (lanes 3–5). Transcription products were resolved by denaturing PAGE and visualized by autoradiography. (B) The same gel was scanned by a PhosphorImager (Molecular Dynamics). The calculated integrated volumes were proportional to levels of the transcripts.

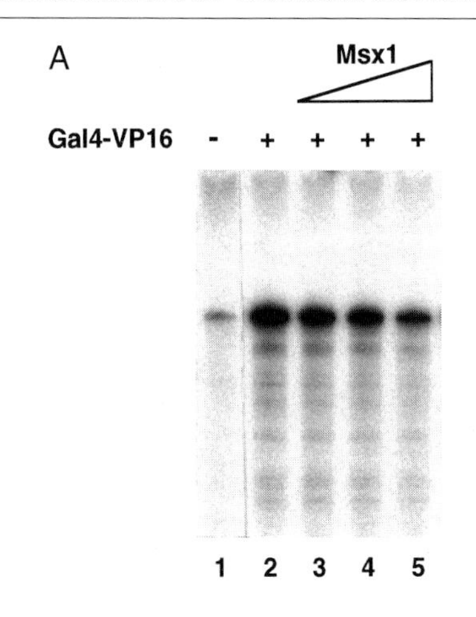

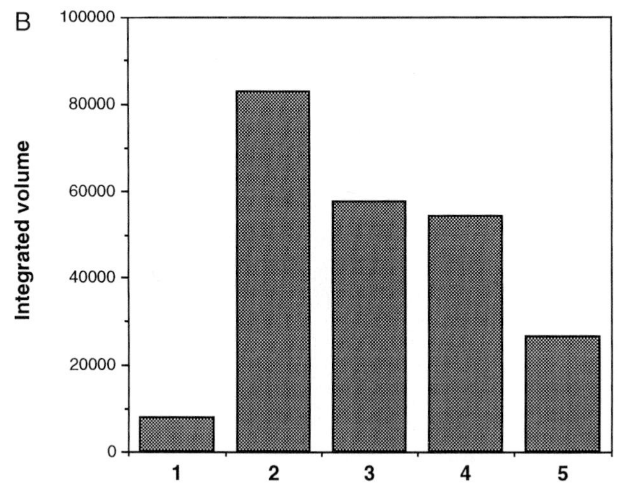

FIG. 4. Quantitation of the transcriptional levels in the reconstituted system. (A) The reporter plasmid (0.15 μg), Gal4-VP16, and Msx1 were tested in the system reconstituted from purified GTFs and RNA polymerase II. Transcription products were resolved by denaturing PAGE and visualized by autoradiography. Lane 1, the basal transcription; lane 2, activated transcription by Gal4-VP16; lanes 3–5, transcriptional repression by increasing concentrations of Msx1. (B) The same gel was scanned by a PhosphorImager (Molecular Dynamics). The calculated integrated volumes were proportional to levels of the transcripts.

systems, there are several systems prepared from other sources. Regardless of the *in vitro* transcription system used, it should be noted that *in vitro* assays are not a substitute for those assays performed in living cells or organisms. Although *in vitro* assays can be extremely informative, the results of such assays must be confirmed or corroborated by the use of *in vivo* systems that assay functions of transcription factors and DNA elements under physiological conditions. In summary, *in vitro* and *in vivo* transcription systems are complementary assays for studies investigating transcriptional regulation and, by employing both systems, solid conclusions may be drawn.

Acknowledgments

We are grateful to Dr. Danny Reinberg and Juan Inostroza for assisting us in establishing the reconstituted system, and for generously providing the purified GTFs and RNA polymerase II. This work was partially supported by an NIH grant (#5-RO1-HD29446-01) to C. A.; H. Z. was a recipient of the CABM Award for Graduate Studies, 1994; N. I. was supported by an NIH-Biotechnology training grant (#2-T32-GM08339-06).

[7] Global Steps during Initiation by RNA Polymerase II

By JAY D. GRALLA

The number of polypeptides known to be associated with an RNA polymerase II transcription complex is now more than halfway to triple digits (see [15] in this volume[1]). This enormous complexity makes the problem of working out the mechanism of initiation extremely challenging. It is helpful to attempt to simplify this complex problem initially by breaking the pathway down into substeps, leaving the roles of each of the individual polypeptides to be established later. This chapter discusses a conceptual framework and a range of assays that can be used.

The minimal conceptual framework derives from studies of the simpler prokaryotic transcription systems.[1a] For example, transcription initiation from a typical *Escherichia coli* promoter uses only five polypeptides: the α, β, β', and σ components of RNA polymerase and either an activator or a repressor.[2] However, even in this simple system, there is considerable diversity in how transcription rates are regulated. It has been found useful to subdivide the overall initiation process into three broad steps: closed complex formation, open complex formation, and abortive initiation and promoter clearance. The centerpiece of this pathway is the open complex,

[1] Y. Li, S. Bjorklund, Y.-J. Kim, and R. D. Kornberg, *Methods Enzymol.* **273,** Chap. 15, 1996 (this volume).

[1a] J. D. Gralla, *Methods Enzymol.* **185,** 37 (1990).

[2] J. Collado-Vides, B. Magasanik, and J. D. Gralla, *Microbiol. Rev.* **66,** 371 (1991).

which is a required intermediate in which about 12 to 15 base pairs of DNA are in the form of a single-stranded bubble. Transcription initiation requires that formation of this open complex be preceded by a series of reactions involving closed complexes. Initiation is completed by a series of reactions that convert the open complex to an elongation complex. Finally, in a fourth step, reinitiation occurs, which is mainly an issue for polymerase II transcription, where reinitiation need not precisely recapitulate the initiation pathway.[3,4]

This general way of breaking down the initiation pathway can be applied to RNA polymerase II transcription. However, because the field has been advancing gradually, systematic terminology and standard assays for the various steps are yet to be established. Many assays in the literature involve the formation of what are generically termed "preinitiation complexes." This term does not have a precise meaning but typically refers to what is formed after long incubations in the absence of nucleoside triphosphates (NTPs). As discussed below, fully assembled closed complexes are likely to form under such conditions. If ATP is present during formation of preinitiation complexes, then open complexes are likely to be present. The use of restricted combinations of nucleoside triphosphates in such assays may lead to the presence of abortive initiation complexes.

The detergent sarkosyl has also been used as a challenge reagent to inactivate certain types of RNA polymerase II complexes and therefore to collect specific types of "sarkosyl-resistant" complexes.[4] The composition of these is also not yet defined. To determine what general types of complexes might be present after sarkosyl treatment it is probably best to use the same guidelines: if ATP has been added before sarkosyl, the complexes are likely to be open; if restricted combinations of nucleotides were added, the open complexes have likely initiated; and if no nucleotides were added, the complexes likely remained closed.

This chapter is divided into four sections. In the first two sections, the conditions leading to closed and open complexes are discussed with the focus on the general properties of each kind of complex and on assays for detecting them. The third section presents a similar analysis of the process by which an open complex is converted to an elongation complex. Finally, an overview of the initiation pathway, based on these assays, is presented.

Formation of Closed Complexes

The first overall step, closed complex formation, is basically defined as formation of DNA–protein complexes in which the required DNA unpair-

[3] Y. Jiang and J. D. Gralla, *Mol. Cell. Biol.* **13,** 4572 (1993).
[4] D. K. Hawley and R. G. Roeder, *J. Biol. Chem.* **262,** 3452 (1987).

ing reaction has not yet been accomplished. This complex cannot efficiently initiate transcription because the template strand is not yet exposed and available to be read. In the case of RNA polymerase II, the term refers to a large number of complexes with widely varying composition. The precise composition of closed complexes will no doubt vary with the choice of conditions and experimental system. In experiments involving fractionated systems one can control the composition to a reasonable extent. Studies suggest that formation of a complex containing transcription factors TFIID, TFIIA, and TFIIB is incapable of opening the DNA.[5] It is not known which macromolecular factors added subsequently are necessary to accomplish DNA opening; many of these may add as a single holoenzyme complex containing polymerase (see Ref. 6 for example).

Although not always stated explicitly, closed complex formation is probably the most commonly discussed step as a target for transcriptional regulation. That is, many proposals center on regulators changing the retention or recruitment of factors involved in the steps prior to DNA opening (reviewed in Ref. 7). Thus if a proposal discusses transcription activators "recruiting" via TFIID or TATA-binding protein-associated factors (TAFs) or TFIIA or TFIIB or a holoenzyme containing polymerase one is predominantly discussing altering the stability or composition of closed complexes. In such proposals the role of the activator is essentially to drive closed complex formation.

At this stage of development of the field, the factor requirements for activator-dependent transcription have not been fully established. Because different promoter–activator combinations will have different requirements there is a continuing need for both crude and partially purified transcription systems.[8] Whether the systems are fractionated or crude, the main concern in the study of closed complexes is how to construct the experimental conditions so as to prevent the opening reaction. This mainly involves omitting ATP and, to a lesser extent, other nucleoside triphosphates.[9,10] That is, if an effect of a regulator can be seen under conditions that lack nucleotides then that effect can likely be ascribed to altered closed complex formation. There is some controversy about this nucleotide effect, but this is largely restricted to certain highly purified systems in which certain factor requirements, including that for an activator, can be bypassed (see below and Ref. 11).

[5] W. Wang, J. D. Gralla, and M. Carey, *Genes Dev.* **6,** 1716 (1992).
[6] A. J. Koleske and R. A. Young, *Nature* (*London*) **368,** 466 (1994).
[7] R. Hori and M. Carey, *Curr. Opin. Genet. Dev.* **4,** 236 (1994).
[8] R. Tjian and T. Maniatis, *Cell* **77,** 5 (1994).
[9] Y. Jiang and J. D. Gralla, *J. Biol. Chem.* **270,** 277 (1994).
[10] W. Wang, M. Carey, and J. D. Gralla, *Science* **255,** 450 (1992).
[11] J. A. Goodrich and R. Tjian, *Cell* **77,** 145 (1994).

Currently, there is no perfect way of testing the effect of factors on closed complex assembly. The best procedure is probably the following, which uses the basic HeLa transcription extract but can also be used in fractionated systems.[5] One mixes the basic transcription system with the DNA in the absence of nucleotides. For maximal levels of closed complex formation a long incubation is necessary. This can be determined in a time course but typically 1 hr will suffice; most half-times found so far are in the range of 10 to 30 min.[3,5,12] At this time a large number of closed complexes will have accumulated, which can be assayed by their ability to lead to transcript formation. They are detected by adding a high concentration of NTPs for a short time and then assaying for transcript.

Because opening and initiation appear to be more rapid than closed complex formation, a 2-min NTP pulse may be sufficient.[3] The pulse time should be long enough to allow closed complexes to initiate and produce detectable transcript but not so long as to allow the same DNA to be used twice as a consequence of reinitiation. The pulse time can be determined by a preliminary experiment in which samples are removed at various times after addition of NTPs to closed complexes. One expects a rapid initial signal within 2 min or so, corresponding to a burst of synthesis from the preassembled closed complexes, followed by a more gradual increase, most likely from reinitiation events. If the presence of a factor during the preincubation is observed to lead to altered transcription, then it has likely had its effect during closed complex formation. However, controls in which the factor is added along with the NTPs are necessary to support this point; there should be little effect in this case.

An alternative to the use of an NTP pulse would be treatment with low concentrations of sarkosyl.[4] The complication here is that sarkosyl has varying effects on transcription and these may vary among promoters. Low concentrations can block closed complex assembly, higher amounts block closed complexes from opening and initiating, and still higher amounts have varying effects on elongation complexes. In any given experimental system it may be possible to find a concentration of sarkosyl that inhibits when added before the 1-hr preincubation but does not inhibit after it (for example, 0.025%). In this circumstance, one can add this amount of sarkosyl and stop further formation of closed complexes. The number of closed complexes formed before sarkosyl addition can then be assayed by adding NTPs. This protocol has the advantage of allowing long elongation times, which can maximize the reproducibility of the signal. The protocol also has disadvantages. Primarily, it can be difficult to find a sarkosyl concentration that cleanly distinguishes among the various complexes. The problem is

[12] J. White, C. Brou, J. Wu, Y. Lutz, V. Moncollin, and P. Chambon, *EMBO J.* **11,** 2229 (1992).

made worse in cases where a subset of factors may remain template bound after initiation[3,4]; there may be concentrations of sarkosyl that inactivate free factors but allow reinitiation using bound factors. For these reasons a short NTP pulse that yields a significant level of transcript is probably a superior protocol.

Formation of Open Complexes

A regulator that alters the number of open complexes formed can do this in one of two ways. First, it may directly alter the reactions that open the DNA and keep it opened. Second, it may work indirectly by affecting closed complex formation, which changes the number of complexes available to be opened. The easiest assay is based on the latter property, which determines whether the factor under study alters the ability to reach the open complex stage, but does not further identify the step affected.[13] Separate assay for the effect on closed complex formation can help narrow the possibilities.

The use of permanganate probing is the most straightforward way to detect open complexes.[14] Permanganate oxidizes the DNA bases, primarily thymines, although it reacts with cytosine and to a lesser extent other bases. These reactions are hindered in double-stranded DNA and thus few bases are modified. However, the reaction proceeds readily on single-stranded DNA and thus the bases are readily modified. This is true for many, but not all, thymines that are part of the single-stranded bubble within the open complex. Apparently, permanganate is a small enough reagent to penetrate some protein components of open complexes and react with unpaired nucleotide bases. Strong reactivity of bases in nucleoprotein complexes compared to double-stranded naked DNA controls is indicative of open complex formation.

Assaying whether factors alter open complex formation involves long incubation times and ATP. The long times are necessary to maximize the number of closed complexes that form. ATP is necessary because DNA opening is stimulated by hydrolysis of the beta–gamma (β–γ) bond of ATP.[9,10] Because the ATP-stimulated opening reaction is rapid, one typically performs most of the 1-hr reaction in the absence of ATP and then adds the ATP during the last few minutes. This procedure minimizes ATP-dependent side reactions that may occur during the long preincubation. Permanganate is then added and any of several procedures can be used to

[13] Y. Jiang, S. Triezenberg, and J. D. Gralla, *J. Biol. Chem.* **269,** 5505 (1994).

[14] J. D. Gralla, M. Hsieh, and C. Wong, *in* "Footprinting Techniques for Studying Nucleic Acid–Protein Complexes" (A. Revzin, ed.), p. 107. Academic Press, Orlando, FL, 1993.

detect the reactive bases (see protocol below). The key to reliable interpretation is to have several comparison lanes, including ones without proteins, without activator, and without ATP. The signal in the complete system lane must be significantly greater than in all control lanes to be considered reliable. In that case the region showing high reactivity is taken to be within the melted bubble of the open complex. The lesser reactivity of bases in other regions typically corresponds to the low reactivity with double-stranded DNA.

There should be further controls on the functionality of the open complexes detected.[9,10] The functionality test is simple and uses two comparison samples. In one, all four NTPs are added to allow the open complex to initiate. This should lead to the lessening or the disappearance of the permanganate signal over the start site as the transcription bubble moves downstream during elongation. A second sample should have low amounts of α-amanitin added just prior to NTPs. This should prevent initiation, counteracting the effect of NTPs and thus freezing the transcription bubble in place. If a signal disappears with NTPs and is frozen by amanitin, then the open complex has used polymerase II to initiate transcription and is therefore defined as functional.

There are two primary factors in determining whether open complexes can be detected reliably. First, the promoter should have several thymines overlapping or just prior to the transcription start site region, where melted bases commonly occur in open complexes. Thymines in open complexes are not all highly reactive and thus it helps detection if several are present. This is the reason that open complex formation was detected first at the adenovirus E4 promoter, which contains six consecutive thymines on the nontemplate strand in this region.[10] Second, the signal-to-noise (S/N) ratio will be proportional to the fraction of templates that are in open complexes. Thus one should use low amounts of template. Such low amounts, although they do not maximize the total number of active complexes, do maximize the fraction of templates that are open. In a typical system greater than 90% of templates are not active and increasing the amount of DNA largely raises the background due to reactivity with double-stranded DNA. This lessens the distinction between the 1 to 10% of the DNA that reacts strongly because it is open and the 90 to 99% of the DNA that reacts weakly because it is closed. The use of 2 to 10 ng of DNA is best for open complex assays as is also true for detecting the activator dependence of transcription. Because basal (unactivated) transcription is so much less efficient, the S/N problem has so far precluded detection of opening within basal transcription complexes.

Some transcription systems are contaminated with low levels of nucleotides and this must be taken into account when designing assays. The problem is that long incubations in some cases may allow leaky initiation, thus reducing the number of open complexes that can be detected. This problem is best overcome by including inhibitors of initiation during the incubation. The best choice is α-amanitin, which blocks initiation and thus prevents the loss of open complexes. In addition, it is best to use sources of hydrolyzable ATP that cannot be used efficiently for transcription. These include dATP and cordecypin, both of which can trigger open complex formation but cannot be used for transcription elongation.

The following protocol has been used successfully to detect open complexes at an activated adenovirus E4 promoter variant.

1. Prepare a standard 40-μl reaction containing HeLa extract, 5 or 10 ng of supercoiled DNA template, excess nonpromoter DNA competitor, and appropriate amounts of activator, as determined from a prior transcription titration. Add α-amanitin to 1 μg/ml. Incubate for 30 to 60 min at 30°. Add dATP to 25 μ*M* for the final 2 min. Add potassium permanganate to 4–8 m*M* for 30 sec to 3 min. Stop the reaction by adding 3 μl of 2-mercaptoethanol. Then add 0.1 ml of 0.3 *M* sodium acetate (pH 5.5)–10 m*M* EDTA–0.2% sodium dodecyl sulfate (SDS)–5 μg of carrier RNA.
2. Digest with 0.625 μg/ml of proteinase K for 1 hr at 37°. Extract the modified DNA with hot phenol, phenol–chloroform, and chloroform. Precipitate with ethanol, redissolve in 40–60 μl of water, and pass it through a 1-ml Sephadex G-50 spin column equilibrated in water.
3. Adjust the volume with water to 100 μl including end-labeled primer (the sequence in this case is GCGGCAGCCTAACAGTCAGCCTTACCAGTA) and buffer with 2 m*M* $MgCl_2$ and 0.2 m*M* dNTP. Add 2.5 units of *Taq* polymerase. Run a thermocycler program as follows: 95° for 0.5–1 min, 60° for 1 min, 72° for 1.5 min. Repeat for 30 to 35 cycles.
4. Process the reaction product for electrophoresis as in DNA sequencing. For example, extract with chloroform and precipitate and rinse with ethanol. Run parallel lanes with sequencing markers and control lanes with products of reactions: lacking dATP; with all four NTPs added for 5 min; with all four NTPs but without α-amanitin; and lacking nuclear extract, which will require less vigorous permanganate reactivity conditions, including shorter times and lower concentrations at the lowest edge of the range indicated above.

As mentioned above, studies of certain purified systems that lack an activator requirement have led to uncertainty about the role of ATP in the opening reaction (see Refs. 11 and 15, for example). These studies have not assayed opening directly, but have shown that either productive or

[15] H. T. Timmers, *EMBO J.* **13,** 391 (1994).

abortive initiation may occur in the absence of hydrolyzable ATP. Some, but possibly not all, of this can be explained by the observation that other nucleotides can substitute for ATP when they are present in higher concentration.[9] It is possible that the use of supercoiled DNA[16] and high concentrations of factors may lead to transient opening that is sufficient to support transcription. This issue will not be resolved until the permanganate assay is applied directly to such systems.

Abortive Initiation and Promoter Clearance

After an open complex forms, it must use NTPs to read the open template and begin transcription. There are few mechanistic studies of this complex process. Most assays can be assigned to one of two categories: appearance of short, possibly abortive, transcripts and appearance of long transcripts. The former is a direct assay for the polymerase reaching various positions in the initially transcribed region. The latter is an assay for whether the tested conditions are sufficient to allow the polymerase to leave the promoter and make a functional transcript.

The first assay is typically termed "abortive initiation," modeled after similar reactions developed in prokaryotic systems.[17] In this case a primer, usually a ribodinucleotide complementary to positions -1 and $+1$, is added along with the NTP complementary to position $+2$. Restricted combinations of NTPs that should move the polymerase step by step into the initial transcribed region may also be used.[18] The goal is to learn what factors and conditions are necessary to allow the polymerase to begin transcription.

Because there are so few published examples[11,19–22] it is too soon to provide generalizations or protocols for polymerase II abortive initiation assays. Typically one detects the radioactive trinucleotide that corresponds to the condensation of the dinucleotide primer and the subsequently encoded NTP. In crude systems one uses gel-exclusion columns to enrich transcription complexes or separate them from released abortive products. On the basis of early studies and with guidance from prokaryotic studies, there are useful considerations that can be discussed. So far it is not clear that a primary criterion for a prokaryotic abortive initiation product has been met for polymerase II transcription: the accumulation of excess prod-

[16] J. D. Parvin and P. A. Sharp, *Cell* **73,** 533 (1993).

[17] W. R. McClure, *Annu. Rev. Biochem.* **54,** 171 (1985).

[18] A. J. Carpousis and J. D. Gralla, *J. Mol. Biol.* **183,** 165 (1985).

[19] C. A. Jacob, S. W. Luse, and D. S. Luse, *J. Biol. Chem.* **266,** 22537 (1991).

[20] D. S. Luse and G. A. Jacob, *J. Biol. Chem.* **262,** 14990 (1987).

[21] G. A. Jacob, J. A. Kitzmiller, and D. S. Luse, *J. Biol. Chem.* **269,** 3655 (1994).

[22] A. J. Carpousis and J. D. Gralla, *Biochemistry* **19,** 3245 (1980).

uct over the amount of template present (see Ref. 21 for a discussion). This is what one expects in a true abortive initiation reaction where a single open complex synthesizes excess RNA reiteratively. The phenomenon is difficult to demonstrate convincingly because of the low efficiency of the polymerase II systems currently available. Therefore, alternative criteria have been used, leading to some uncertainty about the meaning of the results.

Because there are typically numerous background bands in abortive initiation assays there must be criteria for specificity of the product assayed. Chief among these are requirements for the correct dinucleotide primer, NTP, and template. An additional requirement should be that the reaction is inhibited by low levels of α-amanitin. However, apparently specific abortive initiation products can be produced in the presence of amanitin.[20,22a] It is not known what type of reaction produces such products and thus extreme caution should be exercised in interpreting the results of reactions that proceed in the presence of α-amanitin. In one case we have found that such reactions proceed under closed complex conditions whereas under open complex conditions (that is, with added dATP) the reaction follows the predicted α-amanitin sensitivity for RNA polymerase II transcription.[22a]

When specific products are detected, the question then arises as to whether they result from stalled elongation complexes or from reiterative abortive initiation reactions. The best criterion is probably to assay if the short RNA species can be chased into longer RNA when the NTPs required for elongation are added. We have used a protocol[22a] in which the incubation begins and is interrupted part way through with the extra NTPs. A true reiterative abortive initiation reaction should behave distinctively in this assay. The addition of all missing NTPs should approximately freeze the amount of a true abortive product. No reduction in amount should be seen; the RNA is not polymerase bound because it has been released during a reiterative reaction. Increases should be modest because the next time an RNA is initiated, it is subject to elongation by the newly added NTPs. By analogy with prokaryotic studies, one expects that polymerase will lose the ability to do reiterative abortive initiation after a few bonds are formed in the mRNA.[22] Preliminary studies with polymerase II support this expectation.[20,22a] In assays of this type any RNA that is lessened in amount after addition of missing NTPs is probably associated with a stalled elongation-type complex.

A separate issue involves promoter clearance, which is not as well defined. In its original form, promoter clearance was said to occur when

[22a] Y. Jiang, M. Yan, and J. D. Gralla, *J. Biol. Chem.*, **270**, 27332 (1996).

the promoter became sufficiently cleared after initiation to allow a new polymerase to bind and reinitiate.[17] Unfortunately, promoter clearance has also become a generic term defining the transition from preinitiation complex to elongation complex. The generic term might include abortive initiation, the downstream movement of the transcription bubble, and the eventual movement of polymerase away from the promoter; only when the latter has proceeded to a certain distance will the promoter truly be cleared.

Assays for the generically defined promoter clearance can be simple transcription assays (see Ref. 11 for example). One forms some type of preinitiation complex and then determines what additional factors are needed to produce transcripts. If a factor is not needed to form a preinitiation complex but is needed to produce transcript it can be said to be acting at the promoter clearance step. This has the virtue of involving a simple assay and the disadvantage that many substeps are included in this somewhat vague definition.

It is much more difficult to assay the actual physical clearance of the promoter because footprinting is hindered by the low efficiency of the systems available. One assay[23] starts with formation of open complexes followed by NTP addition to allow transcription to begin. As discussed above, this leads to the disappearance of the transcription bubble over the start site because it has moved downstream with the elongating polymerase. The next polymerase to enter the cleared promoter can then be trapped by addition of α-amanitin and detected by permanganate assay of the reopened start site. This is an assay for true promoter clearance as it measures what it takes to make the initiated promoter available for a new polymerase. The issue is more than one of semantics because transcription levels are set by how frequently reinitiation occurs, which cannot begin until the promoter is cleared of the first polymerase.

Studies have indicated that a number of factors influence the generically defined promoter clearance step.[11,21,23–27] These include basal transcription factors, especially TFIIE, TFIIF, and TFIIH, ATP, the C-terminal domain of the polymerase, some activators, and the DNA sequence of the promoter. Just as in the case of the opening reaction, certain requirements (for "clearance" in this case) may be partly bypassed, especially in basal systems using supercoiled DNA. The following section attempts to place these various

[23] Y. Jiang, M. Yan, and J. D. Gralla, *Mol. Cell Biol.*, **16,** 1614 (1996).
[24] C. Chang, C. F. Kostrub, and Z. F. Burton, *J. Biol. Chem.* **268,** 20482 (1993).
[25] M. E. Maxon, J. A. Goodrich, and R. Tjian, *Genes Dev.* **8,** 515 (1994).
[26] M. E. Dahmus, *Biochim. Biophys. Acta* **1261,** 171 (1995).
[27] S. Narayan, W. A. Beard, and S. H. Wilson, *Biochemistry* **34,** 73 (1994).

steps and assays in the overall context of the pathway leading to formation of transcript.

Overview of Global Steps in Initiation Pathway

It is useful to review some key steps in the global pathway and place them in perspective. One can consider transcription to occur in a series of four consecutive steps. The first global step begins with the promoter being marked by the binding of TFIID and activators. Other activators, TFIIA, TFIIB, other general factors, and polymerase may join. Eventually a type of closed complex forms that can be a substrate for the DNA opening reaction. The composition of this full closed complex is not known. It can be detected by its ability to open the DNA rapidly and begin transcription on addition of nucleotide substrates. Studies have suggested that activators can assist transcription by stabilizing precursors to these full closed complexes (reviewed in Ref. 7).

In a second global step, a region within the promoter is opened to allow the template strand to be read. The DNA opening reaction is stimulated by hydrolysis of the β–γ bond of ATP, but other nucleotides can substitute *in vitro* if present in high concentration.[9] It is not known which basal factors are required to open the DNA. Phosphorylation of the polymerase C-terminal domain (CTD) seems not to be required.[28] Open complexes are detected using the permanganate assay. Activators have been shown to act at a stage leading to open complex formation but have not yet been shown to affect the opening reaction directly.[13]

In a third global step,[23] the polymerase begins initiation and clears the promoter for reinitiation. This step apparently begins with an abortive initiation phase in which short RNAs are released after they are made. Subsequently, the transcription bubble moves into the initial transcribed region as the 5′ end of the mRNA is extended. The polymerase is then released from the contacts that hold it to the promoter in a reaction that does not require, but is stimulated by, the ATP-dependent phosphorylation of the polymerase C-terminal domain. After the polymerase moves further downstream the promoter becomes cleared. Regulators have also been proposed to act at this third step[29] (reviewed in Ref. 30).

In a fourth global step, a reinitiation transcription complex is assembled over the cleared promoter. The assembly of reinitiation complexes can

[28] Y. Jiang and J. D. Gralla, *Nucleic Acids Res.* **22,** 4958 (1995).
[29] H. Lee, K. W. Kraus, M. F. Wolfner, and J. T. Lis, *Genes Dev.* **6,** 284 (1992).
[30] J. Greenblatt, J. R. Nodwell, and S. W. Mason, *Nature* (*London*) **364,** 401 (1993).

differ from assembly of initiation complexes. This is because certain factors may not be cleared from the promoter in the third step just described.[4] In this circumstance, reinitiation can be faster than initiation because reinitiation need not recapitulate all the steps during assembly of a first-round transcription complex.[3] Because reinitiation produces the bulk of the RNA made, this process has potential importance in uncoupling transcriptional induction regulation from regulation of the levels of RNA produced after induction. The framework established by these four global steps should be useful in designing further studies, as the number of polypeptides identified in transcribing a promoter approaches triple digits.

Acknowledgment

Preparation of this chapter was supported by USHHS Grant GM 49048 and a grant from the NSF.

[8] Purification and Analysis of Functional Preinitiation Complexes

By STEFAN G. E. ROBERTS and MICHAEL R. GREEN

The initiation of transcription in eukaryotes requires the assembly at the promoter of a plethora of proteins in addition to RNA polymerase II (the preinitiation complex).[1] This occurs, at least in higher eukaryotes, in a stepwise manner involving an array of protein–protein and protein–DNA contacts. The pathway of preinitiation complex formation has been elucidated by using purified components and monitoring their assembly on the promoter, primarily by electrophoretic mobility shift assays (Fig. 1). Transcription factor TFIID is the central DNA-binding unit of the preinitiation complex and provides a platform for the assembly of the remaining factors, in the following order: TFIIA, TFIIB, RNA polymerase II/TFIIF, TFIIE, and TFIIH. This assembly is able to perform a basal level of accurately initiated transcription *in vitro*. However, the scheme outlined in Fig. 1 is by no means complete, as nuclear extracts contain several additional factors that participate in transcription. Indeed, a transcription system composed of purified components responds poorly to promoter-specific activator proteins.[1,2] Thus, although the study of transcription using purified

[1] L. Zawel and D. Reinberg, *Annu. Rev. Biochem.* **64,** 533 (1995).
[2] R. E. Kingston and M. R. Green, *Curr. Biol.* **4,** 325 (1994).

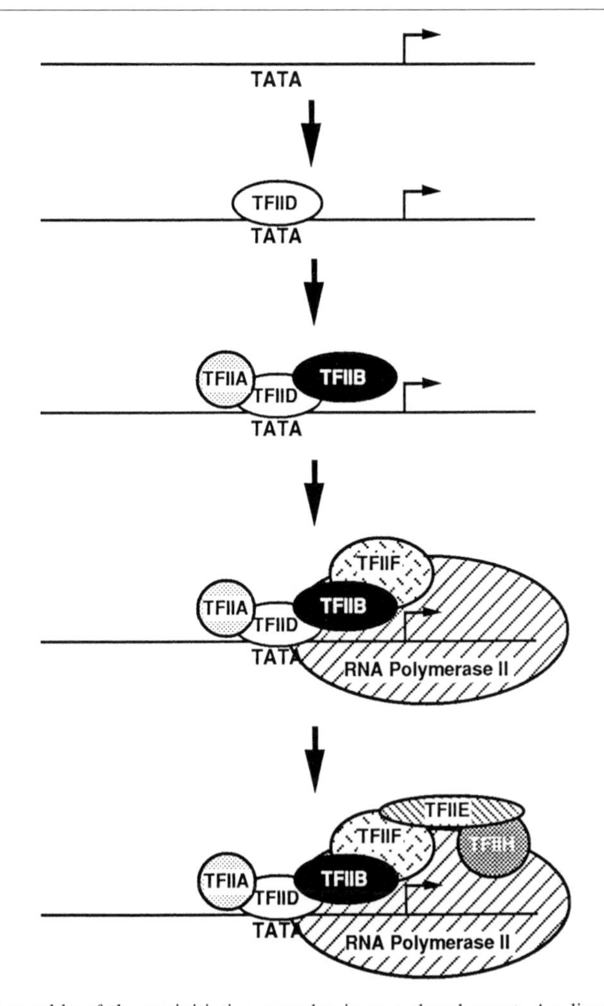

FIG. 1. Assembly of the preinitiation complex is an ordered event. As discussed in text, not all of the factors involved are shown. Some of the protein–protein and protein–DNA contacts are drawn arbitrarily.

proteins yields valuable information, the conclusions are limited because several auxiliary factors are absent. Furthermore, it is questionable whether the ratios of the components in purified transcription systems accurately reflect those found in an untreated nuclear extract. These considerations are crucial for the study of limiting steps in preinitiation complex assembly and the mechanism of transcriptional activation.[2]

A more thorough analysis of transcriptional regulation requires tech-

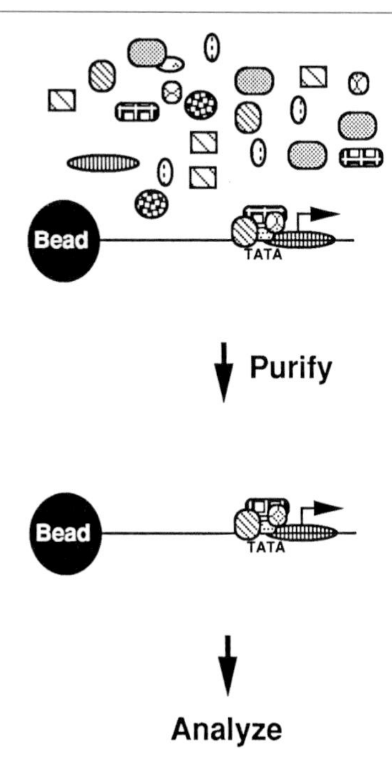

FIG. 2. Preinitiation complexes and intermediates can be purified using an immobilized DNA template. After incubation in a nuclear extract, the DNA is separated from the unbound material and can be analyzed either in a transcription assay or by immunoblotting with antibodies against specific factors.

niques that can be used to monitor preinitiation complex assembly in a crude nuclear extract. In this chapter, we describe methods that were developed for this purpose. Moreover, these methods are linked to functional transcription assays and therefore have additional applications in the study of transcriptional regulation. We describe two different techniques by which this can be achieved. The first method utilizes a linear DNA template immobilized on a solid support that facilitates separation of preinitiation complexes from unbound material (outlined in Fig. 2).[3,4] The second method involves the formation of transcription complexes on a supercoiled DNA template that are resolved from unbound factors by gel-filtration chromatography.[4,5] The latter method is more labor intensive and is not generally appropriate for batch processing of samples or staged incubations involving

[3] Y.-S. Lin and M. R. Green, *Cell* **64,** 971 (1991).
[4] B. Choy and M. R. Green, *Nature* (*London*) **366,** 531 (1993).
[5] M. F. Carey, S. P. Gerard, and N. R. Cozzarelli, *J. Biol. Chem.* **261,** 4309 (1986).

more than one purification step. For these reasons, most of our experiments have used the immobilized DNA template method.

Materials

Buffers

Buffer D: 20 m*M* *N*-2-hydroxyethylpiperazine-*N*′-2-ethanesulfonic acid (HEPES; pH 8.0), 20% (v/v) glycerol, 100 m*M* KCl, 0.2 m*M* EDTA, 1 m*M* dithiothreitol (DTT), 0.5 m*M* phenylmethylsulfonyl fluoride (PMSF).

Transcription buffer: 12 m*M* HEPES (pH 8.0), 12% (v/v) glycerol, 60 m*M* KCl, 0.12 m*M* EDTA, 7.5 m*M* $MgCl_2$, 0.5 m*M* DTT, 0.5 m*M* PMSF

Specific Components

Streptavidin–agarose (Sigma, St. Louis, MO)
Dynabeads M280, metal beads (Dynal, Lake Success, NY)
Magnetic bead collector
Biotin-dNTP
Sepharose CL-2B
Gravity flow column (1 × 30 cm)

Purification of Preinitiation Complexes Using Immobilized DNA Template

To immobilize promoter DNA we exploit the high affinity of streptavidin for biotin. A biotin moiety is incorporated into promoter DNA, usually by a Klenow "fill-in" reaction of a 5′ overhang with commercially available biotin–dNTP. Alternatively, promoter fragments can be produced by polymerase chain reaction (PCR) using a biotinylated oligonucleotide primer. Commercially available sources of streptavidin linked to solid supports are used to immobilize the DNA template. We have used streptavidin–agarose and also streptavidin-coated metal beads. Streptavidin–agarose can be collected by mild centrifugation (2000 rpm at 4° for 30 sec), while the metal beads are collected by attraction of the beads to a magnetic source.

Immobilization of DNA

Typically, we immobilize the biotinylated DNA such that 10 μl of streptavidin–agarose or 2 μl of metal beads is sufficient for a single *in vitro* transcription reaction. Biotinylated DNA is incubated with the streptavidin–agarose (or metal beads) in a buffer containing 1 *M* NaCl, 10 m*M* Tris–HCl (pH 7.40), and 0.2 m*M* EDTA for 2 hr with agitation at room temperature. The beads are then washed three times with the above buffer and twice with transcription buffer. The DNA–beads are stored in transcription buffer at 4° and are stable for at least 1 month.

Formation of Transcriptionally Competent Complexes on Immobilized DNA

To form preinitiation complexes a standard transcription reaction[6] is mixed in a microcentrifuge tube, but ribonucleotide triphosphates (rNTPs) are omitted to prevent initiation.

Mix:

DNA–streptavidin agarose beads (10 μl; or 2 μl of DNA–metal beads)

$MgCl_2$ (100 m*M*), 3 μl

HeLa cell nuclear extract (10 mg/ml in buffer D), 25 μl

Bring to 40 μl with sterile H_2O.

The sample is incubated at 30° for 30 min with occasional agitation of the tubes to keep the beads suspended. The beads are collected by centrifugation at 2000 rpm for 30 sec (agarose) or by using a magnetic particle collector (metal beads). Wash the beads four times with transcription buffer containing 0.003% v/v Nonidet P-40 (NP-40) (to prevent the beads from sticking to the sides of the tube). The efficacy of the immobilized preinitiation complexes can be analyzed by transcription assay as follows: Resuspend the beads in 40 μl of transcription buffer. Add 1 μl of 25 m*M* rNTP mix and incubate at 30° for 30 min. Transcripts are detected by primer extension as described previously.[6]

Figure 3 shows transcription assays with the immobilized DNA template G5E4T[6] in the absence or presence of the activator protein GAL4-AH. Transcription assays with both agarose and metal beads are shown. Lanes 1 and 2 in Fig. 3 show transcription assays in which the complexes were not purified, but were initiated immediately after formation by the addition of rNTPs. As can be seen, GAL4-AH efficiently stimulates transcription from the immobilized G5E4T template. Lanes 3 and 4 in Fig. 3 show transcription assays in which the complexes were purified prior to initiation. By comparison, it is clear that functional preinitiation complexes are amenable to purification by this method. Also, the agarose and metal beads produce equivalent results.

Analysis of Content of Preinitiation Complexes

The immobilized DNA template can be used to purify preinitiation complexes for analysis by immunoblotting. However, we find that the aga-

[6] Y.-S. Lin, M. F. Carey, M. Ptashne, and M. R. Green, *Cell* **54,** 659 (1988).

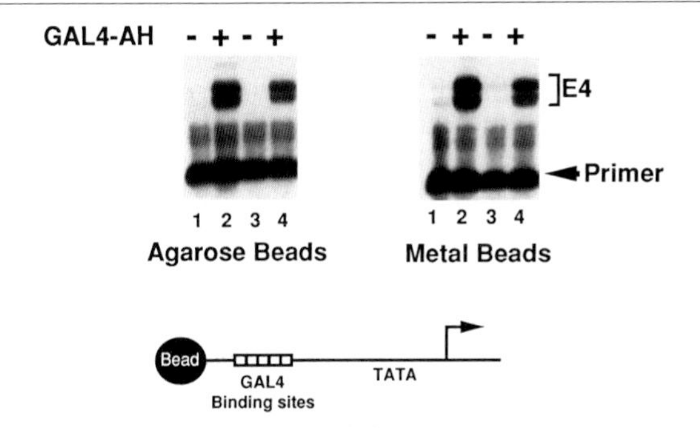

FIG. 3. Transcription of immobilized DNA templates. The DNA template G5E4T was immobilized on either agarose or metal beads coated with streptavidin. After incubating the immobilized DNA template with nuclear extract in the absence or presence of the transcriptional activator GAL4-AH, ribonucleotides were added immediately (lanes 1 and 2), or following purification from unbound material (lanes 3 and 4). Transcripts (E4) were detected by primer extension. The immobilized DNA template G5E4T is shown below.

rose beads bind proteins nonspecifically, resulting in high background levels when individual transcription factors are detected by immunoblotting. Thus, in the following method we use only the magnetic beads. Complexes are purified as described above, but instead of initiating transcription by the addition of rNTPs, the beads are resuspended in sodium dodecyl sulfate-polyacrylamide gel electrophoresis (SDS–PAGE) loading dye and then subjected to SDS–PAGE and immunoblotting with antibodies against components of the transcriptional machinery. Figure 4 shows immunoblots detecting the presence of the general transcription factor TFIIB, which has

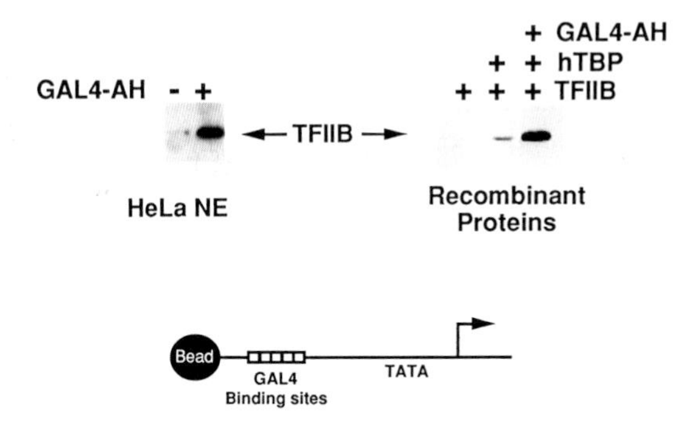

FIG. 4. The DNA template G5E4T immobilized on metal beads (shown at bottom) was incubated with either a HeLa cell nuclear extract or recombinant proteins in the absence or presence of GAL4-AH. Complexes were then purified and analyzed by Western blotting with an anti-TFIIB antibody.

been shown to be recruited to the promoter by activator proteins.[3,4,7,8] The left-hand side in Fig. 4 shows a recruitment assay using HeLa cell nuclear extract. The right-hand side in Fig. 4 uses purified components to show that in a three-protein system (GAL4-AH, TATA-binding protein, and TFIIB) an activator can recruit TFIIB. Thus, this method is applicable to assays using both crude nuclear extracts and purified proteins.

Troubleshooting

Some proteins may nonspecifically bind to the metal beads in the absence or presence of the DNA template. The following modifications can be used to determine empirically the conditions required to minimize background. (1) The beads can be preincubated in transcription buffer containing bovine serum albumin (BSA) to block nonspecific binding sites; (2) the concentration of KCl in the incubations and washes can be increased up to 150 mM; (3) competitor DNA can be added to the incubation [we have used both plasmid DNA fragments and poly(dG · dC)].

Additional Applications

Because the immobilized DNA template can be washed of unbound factors several times, staged complex reactions can be performed. This has been of invaluable use in the study of transcriptional activation.[3,4,7–9] Also, we have used this assay to determine which factors need to reassemble at the promoter during multiple rounds of transcription initiation.[7] Immobilized DNA templates have also been effectively utilized to study the fate of general transcription factors during the transition from initiation to elongation.[10]

Purification of Preinitiation Complexes by Gel Filtration

Alternatively, we have used a method to purify preinitiation complexes that was originally applied to the purification of RNA polymerase III transcription units.[5] However, this method is more time consuming than the immobilized DNA template method and is not appropriate for batch examination of samples. The method employs gel-filtration chromatography to separate the macromolecular preinitiation complex from free proteins. The DNA template is incubated in supercoiled form with HeLa cell nuclear

[7] S. G. E. Roberts, B. Choy, S. S. Walker, Y.-S. Lin, and M. R. Green, *Curr. Biol.* **5,** 508 (1995).
[8] T. K. Kim and R. G. Roeder, *Proc. Natl. Acad. Sci. U.S.A.* **91,** 4170 (1994).
[9] B. M. Shykind, J. Kim, and P. A. Sharp, *Genes Dev.* **9,** 1354 (1995).
[10] L. Zawel, K. P. Kumar, and D. Reinberg, *Genes Dev.* **9,** 1479 (1995).

extract and then layered over a Sepharose CL-2B Column. The preinitiation complexes are present in the early flowthrough fractions of the column. Small DNA fragments are included in the incubation, to adsorb nonspecific interactions and ensure that they are retained by the column (we routinely use *Hae*III/*Hpa*II-digested pGEM3). The fractions are of suitable volume for the initiation of transcription, achieved simply by the addition of ribonucleotides. However, for immunoblotting, the complexes need to be precipitated prior to SDS–PAGE.

Method

Pack a 1 × 30 cm gravity flow column with Sepharose CL-2B preequilibrated with transcription buffer. The column can be calibrated by fractionating a supercoiled DNA template–DNA fragments mix in 200 μl of transcription buffer. Once this has entered the column, transcription buffer is added. Fractions (0.5 ml) are collected to a total of two column volumes and a sample of each is analyzed by agarose gel electrophoresis. The supercoiled DNA template is in the early flowthrough fractions, while the short DNA fragments elute much later. Because transcription complexes elute close to the supercoiled DNA template alone, this information can be used to choose fractions for transcription assays/immunoblotting.

1. Assemble transcription complexes as follows:

HeLa cell nuclear extract	100 μl (10 mg/ml in buffer D)
100 m*M* $MgCl_2$	12 μl
Supercoiled DNA template	1 μg
DNA fragments	10 μg

Add sterile H_2O to a final volume of 160 μl.

2. Incubate at 30° for 30 min.
3. Place the sample over a preequilibrated column and allow the sample to enter the column; then add the transcription buffer. Collect and keep 0.5-ml fractions according to the calibration (above).
4. Fractions can be directly used for transcription assays by adding ribonucleotides. However, for immunoblot analysis, precipitate the complexes with 10% (w/v) trichloroacetic acid–0.1% (w/v) deoxycholate, leaving on ice for 30 min. Wash the pellet with acetone, dry, and resuspend in SDS–PAGE loading buffer.

As stated above, this method is more labor intensive and also requires regular calibration of the column. In addition, a column is needed for the processing of each sample. Thus, for most applications the immobilized DNA template would be the more appropriate choice. However, the gel-filtration method may find a niche in instances where a supercoiled DNA template is required, for example, in some minimal transcription systems.

Acknowledgments

We thank Joe Reese and Alan Whitmarsh for comments on the manuscript. S.G.E.R. is supported by a senior postdoctoral fellowship from the Massachusetts division of the American Cancer Society. This work was supported by a grant from the NIH to M.R.G.

Section III

RNA Polymerase and Its Subunits in Prokaryotes

[9] Reconstitution of RNA Polymerase

By NOBUYUKI FUJITA and AKIRA ISHIHAMA

The DNA-dependent RNA polymerase (EC 2.7.7.6) of *Escherichia coli* is composed of core enzyme (subunit composition $\alpha_2\beta\beta'$) and one of at least six molecular species of the σ subunit (σ^D, σ^N, σ^S, σ^H, σ^F, and σ^E). Despite the complexity of its subunit composition and its extraordinarily large molecular size, RNA polymerase is one of the few large molecular assemblies that can be successfully reconstituted from isolated individual subunits. The first successful attempts were made in the early 1970s,[1,2] soon after the discovery of its subunit composition.[3] In these classic procedures, the subunits were isolated from purified RNA polymerase by cellulose-acetate gel electrophoresis[1] or by successive column chromatography,[2] after dissociation of purified RNA polymerase with high concentrations (typically 6*M*) of urea. These reconstitution systems, in combination with genetic studies of RNA polymerase mutants, were successfully applied to identify functions of each subunit (reviewed in Ref. 4). The RNA polymerase core enzyme is assembled in the order $2\alpha \rightarrow \alpha_2 \rightarrow \alpha_2\beta \rightarrow \alpha_2\beta\beta'$ (premature core) $\rightarrow$ E (active core enzyme) (reviewed in Refs. 5 and 6). The pathway of subunit assembly was established first by use of the *in vitro* reconstitution system[5,6] and then confirmed *in vivo* by both kinetic studies and analysis of assembly-defective mutants.[7]

The second breakthrough was brought about two decades later. Establishment of transient overexpression and purification procedures for the individual subunits[8,9] allowed ready production of 10-mg quantities of each subunit. By employing these systems, it is now possible to prepare highly purified reconstituted enzyme in a few days. The *in vitro* reconstitution is now a powerful tool for the molecular anatomy of RNA polymerase. One

[1] A. Heil and W. Zillig, *FEBS Lett.* **11,** 165 (1970).
[2] A. Ishihama and K. Ito, *J. Mol. Biol.* **72,** 111 (1972).
[3] R. R. Burgess, *J. Biol. Chem.* **244,** 6168 (1969).
[4] T. Yura and A. Ishihama, *Annu. Rev. Genet.* **13,** 59 (1979).
[5] A. Ishihama, *Adv. Biophys.* **14,** 1 (1981).
[6] A. Ishihama, *Adv. Biophys.* **26,** 19 (1990).
[7] A. Ishihama, M. Taketo, T. Saitoh, and A. Ishihama, *in* "RNA Polymerase" (R. Losick and M. Chamberlin, eds.), pp. 475–502. Cold Spring Harbor Laboratory Press, Cold Spring Harbor, NY, (1976).
[8] K. Zalenskaya, J. Lee, C. N. Gujuluva, Y. K. Shin, M. Slutsky, and A. Goldfarb, *Gene* **89,** 7 (1990).
[9] K. Igarashi and A. Ishihama, *Cell* **65,** 1015 (1991).

can easily introduce mutations in one of the cloned subunit genes, overexpress mutant subunits, reconstitute mutant enzymes, and analyze the altered functions associated with the mutant enzymes. This *in vitro* approach is particularly useful for the analysis of structure–function relationships in such essential enzymes as RNA polymerase because even a lethal mutation can be analyzed, and it has been successfully applied (for examples, see Refs. 8–14).

Subunit Preparation

All the core subunits (α, β, and β') and the major σ subunit (σ^{70} or σ^{D}) can be expressed to high levels in *E. coli* cells by using the T7 RNA polymerase–T7 promoter system. After cell disruption, β, β', σ^{70}, and most of their derivatives so far examined are recovered in insoluble inclusion body fractions, but can be solubilized using high concentrations of protein denaturants such as urea and guanidine hydrochloride. The solubilized proteins can be used for reconstitution directly or after purification by passing them through a DEAE column. In contrast, the α subunit and most of its derivatives are recovered in native form either in the soluble fraction or in the particulate fraction, from which the protein can be extracted by increasing the salt concentration without using protein denaturants.

Wild-Type α Subunit

Escherichia coli strain BL21(DE3) transformed with the expression plasmid pGEMAX185[9] is grown at 37° with vigorous shaking in 200 ml of LB medium containing ampicillin (200 μg/ml). At a density of 1×10^8 cells/ml (30 Klett units), expression of the α gene is induced by the addition of isopropyl-β-D-thiogalactopyranoside (IPTG) at a final concentration of 0.4 m*M*, and growth is continued for another 3 hr. Under this induction condition, subunit α reaches nearly 50% of the total cellular proteins (the intracellular content of α in wild-type *E. coli* is 0.2–0.3% the level of total proteins). The level of plasmid-encoded α is more than 100-fold higher than that of chromosome-encoded α (as measured using plasmid-encoded mutant α with different migration rates from wild-type α). Cells are harvested, washed with 10 m*M* Tris–HCl (pH 8.0 at 4°), 1 m*M* EDTA, and 150 m*M* NaCl, and stored frozen at −80° until use.

[10] J. Lee, M. Kashlev, S. Borokhov, and A. Goldfarb, *Proc. Natl. Acad. Sci. U.S.A.* **88,** 6018 (1991).

[11] C. Zou, N. Fujita, and A. Ishihama, *Mol. Microbiol.* **6,** 2599 (1992).

[12] M. Kimura, N. Fujita, and A. Ishihama, *J. Mol. Biol.* **242,** 107 (1994).

[13] H. Tang, A. Severinov, A. Goldfarb, D. Fenyo, B. Chait, and R. Ebright, *Genes Dev.* **8,** 3058 (1994).

[14] M. Kimura and A. Ishihama, *J. Mol. Biol.* **254,** 342 (1995).

Frozen cells are suspended in 3 ml of lysis buffer A [50 m*M* Tris–HCl (pH 8.0 at 4°) and 1 m*M* EDTA]. The cell suspension is transferred to a 15-ml conical disposable tube and, after adding 8 μl of 100 m*M* phenylmethylsulfonyl fluoride (PMSF) in 2-propanol and 80 μl of 10-mg/ml lysozyme in the lysis buffer, is incubated for 20 min on ice with occasional agitation. The cells are lysed by sonication and centrifuged at 10,000 *g* for 10 min at 4°. The precipitates are suspended in 1 ml of extraction buffer A (lysis buffer plus 0.5 *M* NaCl), and after a 20-min incubation on ice the α subunit is recovered in the supernatant by centrifugation at 10,000 *g* for 10 min at 4°. The extraction with extraction buffer A is repeated twice more. To the pooled extract (total volume, 3 ml), 1.5 vol of saturated ammonium sulfate is added. The α subunit is recovered in the precipitate after centrifugation at 10,000 *g* for 20 min at 4°. For maximum recovery of the α subunit, an overnight incubation at 4° is essential before the centrifugation. The precipitates are dissolved in 2 ml of TGED buffer [10 m*M* Tris–HCl (pH 7.6 at 4°), 5% (v/v) glycerol, 0.1 m*M* EDTA, and 0.1 m*M* dithiothreitol (DTT)] and dialyzed against TGED buffer containing 0.1 *M* NaCl. Precipitates formed during the dialysis are removed by centrifugation. Nearly half the initial amount of α subunit is lost in this step because of coprecipitation with the impurities. If a higher yield is desired, the precipitates are resolubilized in the extraction buffer, dialyzed again, and the soluble fraction is pooled. Alternatively, before adding ammonium sulfate the 0.5 *M* NaCl extract is treated with Polymin P at a final concentration of 0.1% (v/v), incubated on ice for 1 hr, and the precipitates are removed by centrifugation.

The fraction remaining soluble after dialysis is applied to a DEAE column [Protein Pak G-DEAE (Waters, Milford, MA) or an equivalent] attached to a high-performance liquid chromatography (HPLC) system. An open column (1.5 × 10 cm) of DE52 (Whatman, Maidstone, England) or DEAE-Toyopearl 650M (Tosoh, Tokyo, Japan) can also be used with essentially the same efficiency. Proteins are eluted with a linear gradient of 0.1 to 0.5 *M* NaCl in TGED buffer. Peak fractions containing α subunit are pooled, dialyzed against a storage buffer [10 m*M* Tris–HCl (pH 7.6 at 4°), 10 m*M* $MgCl_2$, 0.1 m*M* EDTA, 1 m*M* DTT, 50% (v/v) glycerol, and 0.2 *M* KCl], and stored at either −30 or −80° in small aliquots. The yield of α is about 5 mg, and the purity is more than 95%.

Mutant α Derivatives

Most of the mutant derivatives of α, including those with deletions, insertions, and amino acid substitutions, can be expressed to high level from the corresponding plasmids derived from pGEMAX185. In some particular cases, however, the significant basal expression of mutant derivatives occurring even in the absence of IPTG is highly toxic to the bacterial cells, and the plasmid cannot be stably maintained in the transformed cells. In such cases, for example, with amino-terminal and internal deletions,[12] stable transformation and high-level expression can be achieved by introduction of a *lac* operator sequence downstream of the T7 promoter and the use of a *lacI*^{q} host strain such as JM109(DE3), in order to minimize the level of basal expression.

Most, if not all, of the mutant derivatives with amino acid substitutions[11,14] or carboxy-terminal and short amino-terminal deletions[9,12] are recovered either in a salt-extractable precipitate, as for the wild-type α subunit, or in the soluble fraction. For purification of the soluble α derivatives, the cell lysate obtained by the procedure already described is centrifuged at 10,000 *g* for 10 min at 4° to remove cell debris. The supernatant is further centrifuged at 100,000 *g* for 1.5 hr at 4° to remove the ribosomal fraction. Proteins are then precipitated by adding 1.5 vol of saturated ammonium sulfate to the supernatant, and after more than 1 hr of incubation on ice, centrifuged at 10,000 *g* for 20 min. The precipitates are dissolved in 2 ml of TGED buffer and dialyzed against TGED buffer containing 0.1 *M* NaCl. After precipitates formed during dialysis are removed by centrifugation at 10,000 *g* for 10 min at 4°, the supernatant is applied to a column (1.5 × 10 cm) of DE52 or DEAE-Toyopearl 650M that has been equilibrated with TGED buffer containing 0.1 *M* NaCl. The column is eluted with a linear gradient of 0.1 to 0.5 *M* NaCl in TGED buffer. Proteins in the peak fractions are concentrated, if necessary, by precipitating with ammonium sulfate and redissolving in a small volume of TGED buffer, dialyzed against the storage buffer, and stored at −30 or at −80° in small aliquots. Mutant α derivatives thus obtained contain some minor impurities, but are usually pure enough for the following reconstitution experiments. Impurities can be completely removed by HPLC purification of the reconstituted core enzyme. When further purification of α derivatives alone is necessary at this step, one can use either gel-permeation HPLC (Protein Pak 300 from Waters, TSK gel G-3000SW from Tosoh, or an equivalent) or heparin–agarose column chromatography, except that some α derivatives cannot be retained on a heparin–agarose column.

Some α derivatives, including large amino-terminal and internal deletions.[12] form tight inclusion bodies. For purification of these derivatives, the frozen cells are suspended in 3 ml of lysis buffer B (lysis buffer A plus 0.1 *M* NaCl), and treated with PMSF and lysozyme as described above. After a 20-min incubation on ice, the cells are lysed by adding 40 μl of 8% (w/v) sodium deoxycholate. After a 20-min incubation on ice with occasional mixing, the lysate is homogenized by gentle sonication. The inclusion bodies are recovered by centrifugation at 10,000 *g* for 10 min at 4°, and resuspended in 1 ml of extraction buffer B [50 m*M* Tris–HCl (pH 8.0 at 4°), 10 m*M* EDTA, 0.1 *M* NaCl, and 0.5% (v/v) Triton X-100]. After a 5-min incubation at room temperature, the suspension is centrifuged at 10,000 *g* for 10 min at 4°. The precipitates are washed again by resuspending in 1 ml of the same buffer, followed by centrifugation. The washed precipitates are resuspended in 1 ml of dissociation buffer [50 m*M* Tris–HCl (pH 8.0 at 4°), 1 m*M* EDTA, 10 m*M* DTT, 0.2 *M* KCl, 10 m*M* $MgCl_2$, 20% (v/v) glycerol, and 6 *M* urea], and centrifuged at 100,000 *g* for 2 hr at 4°. The supernatant is diluted 10-fold with TGED buffer containing 6 *M* urea to reduce the salt concentration, and applied to a DEAE column (Protein Pak G-DEAE or an equivalent) in an HPLC system. Proteins are eluted with a linear gradient of 0.02 to 0.25 *M* NaCl in TGED buffer containing 6 *M* urea. Peak fractions are pooled and stored in small aliquots at −80°.

β and β′ Subunits

The β and β′ subunit genes are organized in a single operon and are expressed coordinately. It is therefore convenient to overexpress both β and β′ from a single expression plasmid. The plasmid pGEMBC[9] contains both the β and β′ genes under the control of the T7 promoter, and can be used for their high-level and coordinate expression. Because this plasmid is unstable in BL21(DE3) cells, care should be taken in storage and propagation of the transformed cells. For reproducible and high-level expression, it is best to store pGEMBC in the DNA form and use it to transform BL21(DE3) cells *de novo* in each experiment. A fresh, small colony of transformants (not more than a few days old) is picked up in the morning, inoculated into 200 ml of LB medium containing ampicillin (200 μg/ml), grown at 37° with vigorous shaking to a cell density of 1×10^8 cells/ml (30 Klett units), and induced by adding IPTG at a final concentration of 0.4 m*M*. Growth is then continued for another 1.5 hr. All these procedures should be done in 1 day. Alternatively, the growth in liquid medium is suspended at early logarithmic phase, the culture is stored overnight at 4°, and then growth is resumed on the next day by adding fresh medium. Cells are harvested, washed with 10 m*M* Tris–HCl (pH 8.0 at 4°), 1 m*M* EDTA, and 150 m*M* NaCl, and stored at −80°.

For purification of the mixture of β and β′ subunits, the frozen cells are suspended in 1.5 ml of lysis buffer B, transferred to a 15-ml disposable tube, and treated with 4 μl of 100 m*M* PMSF and 40 μl of lysozyme (10 mg/ml). After a 20-min incubation on ice, the cells are lysed by adding 20 μl of 8% (w/v) sodium deoxycholate and incubating for 20 min on ice with occasional mixing. The lysate is homogenized by gentle sonication, transferred to an Eppendorf tube, and centrifuged at 10,000 *g* for 10 min at 4°. The inclusion bodies are suspended in 1 ml of extraction buffer B, incubated for 5 min at room temperature, and then centrifuged at 10,000 *g* for 10 min at 4°. The precipitates are washed again by resuspending in 1 ml of the same buffer, followed by centrifugation. The β and β' subunits are extracted by resuspending the precipitates in 0.4 ml of dissociation buffer, incubated for 30 min on ice, and then centrifuged at 10,000 *g* for 10 min at 4°. The extraction is repeated once more with 0.4 ml of the same buffer, and the pooled supernatant (0.8 ml) is further centrifuged at 100,000 *g* for 2 hr at 4°. The supernatant is stored in small aliquots at $-80°$. This preparation contains slightly less of the β' subunit than of the β subunit. If necessary, the two subunits can be separated from each other by either DEAE column chromatography[2] or phosphocellulose column chromatography[15] in the presence of 6 *M* urea. The yield of $\beta\beta'$ subunits is at least 2.5 mg at this step.

Because the mixture of β and β' subunits thus obtained (more than 90% pure as judged by sodium dodecyl sulfate-polyacrylamide gel electrophoresis followed by Coomassie blue staining) is used for reconstitution experiments without further purification, the procedure described above should be carried out carefully. The purity and integrity of the β and β' subunits at this step (and hence the efficiency of reconstitution of active enzyme complex) are affected by several factors. A high-level induction of β and β' subunits is a prerequisite for minimizing contamination of other cellular components in the final preparation. Although the induction level is considerably lower than is attainable with the other subunits (α and σ), reproducible results are obtained by carefully following the procedure described above. Thorough washing of inclusion bodies with the buffer containing Triton X-100 is essential for high reconstitution efficiency. Cyanate, which accumulates in urea solutions, inactivates RNA polymerase subunits.[16] The β' sununit is particularly sensitive to cyanate, and the use of aged urea solution typically results in the accumulation of $\alpha_2\beta$ subassembly, which is a major by-product of the reconstitution reaction. For this reason, urea solution used for the extraction of β and β' subunits should be deionized prior to use. This can be conveniently done by storing 8 *M* urea solution at 4° over a layer of ion-exchange resin (AG501-X8; Bio-Rad, Richmond, CA). For maximum recovery of enzyme activity, solutions containing urea should always be kept cool because high temperature accelerates the formation of cyanate.

Another expression system for the β and β' subunits has been developed by Zalenskaya and co-workers[8] (see also [10] in this volume[16a]). In this

[15] L. R. Yarbrough and J. Hurwitz, *J. Biol. Chem.* **249,** 5400 (1974).

[16] K. Ito and A. Ishihama, *J. Mol. Biol.* **79,** 115 (1973).

[16a] H. Tang, Y. Kim, K. Severinov, A. Goldfarb, and R. H. Ebright, *Methods Enzymol.* **273,** Chap. 10, 1996 (this volume).

case the β and β' genes are separately expressed from two expression plasmids, pXT7β and pT7β', respectively. After extraction from inclusion bodies with a buffer containing 6 *M* guanidine hydrochloride, the β and β' subunit preparations are mixed and further purified by chromatography on a Sephacryl 300 column in the presence of 6 *M* guanidine hydrochloride. Although it would take longer for the expression and purification steps, this system is a better choice if the β or β' subunit gene is to be analyzed.

σ^{70} *Subunit*

The σ^{70} subunit can be easily overexpressed and purified even using a conventional expression system.[17] By the use of the T7 expression system, however, the σ^{70} subunit after a few hours of induction accounts for more than 50% of total cellular proteins, and a single-column chromatography gives a preparation pure enough for most experiments.

Escherichia coli strain BL21(DE3) transformed with the expression plasmid pGEMD[9] or pGRM70[18] is grown, induced, harvested, washed, and stored at −80° as described for the α subunit. Frozen cells from 200 ml of the induced culture are suspended in 3 ml of lysis buffer B, transferred to a 15-ml disposable tube, and treated with 8 μl of 100 m*M* PMSF and 80 μl of lysozyme (10 mg/ml). After a 20-min incubation on ice, the cells are lysed by adding 40 μl of 8% (w/v) sodium deoxycholate and incubating for 20 min on ice with occasional mixing. The lysate is homogenized by gentle sonication, transferred to two Eppendorf tubes, and centrifuged at 10,000 *g* for 10 min at 4°. Inclusion bodies in each tube are suspended in 1 ml of extraction buffer B, incubated for 5 min at room temperature, and then centrifuged at 10,000 *g* for 10 min at 4°. The precipitates are washed again by resuspending in 1 ml each of the same buffer followed by centrifugation. The precipitates containing σ^{70} subumit are dissolved in 1 ml each of TGED buffer containing 6 *M* guanidine hydrochloride, and then dialyzed against TGED buffer containing 0.2 *M* NaCl at 4°. σ^{70} subunit is known to renature instantaneously and quantitatively on removal of guanidine hydrochloride by either dialysis or dilution.

The solution containing renatured σ^{70} is centrifuged at 10,000 *g* for 10 min at 4° to remove precipitates formed during dialysis, diluted with an equal volume of TGED buffer to reduce salt concentration, and then applied to a column (1.5 × 10 cm) of DE52 or DEAE-Toyopearl 650M that has been equilibrated with TGED buffer containing 0.1 *M* NaCl. Proteins are eluted with a linear gradient of 0.1 to 0.5 *M* NaCl in TGED buffer. Peak fractions containing σ^{70} subunit are pooled, concentrated by precipi-

[17] M. Gribskov and R. R. Burgess, *Gene* **26,** 109 (1983).

[18] A. Kumar, R. A. Malloch, N. Fujita, D. A. Smillie, A. Ishihama, and R. S. Hayward, *J. Mol. Biol.* **232,** 406 (1993).

tating with 60% saturation of ammonium sulfate if necessary, dialyzed against the storage buffer, and stored at −30 or at −80° in small aliquots. The yield of σ^{70} is approximately 10 mg.

All derivatives of the σ^{70} subunit so far examined, except for one extensive carboxy-terminal deletion, form inclusion bodies on induction.[19] They can be purified by exactly the same procedure as described for the wild-type σ subunit. In the case of the long carboxy-terminal deletion (σ^{70}-529), most of the σ derivative is recovered in the soluble fraction. It can be purified from the supernatant fraction after cell lysis by ultracentrifugation, ammonium sulfate precipitation, DEAE column chromatography, and gel-permeation HPLC, as described for the soluble α derivatives.

Reconstitution

Core enzyme can be reconstituted by mixing the α, β, and β' subunits under denaturing conditions (in the presence of either 6 *M* urea or guanidine hydrochloride), and then removing the denaturant by dialysis against the appropriate reconsitution buffer. Reconstitution buffer typically contains 10–20% (v/v) glycerol and 0.2–0.3 *M* KCl, which are essential for preventing nonspecific aggregation of the subunits and for achieving a high efficiency of reconstitution. Protein concentration is another important factor affecting the reconstitution efficiency. At protein concentrations higher than 0.5 mg/ml, large subunits (β' in particular) tend to aggregate by themselves on removal of the denaturant, resulting in the accumulation of assembly intermediates, i.e., α dimer and $\alpha_2\beta$ subassembly. Low concentrations of the proteins are preferable especially when guanidine hydrochloride is used as the denaturant.

The core enzyme reconstituted at low temperature is known to exist as an inactive premature from.[20] It can be activated by heating at 30° for 30–60 min. Although the activated core enzyme can be used directly for many purposes, it contains considerable amounts of free subunits and $\alpha_2\beta$ subassembly (and, in some cases, nucleic acids derived from the inclusion bodies). For quantitative experiments, therefore, further purification of the reconstituted enzyme is essential. For example, DEAE column chromatography on an HPLC system can be used to separate core enzyme from $\alpha_2\beta$ subassembly, free subunits and other contaminating proteins. Free β and β' subunits, if present, form a broad peak that is partially overlapping with the core enzyme peak. Addition of a slightly excess amount of α subunit

[19] A. Kumar, B. Grimes, N. Fujita, K. Makino, R. A. Malloch, R. S. Hayward, and A. Ishihama, *J. Mol. Biol.* **235,** 405 (1994).

[20] A. Ishihama, R. Fukuda, and K. Ito, *J. Mol. Biol.* **79,** 127 (1973).

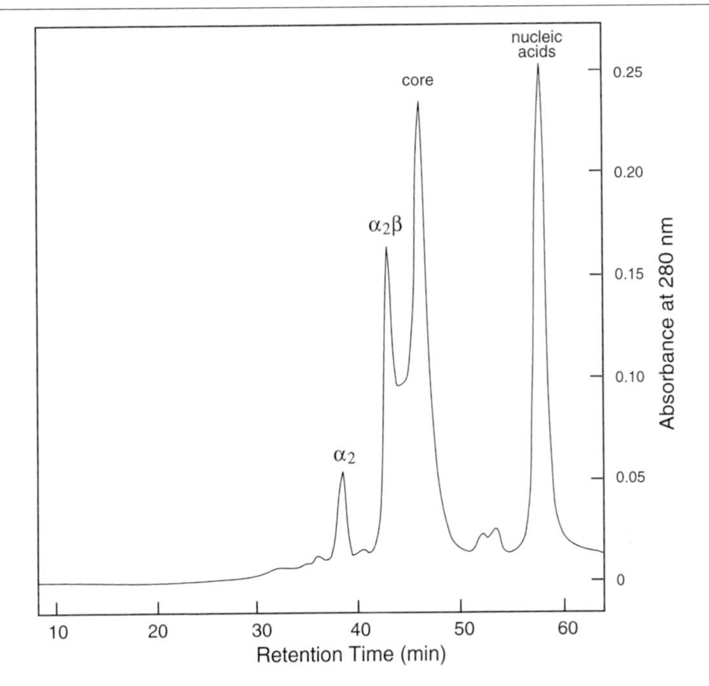

FIG. 1. HPLC purification of the reconstituted core enzyme.

in the reconstitution reaction largely prevents the appearance of free β and β' subunit.

Holoenzyme can be reconstituted easily by adding a two- to fourfold molar excess of σ^{70} subunit to the purified core enzyme in appropriate buffer, e.g., the storage buffer, and incubating for a few minutes at 30°. Because σ^{70} subunit is known to facilitate the maturation process of the inactive core complex,[20,21] it is a good alternative[8] to add σ^{70} subunit in the maturation step described above.

In a typical reconstitution experiment, 1 mg of the $\beta\beta'$ mixture, which contains about 3 nmol each of β and β' subunits, is mixed with 0.24 mg (6.6 nmol) of the α subunit, and diluted to a final volume of 2.5 ml with the dissociation buffer. The mixture is dialyzed at 4° against reconstitution buffer [50 m*M* Tris–HCl (pH 7.8 at 4°), 0.1 m*M* EDTA, 1 m*M* DTT, 0.3 *M* KCl, and 20% (v/v) glycerol]. After dialysis, the mixture is incubated for 30 min at 30° to activate the premature core enzyme. After dilution by adding 2 vol of TGED buffer to reduce the salt concentration, the sample is applied to a DEAE column (Protein Pak G-DEAE from Waters or an

[21] R. Fukuda and A. Ishihama, *J. Mol. Biol.* **87,** 523 (1974).

equivalent) in an HPLC system (Fig. 1). Proteins are eluted with a linear gradient of 0.1 to 0.7 *M* NaCl in TGED buffer. Three protein peaks (α dimer, $\alpha_2\beta$ subassembly, and core enzyme in this order) are usually observed, followed by a peak of contaminating nucleic acids. Peak fractions containing the core enzyme are pooled, dialyzed against the storage buffer, and stored at either -30 or at $-80°$ in small aliquots. The recovery of the core enzyme after the purification step is typically 20–30%.

Acknowledgment

We thank Richard S. Hayward for critical reading of the manuscript.

[10] *Escherichia coli* RNA Polymerase Holoenzyme: Rapid Reconstitution from Recombinant α, β, β', and σ Subunits

By Hong Tang, Younggyu Kim, Konstantine Severinov, Alex Goldfarb, and Richard H. Ebright

Escherichia coli RNA polymerase holoenzyme (RNAP) has the subunit composition $(\alpha)_2\beta\beta'\sigma$.[1,2] It is possible to reconstitute RNAP from individual recombinant α, β, β', and σ^{70} subunits.[3] Reconstitution of RNAP from recombinant subunits permits construction of mutant RNAP derivatives, including lethal-mutant RNAP derivatives, and therefore is a powerful tool for genetic analysis of RNAP structure and function.[4–14]

[1] M. Chamberlin, *in* "RNA Polymerase" (R. Losick and M. Chamberlin, eds.), p. 17. Cold Spring Harbor Laboratory, Cold Spring Harbor, NY, 1976.

[2] R. Burgess, *in* "RNA Polymerase" (R. Losick and M. Chamberlin, eds.), p. 69. Cold Spring Harbor Laboratory, Cold Spring Harbor, NY, 1976.

[3] K. Zalenskaya, J. Lee, N. G. Chandrasekhar, Y. Shin, M. Slutsky, and A. Goldfarb, *Gene* **89,** 7 (1990).

[4] M. Kashlev, J. Lee, K. Zalenskaya, V. Nikiforov, and A. Goldfarb, *Science* **248,** 1006 (1990).

[5] A. Mustaev, M. Kashlev, J. Lee, A. Polyakov, A. Lebedev, K. Zalenskaya, M. Grachev, A. Goldfarb, and V. Nikiforov, *J. Biol. Chem.* **266,** 23927 (1991).

[6] J. Lee, M. Kashlev, S. Borokhov, and A. Goldfarb, *Proc. Natl. Acad. Sci. U.S.A.* **88,** 6018 (1991).

[7] K. Igarashi and A. Ishihama, *Cell* **65,** 1015 (1991).

[8] C. Zou, N. Fujita, K. Igarashi, and A. Ishihama, *Mol. Microbiol.* **6,** 2599 (1992).

[9] E. Martin, V. Sagitov, E. Burova, V. Nikiforov, and A. Goldfarb, *J. Biol. Chem.* **267,** 20175 (1992).

[10] K. Severinov, M. Soushko, A. Goldfarb, and V. Nikiforov, *J. Biol. Chem.* **268,** 14820 (1993).

Conventional procedures for reconstitution of RNAP from recombinant subunits have been extremely labor intensive, requiring prereconstitution column purification of α, β, β', and σ, and postreconstitution column purification of RNAP (up to eight chromatographic columns).[15] We have reported a 1-day, no-column procedure for reconstitution of RNAP from recombinant subunits.[16] In this improved procedure, hexahistidine-tagged recombinant α subunit purified by batch-mode metal ion-affinity chromatography is incubated with crude recombinant β, β', and σ^{70} subunits from inclusion bodies, and the resulting RNAP is purified by batch-mode metal ion-affinity chromatography.[16] This improved procedure permits reconstitution of RNAP within 21–24 hr, starting with cell pellets of bacterial cultures producing α, β, β', and σ^{70}. (Preparation of near-homogeneous recombinant α requires 1–2 hr from harvesting of bacterial cultures. Preparation of crude recombinant β, β', and σ^{70} requires a total of 1–2 hr from harvesting of bacterial cultures. Reconstitution and postreconstitution purification require 19–20 hr.) Because this improved procedure has no column chromatography steps, it permits parallel processing of multiple samples (e.g., multiple mutants), and it can be scaled up easily for production of multimilligram or gram quantities of RNAP.

Methods

Preparation of Hexahistidine-Tagged Recombinant α

N-Terminally hexahistidine-tagged α (encoded by plasmid pHTT7f1-NHα; Table I) and C-terminally hexahistidine-tagged α (encoded by plasmid pHTT7f1-CHα; Table I) are prepared by identical procedures and function equally well in reconstitution.[16]

Escherichia coli strain BL21(DE3)[17] (Novagen, Inc.) transformed with plasmid pHTT7f1-NHα or plasmid pHTT7f1-CHα is shaken at 37° in 5 ml

[11] K. Severinov, M. Kashlev, E. Severinova, I. Bass, K. McWilliams, E. Kutter, V. Nikiforov, L. Snyder, and A. Goldfarb, *J. Biol. Chem.* **269,** 14254 (1994).

[12] H. Tang, K. Severinov, A. Goldfarb, D. Fenyo, B. Chait, and R. Ebright, *Genes Dev.* **8,** 3058 (1994).

[13] M. Kimura, F. Nobuyuki, and A. Ishihama, *J. Mol. Biol.* **242,** 107 (1994).

[14] K. Severinov, W. Ross, H. Tang, D. Fenyo, B. Chait, R. Gourse, R. Ebright, and A. Goldfarb, (submitted for publication).

[15] S. Borukhov and A. Goldfarb, *Protein Exp. Purif.* **4,** 503 (1993).

[16] H. Tang, K. Severinov, A. Goldfarb, and R. Ebright, *Proc. Natl. Acad. Sci. U.S.A.* **92,** 4902 (1995).

[17] F. W. Studier, A. Rosenberg, J. Dunn, and J. Dubendorff, *Methods Enzymol.* **185,** 60 (1990).

TABLE I
PLASMIDS

Plasmid	Relevant characteristics	Ref.
pHTT7f1-NHα	Ap^R; ori-pBR322; ori-f1; *ϕ10P-rpoA(H6,Nter)*[a]	16
pHTT7f1-CHα	Ap^R; ori-pBR322; ori-f1; *ϕ10P-rpoA(H6,Cter)*[b]	16
pMKSe2	Ap^R; ori-pBR322; *lacP-rpoB*	10
pT7β'	Ap^R; ori-pBR322; *ϕ10P-rpoC*	3
pHTT7f1-σ	Ap^R; ori-pBR322; ori-f1; *ϕ10P-rpoD*	16

[a] *rpoA(H6,Nter)* is a derivative of *rpoA* having a nonnative hexahistidine coding sequence immediately after the *rpoA* start codon.

[b] *rpoA(H6,Cter)* is a derivative of *rpoA* having a nonnative hexahistidine coding sequence immediately before the *rpoA* stop codon.

of LB[18] plus ampicillin (200 μg/ml) until OD_{600} 0.7, induced by addition of isopropylthio-β-D-galactoside to 1 m*M*, and shaken for an additional 3 hr at 37°. The culture is harvested by centrifugation (16,000 *g*; 2 min at 4°), the cell pellet is resuspended in 0.4 ml of buffer A [20 m*M* Tris–HCl (pH 7.9), 500 m*M* NaCl, 5 m*M* imidazole], cells are lysed by sonication, and the lysate is cleared by centrifugation (16,000 *g*; 15 min at 4°). The sample is adjusted to 1 ml with buffer A, and hexahistidine-tagged α is precipitated by addition of $(NH_4)_2SO_4$ to 60%, collected by centrifugation (16,000 *g*; 20 min at 4°), and redissolved in buffer B [6 *M* guanidine hydrochloride, 20 m*M* Tris–HCl (pH 7.9), 500 m*M* NaCl, 5 m*M* imidazole]. The sample is adsorbed onto 0.2 ml of Ni^{2+}-NTA agarose (Qiagen, Chatsworth, CA) in buffer B, washed twice with 1 ml of buffer B, washed twice with 0.6 ml of buffer B plus 30 m*M* imidazole, and eluted with 0.6 ml of buffer B plus 500 m*M* imidazole. Adsorption, washes, and elution are performed in a siliconized 1.7-ml microcentrifuge tube with incubations of 1 min at 4° with gentle mixing, and with removal of supernatants after centrifugation (1500 *g*; 1 min at 22°). The typical yield is 100–300 μg (protein estimated by Bradford assay using bovine serum albumin as standard[19]), and the typical purity is >95%.

Preparation of Crude Recombinant β, β', and σ^{70}

Inclusion bodies containing β, β', and σ^{70} are isolated from induced cultures of transformants of plasmid pMKSe2, plasmid pT7β', and plasmid pHTT7f1-σ, respectively (Table I), using a modification of the procedure of ref. 15.

[18] J. Miller, "Experiments in Molecular Genetics." Cold Spring Harbor Laboratory, Cold Spring Harbor, NY, 1972.

[19] M. Bradford, *Anal. Biochem.* **72,** 248 (1976).

Escherichia coli strain XL1-Blue (Stratagene, La Jolla, CA) transformed with plasmid pMKSe2, *E. coli* strain BL21(DE3)[17] transformed with plasmid pT7β', and *E. coli* strain BL21(DE3)[17] transformed with plasmid pHTT7f1-σ are shaken at 37° in 1 liter of LB[18] plus ampicillin (200 μg/ml) until OD_{600} 0.5–0.6, induced by addition of isopropylthio-β-D-galactoside to 1 m*M*, and shaken for an additional 3 hr at 37°. Cultures are harvested by centrifugation (3000 *g*; 15 min at 4°), and cell pellets are resuspended in 16 ml of buffer C [40 m*M* Tris–HCl (pH 7.9), 300 m*M* KCl, 10 m*M* EDTA, 1 m*M* dithiothreitol, 1 m*M* phenylmethylsulfonyl fluoride] plus lysozyme (0.2 mg/ml) and 0.2% (w/v) sodium deoxycholate. After 20 min at 4°, cells are lysed by sonication. Inclusion bodies containing β, β', and σ^{70} are isolated by centrifugation (38,000 *g*; 30 min at 4°), washed once with 16 ml of buffer C plus 0.2% (w/v) *n*-octyl-β-D-glucoside, and washed once with 12 ml of buffer C. Washes are performed with incubations of 10 min at 4°, followed by sonication, followed by centrifugation (38,000 *g*; 30 min at 4°). Washed inclusion bodies containing β, β', and σ^{70} are solubilized in buffer D [6 *M* guanidine-HCl, 50 m*M* Tris–HCl (pH 7.9), 10 m*M* $MgCl_2$, 10 μ*M* $ZnCl_2$, 1 m*M* EDTA, 10 m*M* dithiothreitol, 10% (v/v) glycerol], and protein concentrations are adjusted to ~10 mg/ml (protein estimated by Bradford assay using bovine serum albumin as standard[19]). The typical yields are 30–40 mg of β, 50–60 mg of β', and 60–70 mg of σ^{70}, and the typical purities are 50–90%.

If desired, cell pellets can be stored at −80° (stable for at least 1 week). In addition, if desired, washed inclusion bodies can be resuspended in buffer C plus 10% (v/v) glycerol and stored at −80° (stable for at least 1 month). However, after solubilization in buffer D, crude β, β', an σ^{70} are not stable and must be used immediately.

Preparation of RNA Polymerase

Two reconstitution mixtures are prepared. The first reconstitution mixture (2 ml) contain 60 μg of N- or C-terminally hexahistidine-tagged α, 300 μg of crude β, and 600 μg of crude β' in buffer D. The second reconstitution mixture (0.3 ml) contains 450 μg of crude σ^{70} in buffer D. The reconstitution mixtures are dialyzed separately for 16 hr at 4° against two 500-ml changes of buffer E [50 m*M* Tris–HCl (pH 7.9), 200 m*M* KCl, 10 m*M* $MgCl_2$, 10 μ*M* $ZnCl_2$, 1 m*M* EDTA, 5 m*M*-2-mercaptoethanol, 20% (v/v) glycerol] (UH020/25 collodion membranes; Schleicher & Schuell, Keene, NH). Following dialysis, the reconstitution mixtures are combined, incubated for 45 min at 30°, and cleared by centrifugation (16,000 *g*; 10 min at 4°). The sample is adsorbed onto 0.1 ml of Ni^{2+}-NTA agarose (Qiagen) in buffer F [50 m*M* Tris–HCl (pH 7.9), 0.5 m*M* EDTA, 5% (v/v) glycerol], washed three times with 1.5 ml of buffer F plus 5 m*M* imidazole, and eluted in 0.25 ml of buffer F plus 150 m*M* imidazole. Adsorption, washes, and elution are performed in a siliconized 2.0-ml microcentrifuge tube with incubations of 45 min at 4° with gentle mixing (15 sec for washes), and with removal of supernatants after centrifugation (16,000 *g*; 2 min at 4°). The resulting sample is concentrated to 50 μl by centrifugal ultrafiltration [Centricon-100 filter units (Amicon, Danvers, MA); 1000 *g*; 30 min at 4°], mixed with 50 μl of glycerol, and stored at −20°. The typical yield is 100–300 μg, and the typical purity is >95%.

General Comments

With this procedure, the typical efficiency of reconstitution and postreconstitution purification is fully 30–90% (i.e., fully 30–90% of α added to the reconstitution mixture is recovered in RNAP). The typical overall yield is 3–10 mg of RNAP per liter of bacterial cultures producing α, β, β', and σ^{70}. RNA polymerase prepared by this procedure is indistinguishable from RNAP reconstituted by conventional methods with respect to subunit stoichiometry, basal transcription, UP-element-dependent transcription, and CAP-dependent transcription.[16]

Acknowledgments

This work was supported by National Institutes of Health Grants GM41376 and GM51527 to R.H.E. and GM30717 to A.G. We thank Drs. T. Gaal, W. Ross, and R. Gourse for discussion.

[11] σ Factors: Purification and DNA Binding

By ALICIA J. DOMBROSKI

The sigma (σ) subunit of prokaryotic RNA polymerases provides specificity for transcription initiation by directing polymerase to the appropriate promoter DNA sequences.[1,2] A large number of different σ subunits, which are related to the *Escherichia coli* primary σ (σ^{70}), have been described.[3] Members of this group are characterized by four regions of amino acid

[1] R. R. Burgess and A. A. Travers, *Fed. Proc. Fed. Am. Soc. Exp. Biol.* **29,** 1164 (1970).
[2] R. R. Burgess, A. A. Travers, J. J. Dunn, and E. K. F. Bautz, *Nature* (*London*) **221,** 43 (1969).
[3] M. Lonetto, M. Gribskov, and C. A. Gross, *J. Bacteriol.* **174,** 3843 (1992).

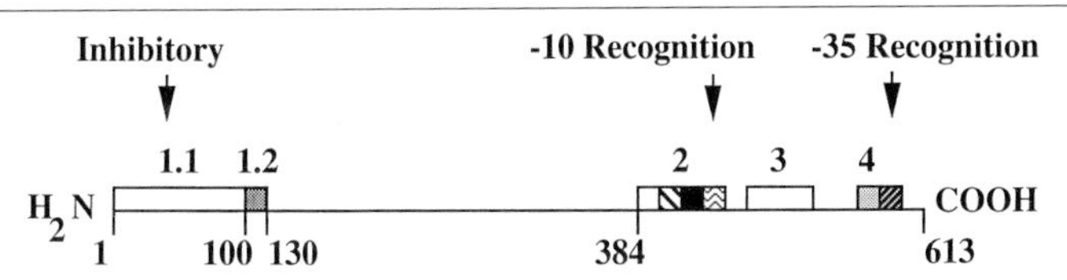

FIG. 1. Linear diagram of σ^{70}. The most highly conserved regions within the σ^{70} family of proteins are designated 1.1, 1.2, 2, 3, and 4. Subregions are indicated by variations in shading.

similarity (Fig. 1). Region 2 interacts with the promoter at the conserved −10 hexamer, while region 4 interacts at the −35 hexamer.[4–9] Interestingly, the ability of σ to bind to DNA *in vitro,* in the absence of the core polymerase subunits, varies depending on the structure of the σ polypeptide. For example, DNA binding by σ^{70} is not observed unless the amino-terminal inhibitory domain (approximately the first 50 amino acids of region 1.1) is removed.[9,10] Alternative σ factors, such as *Bacillus subtilis* σ^{K} and *Salmonella typhimurium* σ^{28}, which lack the inhibitory domain, bind to DNA in full-length form.[10] Another family of σ factors, related to *E. coli* σ^{54}, contains members that are quite dissimilar to the σ^{70} family in both structure and function. Unlike σ^{70}, full-length σ^{54} is capable of binding to DNA.[11,12] This chapter focuses on purification of σ polypeptides using affinity tags, and analysis of DNA binding *in vitro.*

Plasmids for Overexpression of σ Polypeptides

Traditional protein purification protocols have been used for separating σ from the rest of the RNA polymerase complex.[13–16] These methods are satisfactory for purification of large quantities of a single protein, but are

[4] T. Gardella, H. Moyle, and M. M. Susskind, *J. Mol. Biol.* **206,** 579 (1989).
[5] D. A. Siegele, J. C. Hu, W. A. Walter, and C. A. Gross, *J. Mol. Biol.* **206,** 591 (1989).
[6] P. Zuber, J. Healy, H. L. Carter III, S. Cutting, C. P. Moran, Jr., and R. Losick, *J. Mol. Biol.* **206,** 605 (1989).
[7] D. Daniels, R. Zuber, and R. Losick, *Proc. Natl. Acad. Sci. U.S.A.* **80,** 8075 (1990).
[8] C. Waldburger, T. Gardella, R. Wong, and M. M. Susskind, *J. Mol. Biol.* **215,** 267 (1990).
[9] A. J. Dombroski, W. A. Walter, M. T. Record, Jr., D. A. Siegele, and C. A. Gross, *Cell* **70,** 501 (1992).
[10] A. J. Dombroski, W. A. Walter, and C. A. Gross, *Genes Dev.* **7,** 2446 (1993).
[11] M. Buck and W. Cannon, *Nature* (*London*) **358,** 422 (1992).
[12] W. Cannon, F. Claverie-Martin, S. Austin, and M. Buck, *Mol. Microbiol.* **8,** 287 (1993).
[13] M. Gribskov and R. R. Burgess, *Gene* **25,** 167 (1983).
[14] B.-Y. Chang and R. H. Doi, *J. Bacteriol.* **172,** 3257 (1990).
[15] L. Duncan and R. Losick, *Proc. Natl. Acad. Sci. U.S.A.* **90,** 2325 (1993).
[16] L. H. Nguyen, D. B. Jensen, and R. R. Burgess, *Protein Expression Purif.* **4,** 425 (1993).

less practical if rapid purification of several σ derivatives is required. Methods that utilize tagging and overexpression of σ, followed by affinity resin purification, have allowed rapid preparation of a large number of different derivatives of σ.[9,10]

Commercial vector systems are now commonly used to attach affinity "tags" to proteins. The two systems discussed here result in the attachment of either glutathione S-transferase (GST) or several histidine residues (His-tag) to the σ polypeptide at the amino terminus. Vectors that introduce a maltose-binding protein tag (New England Biolabs, Beverly, MA) or a thioredoxin tag (Invitrogen, San Diego, CA) are also available.

The pGEX plasmid vectors (Pharmacia, Piscataway, NJ) are designed to generate an in-frame fusion between GST and the protein to be purified.[17] Expression of the GST fusions is driven by the *E. coli tac* promoter. A copy of the *lacI^q* gene on the same plasmid keeps the promoter repressed until inducer is added. The fusions are induced by addition of isopropyl-β-D-thiogalactopyranoside (IPTG), and then purified in a batch process using glutathione–agarose affinity beads (sulfur linkage; Sigma, St. Louis, MO). The fusion proteins typically contain a specific protease cleavage site between GST and σ, to permit removal of the GST moiety. The advantages of this system are as follows: (1) many different *E. coli* strains can be used as the host because *lacI* is carried on the same plasmid as the fusion protein; (2) σ variants that are rapidly degraded can often be stabilized during purification by the presence of GST; (3) GST can aid in maintaining protein solubility during purification; and (4) GST often does not interfere with the function of the fused protein. The disadvantages are that (1) for certain applications, the 26-kDa GST moiety must be removed because it interferes with subsequent analysis, and (2) in some cases, when σ and GST are separated, the released σ is less soluble than the original GST–σ fusion.

One alternative to GST fusions is addition of several histidine residues to the σ polypeptide. The pET vectors (Novagen, Madison, WI) are designed with expression of the histidine-tagged protein under the control of a bacteriophage T7 promoter. T7 RNA polymerase is provided from an IPTG-inducible construction that is carried within the bacterial chromosome on a lambda (λ) phage lysogen. A protease cleavage site is situated between the His-tag and the σ polypeptide. The advantages of this system are that (1) toxic proteins can be kept effectively repressed by inclusion of a second plasmid carrying the T7 lysozyme gene (T7 lysozyme inhibits any T7 polymerase that is present due to leaky expression), and (2) the His-tag usually does not interfere with protein function. The disadvantages are that (1) some of the vectors have a limited selection of restriction

[17] D. B. Smith and K. S. Johnson, *Gene* **67,** 31 (1988).

enzyme sites available for cloning, and (2) T7 RNA polymerase must be provided *in trans.*

Another set of plasmid vectors that can be used to generate histidine-tagged σ proteins is the pQE series from the QIAexpress system (Qiagen, Chatsworth, CA). In this case the fusion proteins are under the control of the bacteriophage T5 promoter and a *lac* operator. Induction is achieved by addition of IPTG. The advantage of this system are as follows: (1) there is a large number of restriction sites available for cloning, and (2) translational stop codons are present in every reading frame. The disadvantages are that (1) no protease cleavage sites are provided for removal of the histidines (for some applications, the presence of the His-tag interferes with analysis of protein structure and/or function, (2) *lacI* must be provided *in trans,* and (3) leaky expression from this promoter, even in the presence of the *lacI^q* allele, can be substantial. This leaky expression can usually be controlled by including 2% (w/v) glucose in the medium. We have added a thrombin protease cleavage site between the *Bam*HI and *Sac*I restriction sites of pQE30 to generate pQE30T, allowing cleavage of the fusion polypeptide between the His-tag and the σ polypeptide.

With any of these overexpression systems, it is possible that the fusion proteins will be unstable *in vivo.* The BL21 family of *E. coli* strains (Novagen), which are deficient for two proteases (lon and ompT), have been useful for stabilizing several σ fusion proteins.

Overexpression and Purification of σ Polypeptides

GST–σ Fusion Proteins

First, a small-scale induction (5 ml) with several candidate overexpression plasmids is used to verify that the fusion protein is expressed. An *E. coli* strain, such as XL1-blue {*recA1* [F′,*proAB, lacI^{q}Z* ΔM15, Tn*10*(tet)]} (Stratagene, La Jolla, CA), carrying the GST–σ plasmid, is grown overnight in a 37° shaker in LB medium plus ampicillin (100 μg/ml) and tetracycline (20 μg/ml). The culture is diluted 1:10 in fresh LB plus ampicillin, and grown again until midlog phase (OD_{600} ~0.5). The IPTG is added to a final concentration of 0.1–0.2 m*M* and shaking continued for 3–4 hr. Culture samples are taken before addition of IPTG and at the end point of the induction for gel analysis of the efficiency of induction. This is done by measuring the OD_{600} of the culture at the desired time and removing a volume (μl) that corresponds to $1/OD_{600} \times 100$. The cells are pelleted in a microcentrifuge for 5 min at 25°, and resuspended in 75 μl of sodium dodecyl sulfate (SDS) gel loading buffer [0.25 *M* Tris (pH 6.8), 4%

(w/v) SDS, 20% (v/v) glycerol, 2% (v/v) 2-mercaptoethanol, 0.003% (w/v) bromphenol blue]. After boiling for 3–5 min, a 20-μl aliquot is loaded onto an SDS-polyacrylamide gel (usually 10–12% for σ derivatives). *Note:* σ factors often exhibit anomalous mobility on SDS gels.

Overexpression of σ subunits typically results in the formation of inclusion bodies that localize to the insoluble fraction.[13,14,16] Partial polypeptides of σ are generally also found in the insoluble fraction.[9,10] Thus, prior to purification it is necessary to determine the solubility of the fusion protein. A small aliquot of induced culture is lysed, centrifuged, and samples from both the supernatant and pellet fractions analyzed by SDS-polyacrylamide gel electrophoresis (SDS–PAGE). If the fusion protein is in the soluble fraction, then the subsequent urea denaturation and renaturation steps can be omitted, and the supernatant can be applied directly to the resin.

Large-scale inductions for protein purification are carried out in a 1-liter volume, resulting in 1–2 g of wet cell pellet, which can be stored at −70°. A portion of the cell pellet (0.5–1.0 g) is resuspended in 12 ml of cell lysis buffer [50 m*M* Tris (pH 7.9), 0.1 *M* NaCl, 0.1 m*M* EDTA, 0.01% (v/v) Triton X-100] on ice and lysed using either a cell disruption bomb (Parr Instruments, Moline, IL) or a French press, at 4°. Two passes through the cell lysis step result in more complete lysis. The cell lysate is centrifuged at 12,000 rpm in a Beckman JA-17 rotor (~20,000 *g*), at 4° for 10–20 min. If the fusion protein is soluble, then the supernatant can be used directly in the affinity resin purification step. If the GST–σ is found in the cell pellet it must be extracted first. The best results have been obtained by denaturation of the protein with urea (guanidine hydrochloride also works but some σ polypeptides precipitate on renaturation). The pellet is resuspended in 2 ml of 6 *M* urea in denaturation buffer [50 m*M* Tris (pH 7.9), 1 m*M* EDTA, 8 m*M* dithiothreitol (DTT), 0.01% (v/v) Triton X-100] and vortexed. Centrifugation at 7500 rpm in a Beckman (Palo Alto, CA) JA-17 rotor (~8000 *g*) at 4° for 10 min yields denatured σ in the supernatant. The protein is renatured by dialysis into reconstitution buffer [50 m*M* Tris (pH 7.9), 1 m*M* EDTA, 20% (v/v) glycerol, 1 m*M* DTT, 0.01% (v/v) Triton X-100] at 4° with four changes of buffer (at least 250 ml each), for at least 1 hr each. At this point the volume of the sample is approximately 1 ml.

Glutathione–agarose beads (50% solution) can be prepared in the meantime by resuspending 400 mg of beads in 10 ml of MTPBS buffer [150 m*M* NaCl, 16 m*M* Na_2HPO_4, 4 m*M* NaH_2PO_4 (pH 7.3)]. *Note:* MTPBS is prepared and stored as a fivefold concentrated solution. The beads are pelleted by centrifuging at 1000 rpm in a Beckman JA-17 rotor (~150 *g*) for 5 min at 4° and washed three more times in the same manner, then finally resuspended in 10 ml of MTPBS and stored at 4°. An equal volume of 50% glutathione–agarose beads (Sigma) is added to the renatured protein in a 15-ml centrifuge tube and the mixture incubated on ice for 10 min with occasional gentle mixing. The beads are pelleted by centrifugation at 1000 rpm for 5 min at 4°. Contaminating proteins are removed by washing the beads three times with MTPBS (10 ml each time, followed by centrifugation). The GST–σ fusion protein is eluted with 1 vol of 5 m*M* glutathione in 50 m*M* Tris, pH 8.0, twice. The eluted protein is dialyzed into storage buffer [16 m*M* Na_2HPO_4, 4 m*M* NaH_2PO_4 (pH 7.3), 15% (v/v) glycerol, 0.01% (v/v) Triton X-100]. *Note:* For DNA-binding analysis, the protein is stored at 4° in 15% (v/v) glycerol because higher levels of glycerol have been found to interfere with filter retention assays. For long-term storage at −20 to −70°, the protein is dialyzed into the same buffer, but containing 50% (v/v) glycerol. The protein concentration is determined using the Bio-Rad (Richmond, CA) protein assay, or any other protein quantification method. Usually 10–20 μl of the final preparation is sufficient for the Bio-Rad microassay. Finally, between 0.5 and 2 μg of the protein is visualized using SDS–PAGE and Coomassie blue staining to determine approximate degree of purity. Prior to storage, bovine serum albumin (BSA) should be added to 100 μg/ml. Do not add BSA if the affinity tag is to be removed. Yields vary depending on the particular fusion protein, but can be as high as 2 mg.

His-Tag–σ Fusion Proteins

Induction of protein overexpression is basically the same as for the GST–σ fusions, with two notable exceptions. First, to minimize leaky expression under noninducing conditions for the pQE vectors, an overnight culture is prepared in medium containing 2% (w/v) glucose. Second, overexpression is induced by addition of IPTG to a final concentration of 2 m*M* (10-fold higher than for the GST fusions). A portion of the cell pellet (0.5–1.0 g) is resuspended in 15 ml of binding buffer [5 m*M* imidazole, 0.5 *M* NaCl, 20 m*M* Tris (pH 7.9)] plus 1 m*M* phenylmethylsulfonyl fluoride (PMSF, protease inhibitor) on ice, lysed and centrifuged as described for GST fusion proteins. *Note:* Binding buffer can be prepared and stored at 4° as an eightfold concentrated solution, while PMSF is prepared at 0.2 *M* in ethanol and stored at −20°. While the centrifugation is in progress the affinity resin can be prepared.

A 0.5-ml aliquot of His-Bind resin (Novagen) is placed in a 1.5-ml microcentrifuge tube, centrifuged at 3500 rpm for 1 min, and the supernatant is removed. The resin is washed twice in 0.5 ml of sterile distilled water followed by centrifugation at 3500 rpm for 1 min. The resin is resuspended in 0.5 ml of charge buffer (50 m*M* $NiSO_4$), then centrifuged at 3500 rpm for 1 min, twice. The supernatant is removed and the resin is washed twice with 0.5 ml of binding buffer, or binding buffer–6 *M* urea if the protein fractionates with the membrane fraction (see GST–σ Fusion Proteins for determination of solubility). If the fusion protein is soluble, then the denaturation step is omitted and urea is excluded from all buffers.

The cell pellet is resuspended in 10 ml of binding buffer to wash, and centrifuged again at 12,000 rpm (~20,000 *g*) for 15 min at 4°. To denature, the pellet is resuspended in 4 ml of binding buffer–6 *M* urea and incubated on ice for 1 hr, followed by centrifugation at 12,000 rpm for 15 min at 4°. The supernatant is filtered through a 0.8 μm pore size filter to remove particles and then added to the equilibrated resin in 1.5-ml microfuge tubes. Typically, 1 ml of sample is added to each of four tubes containing resin. The tubes are inverted gently to mix, and incubated on ice for 5–10 min. The resin is then washed twice with 1 ml of binding buffer and four times with 1 ml wash buffer [60 m*M* imidazole, 20 m*M* Tris (pH 7.9), 0.5 *M* NaCl, 6 *M* urea], centrifuging at 3500 rpm for 1 min after each wash. Wash buffer is prepared and stored as an 8× solution. *Note:* If purifying a soluble protein, omit urea from the wash buffer. The protein is eluted by washing the resin twice with 0.4 ml of elution buffer [1 *M* imidazole, 0.5 *M* NaCl, 20 m*M* Tris (pH 7.9)]. Elution buffer can be prepared and stored as a fourfold concentrated solution. The eluate is dialyzed against renaturation buffer [50 m*M* Tris (pH 7.9), 1 m*M* EDTA, 1 m*M* DTT, 20% (v/v) glycerol, 0.05% (v/v) Triton X-100] at 4° with four changes of buffer (at least 250 ml each), for at least 1 hr each. Some His-tag–σ derivatives precipitate during the renaturation dialysis, but this can be minimized by conducting a step dialysis such that the first buffer contains 4 *M* urea, the second 2 *M* urea, the third 1 *M* urea, and the final buffer no urea. Concentration determination and storage are the same as for the GST fusions.

Removal of GST or His-Tag Using Thrombin

Proteolytic cleavage to separate the σ polypeptide from the GST or His-tag can be conducted following purification. However, often the best results are obtained by performing the cleavage during purification, while the fusion proteins are still bound to their affinity resins. For GST–σ fusions, after the last wash, but prior to elution, the beads are resuspended in thrombin cleavage buffer [20 m*M* Tris (pH 8), 150 m*M* NaCl, 2.5 m*M* $CaCl_2$]. Thrombin is added at a ratio of approximately 1:200 (w/w) to the fusion protein. The mixture is incubated for 2 hr at room temperature followed by centrifugation to pellet the beads. The freed σ polypeptide is recovered in the supernatant fraction. Uncleaved GST–σ fusion protein and the GST fragment from the cleavage both remain bound to the column. The σ polypeptide is then dialyzed into storage buffer and quantified as described above. The same method is used to cleave the His-tag when purifying a soluble protein. Some fusion proteins come off the resin during cleavage, or the σ polypeptide precipitates after cleavage. In these cases we have found that alternative cleavage buffers, containing 15% (v/v) glycerol and variations in the concentration of Tris and NaCl, can improve the yield of the final product.

Proteolysis to remove the His-tag requires a modification in the purification protocol, if done under denaturing conditions. In this case the denatured protein is renatured in renaturation buffer and then dialyzed into binding buffer prior to introduction to the resin (prepared without urea). A 1-ml sample is added to each of four tubes containing resin, inverted gently to mix, and incubated on ice for 5–10 min. The resin is then washed four times with 1 ml of wash buffer (without urea) and centrifuged at 3500 rpm for 1 min at 4° after each wash. For insoluble proteins, the thrombin cleavage can be performed under conditions of 3 *M* urea, thus avoiding complete renaturation prior to proteolysis.

Nitrocellulose Filter Retention as Measure of σ–DNA Interactions

While many methods have been developed for characterizing protein–DNA interactions, nitrocellulose filter retention has often been the method of choice for analysis of interactions between RNA polymerase and DNA,[18–21] and between σ and DNA,[9,10] as well as between other DNA-binding proteins and their binding sites.[22–25] Here the use of nitrocellulose filter retention techniques to determine both affinity and specificity of interactions between σ polypeptides and promoter DNA is described.

This technique is based on the fact that under the appropriate conditions, most proteins bind to nitrocellulose membranes, while double-stranded DNA passes through. DNA that is complexed to protein is therefore retained on the filter. The advantages to using nitrocellulose filter retention to assay protein–DNA interactions are as follows: (1) the simplicity of the method allows binding analysis to be performed on a large number of

[18] D. C. Hinkle and M. J. Chamberlin, *J. Mol. Biol.* **70,** 157 (1972).
[19] P. Melacon, R. R. Burgess, and M. T. Record, Jr., *Biochemistry* **21,** 4318 (1982).
[20] H. S. Strauss, R. R. Burgess, and M. T. Record, Jr., *Biochemistry* **19,** 3504 (1980).
[21] J.-H. Roe, R. R. Burgess, and M. T. Record, Jr., *J. Mol. Biol.* **176,** 495 (1984).
[22] A. D. Riggs, H. Suzuki, and S. Bourgeois, *J. Mol. Biol.* **48,** 67 (1970).
[23] G. M. Clore and A. M. Gronenborn, *J. Mol. Biol.* **155,** 447 (1982).
[24] R. H. Ebright, Y. W. Ebright, and A. Gunasekera, *Nucleic Acids Res.* **17,** 10295 (1989).
[25] P. Oertel-Buchheit, D. Porte, M. Schnarr, and M. Granger-Schnarr, *J. Mol. Biol.* **225,** 609 (1992).

samples over a short time; and (2) the filtering step takes only seconds to complete, thus trapping lower stability complexes.

Preparation of DNA

A linear fragment of DNA is labeled with ^{32}P and used as the substrate for binding. The binding site for the protein, in this case the promoter sequence, should be approximately centered within the fragment. A fragment length of ~100 base pairs works well for most σ derivatives. Oligonucleotide primers of 15–20 bases in length, which correspond to the 5′ and 3′ ends of the promoter fragment, are used to amplify the fragment using polymerase chain reaction (PCR). First, one primer is 5′ end labeled in a volume of 100 μl that contains 50 pmol of primer, kinase buffer [50 m*M* Tris–HCl (pH 7.6), 10 m*M* $MgCl_2$, 5 m*M* DTT, 0.1 m*M* EDTA, 0.1 m*M* spermidine], 150 μCi of [γ-^{32}P]ATP (3000–6000 Ci/mmol), and 30 units of T4 polynucleotide kinase in a 1.5-ml microcentrifuge tube. The mixture is incubated at 37° for 30–45 min, followed by addition of 2 μl of 0.5 *M* EDTA, and extraction with 1 vol of phenol–$CHCl_3$–isoamyl alcohol (25 : 24 : 1, v/v). The supernatant is removed to a new microcentrifuge tube and 20 μl of 50 m*M* $MgCl_2$ and 12 μl of 3 *M* sodium acetate are added, and the solution vortexed. The labeled primer is precipitated by addition of 0.35 ml of 100% ethanol followed by chilling at −70° for 30 min, and microcentrifugation for 20 min at 4°. The pellet (not visible) is washed gently with 100 μl of 70% (v/v) ethanol, centrifuged again for 5 min at 4°, air dried briefly, and resuspended in 20 μl of sterile H_2O.

The PCR amplification of the DNA is carried out in a 100-μl volume containing 0.01 *M* Tris (pH 8.3), 0.05 *M* KCl, 1.5 m*M* $MgCl_2$, 0.2 m*M* dNTPs, 50 pmol of ^{32}P-labeled primer 1, 50 pmol of unlabeled primer 2, 10–50 ng of template DNA (usually a plasmid containing the promoter of interest), and 2.5 units of Amplitaq DNA polymerase (Perkin-Elmer, Norwalk, CT). A 1-μl aliquot of this mixture is saved for scintillation counting to determine the counts per minute per picomole of primer. A DNA thermocycler is programmed for 35 cycles with the following profile: 80° for 5 min (1 cycle); 94° for 1 min, 55° for 1 min, 72° for 1 min (35 cycles); 72° for 10 min; 25° to hold. Following amplification, the ^{32}P-labeled DNA fragment is purified using the QIAquick-spin PCR purification kit (Qiagen) with elution using 50 μl of 2.5 m*M* Tris, 0.25 m*M* EDTA. The counts per minute per microliter of the DNA solution are measured and compared to the counts per minute per picomole of primer to determine the approximate picomoles per microliter of DNA. The size, purity, and quantity of DNA are checked by loading 0.5–1.5 pmol onto a 3% (w/v) Metaphor agarose gel (FMC, Philadelphia, PA) followed by ethidium bromide staining.

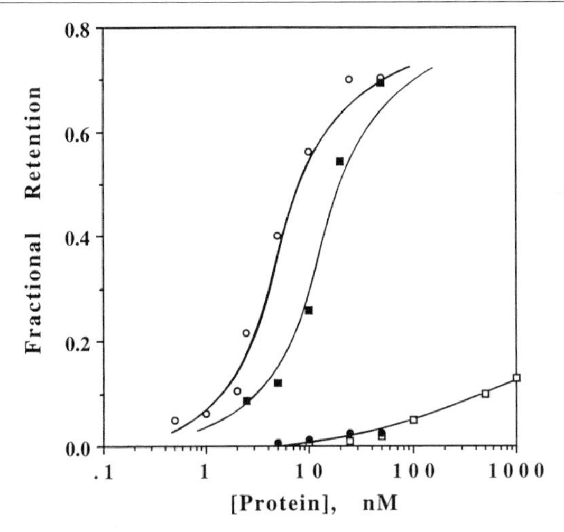

FIG. 2. Affinity of σ^{70} polypeptides for promoter DNA. The retention of ^{32}P-labeled promoter DNA on nitrocellulose as a function of increasing concentration of protein is shown for several different σ^{70} polypeptides. σ^{70} (open squares) and a derivative lacking the first eight amino acids (solid circles) bind DNA poorly. Polypeptides lacking the first 72 amino acids (solid squares) or 360 amino acids (open circles) demonstrate higher affinity for DNA.

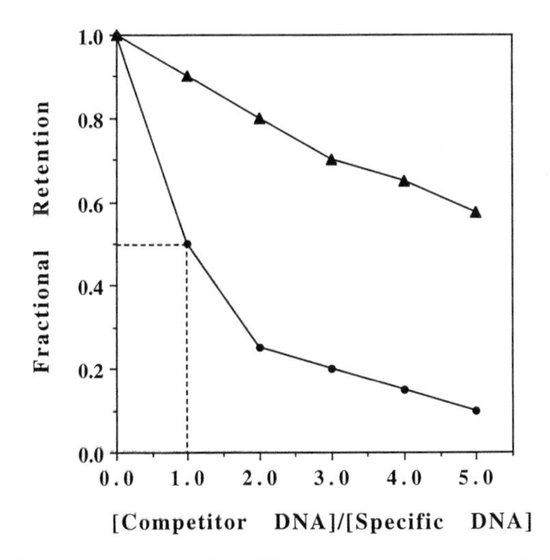

FIG. 3. Equilibrium competition binding. This hypothetical example shows that as the ratio of unlabeled competitor DNA to labeled specific DNA increases, DNA binding decreases. A specific competitor (circles) reduces binding at a lower concentration than a nonspecific competitor (triangles). At a 1:1 ratio of competitor to specific DNA, binding is reduced by one-half.

DNA-Binding Affinity

Standard binding mixtures contain 100 p*M* ^{32}P-labeled promoter DNA in filter-binding buffer [25 m*M* Tris (pH 7.5), 14 m*M* potassium acetate, 0.1 m*M* EDTA, 1 m*M* DTT, 0.01–0.03% (v/v) Triton X-100 (the optimal concentration is determined experimentally to maximize filter retention), and BSA (100 μg/ml)] in a volume of 25–100 μl. The σ polypeptide, diluted in protein dilution buffer [10 m*M* Tris (pH 8.0), 10 m*M* KCl, 10 m*M* 2-mercaptoethanol, 1 m*M* EDTA, 0.1% (v/v) Triton X-100, and BSA (0.4 μg/ml)], is added at various concentrations (0.5 to 500 n*M*, generally) followed by incubation at 37° for 20 min. The mixture is filtered through 24-mm nitrocellulose disks [BA85 (Schleicher & Schuell, Keene, NH) or GSWP02500 (Millipore, Bedford, MA)] using a Hoefer (San Francisco, CA) model FH224V 10-place filter manifold. The filters are washed once with a 10× volume of filter wash buffer [25 m*M* Tris (pH 7.5), 0.1 m*M* EDTA, 1 m*M* DTT]. The filters are dried under an infrared heat source and subjected to liquid scintillation counting using Ultima Gold (Packard, Meriden, CT) as the cocktail. Background DNA retention, in the absence of protein, is typically 5–10% of the total. A representative binding isotherm is shown in Fig. 2. *Note:* σ–DNA interactions are extremely sensitive to solution conditions.

DNA-Binding Specificity

Because most σ subunits exhibit a significant level of nonspecific binding, competition for binding between promoter and nonpromoter DNA has been the most successful method for assessing selectivity. For equilibrium competition binding assays, a fixed amount of σ polypeptide (corresponding to the linear portion of the binding isotherm) is incubated in filter-binding buffer with a constant amount of ^{32}P-labeled promoter DNA and increasing amounts of unlabeled competitor DNA. The competitor DNA, which is equivalent in length to the promoter DNA, can be completely promoter free or can carry a promoter variant. Incubation and filtering procedures are the same as for determining binding affinity. The reduction in binding as a function of increasing competitor DNA is monitored (Fig. 3).

Acknowledgments

Work from this laboratory was supported in part by grants from the American Heart Association–Texas Affiliate (94G-263) and from the American Cancer Society (NP-902). I thank B. Johnson, D. Le, and C. Wilson for contributions to the development of methods and for critical reading of the manuscript.

[12] Purification of Overproduced *Escherichia coli* RNA Polymerase σ Factors by Solubilizing Inclusion Bodies and Refolding from Sarkosyl

By RICHARD R. BURGESS

Many recombinant proteins have been expressed in a variety of hosts. One of the most common occurrences, when overexpression is in *Escherichia coli,* is that the overproduced protein is found in the form of an insoluble inclusion body.[1] To obtain purified, active, monomeric protein, one must isolate the inclusion bodies, solubilize the protein in the inclusion bodies, and then allow the solubilized protein to refold by gradually removing the solubilizing agent. This approach has been used to isolate overproduced components of the *E. coli* transcription machinery.[2–5] Most of the early work involved solubilizing with denaturants such as 8 *M* urea or 6 *M* guanidinium hydrochloride and refolding by slow dilution.[2] It was commonly found that, on dilution to low denaturant, most of the protein aggregated and precipitated. Even when no visible precipitate formed, there was often soluble multimer present in the diluted solution. This was subsequently lost when the apparently soluble protein was bound to an ion-exchange column and subjected to salt gradient elution. The majority of the refolded protein was multimeric, bound tightly to the column, and was never recovered. However, the 5–10% that was monomeric and eluted from the column at the expected salt concentration was fully active and useful in many biochemical studies.[2]

In this chapter, we summarize a method for solubilizing and refolding proteins from inclusion bodies utilizing the mild anionic detergent, Sarkosyl. We have found this method to be helpful in refolding several polymerase subunits and sigma (σ) factors with recoveries of monomeric refolded protein of greater than 50%.[3–6]

[1] C. H. Schein, *Bio/Technology* **7,** 1141 (1989).

[2] M. Gribskov and R. R. Burgess, *Gene* **26,** 109 (1983).

[3] D. R. Gentry and R. R. Burgess, *Protein Expr. Purif.* **1,** 81 (1990).

[4] L. H. Nguyen, D. B. Jensen, and R. R. Burgess, *Protein Expr. Purif.* **4,** 425 (1993).

[5] L. H. Nguyen, D. B. Jensen, N. E. Thompson, D. R. Gentry, and R. R. Burgess, *Biochemistry* **32,** 11112 (1993).

[6] D. Marshak, J. Kadonaga, R. Burgess, M. Knuth, S.-H. Lin, and W. Brennan, "Strategies for Protein Purification and Characterization: A Laboratory Manual." pp. 205–274. Cold Spring Harbor Laboratory Press, Cold Spring Harbor, NY, 1996.

Procedure

The procedure described below for *E. coli* σ^{32} is based on a published procedure,[4] and modified during the Cold Spring Harbor Protein Purification Course in Apil 1994 and 1995.[6] A schematic of this purification procedure is shown in Fig. 1.

Buffers and Solutions

Buffer A contains 50 m*M* Tris–HCl (pH 7.9), 5% (v/v) glycerol, 0.1 m*M* EDTA, 0.1 m*M* dithiothreitol (DTT), and 50 m*M* NaCl. Storage buffer is Buffer A containing 50% (v/v) glycerol. Stock solutions (20%, w/v) of sodium deoxycholate (NaDOC) and Sarkosyl (both from Sigma, St. Louis, MO) are prepared in distilled water.

Purification Procedure

1. *Escherichia coli* cells [BL21(DE3)/pLysS, pLHN16] are grown as described.[4] Briefly, cells are grown on 2× LB medium at 37° to $A_{550\text{ nm}}$ 0.8 and induced by adding 1 m*M* isopropylthio-β-D-galactoside (IPTG). Rifampicin (0.15 mg/ml) is added 30 min later and the cells allowed to grow for an additional 3.5 hr. Cells are harvested by centrifugation at 8000 rpm for 30 min at 4°.

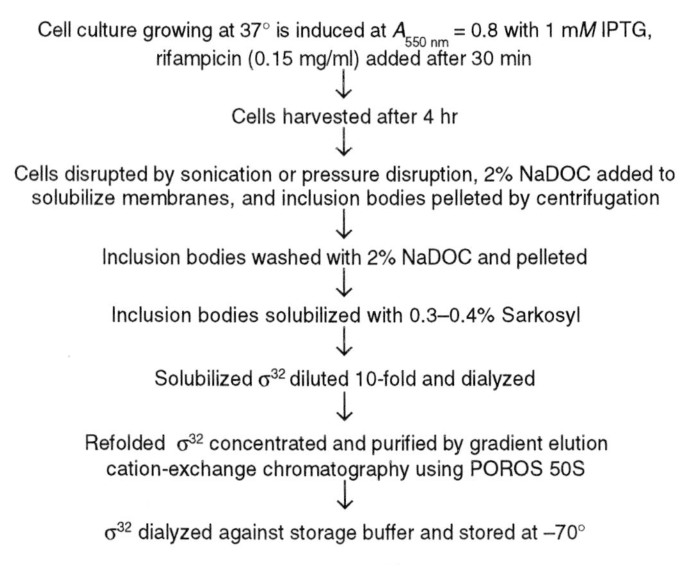

FIG. 1. Schematic for the isolation of *E. coli* σ^{32} from BL21(DE3)/pLysS, pLHN16.

2. Cells (3 g) are resuspended in 20 ml of buffer A and sonicated for 90 sec using a macrotip, or disrupted using a pressure disruption bomb as described.[3] Sodium deoxycholate is added to a final concentration of 2% (w/v), and the solution is mixed well and allowed to incubate at 4° for 10 min before centrifugation (13,000 rpm, 10 min, 4°).

3. The inclusion body pellet is washed once by resuspending in 20 ml of buffer A containing 2% (w/v) NaDOC and centrifuging as described above.

4. The washed inclusion body pellet is dissolved in 40 ml of buffer A containing 0.4% (w/v) Sarkosyl and incubated at 20° for 30 min. This usually gives a solution that contains about 1 mg of σ^{32} protein per milliliter that is, greater than 90% pure. Any insoluble material is removed by centrifugation (13,000 rpm, 10 min, 4°).

5. The dissolved σ^{32} is diluted 10-fold, slowly, by twofold dilutions with buffer A at 4°, waiting 10–15 min after each twofold dilution. The diluted σ^{32} (at about 0.1 mg/ml) is dialyzed twice for 8 hr against 10 vol of buffer A at 4°. The dialyzed material is centrifuged (8000 rpm, 30 min, 4°) to remove any precipitated material.

6. The clarified solution is applied to a 5-ml POROS 50S cation-exchange column (PerSeptive Biosystems, Framingham, MA) at 4 ml/min, washed with buffer A for 15 min, and eluted at 4 ml/min for 60 min with a linear salt gradient from buffer A to buffer A containing 1 *M* NaCl. The monomeric σ^{32} elutes at about 0.35 *M* NaCl.

7. Peak fractions are pooled and dialyzed against storage buffer for 12 hr at 4°. This dialysis against 50% (v/v) glycerol results in a threefold concentration of the protein. The protein can be stored at −20° (at which it will not freeze) or at −70° (at which it will freeze).

General Comments and Precautions

1. Sarkosyl seems to allow refolding of solubilized protein with less aggregation than seen with urea or guanidinium hydrochloride solubilization. A possible explanation is that the partially refolded protein exposes hydrophobic surfaces that result in aggregation. The detergent coats these hydrophobic surfaces and minimizes aggregation. In general, one can refold at higher protein concentrations (as much as 10-fold higher) using Sarkosyl than when using guanidinium hydrochloride.
2. Even when the diluted Sarkosyl-solubilized protein solution is dialyzed, there is still substantial residual Sarkosyl (0.01–0.02%, w/v) in the material applied to the POROS 50S column. The σ^{32} binds to the column and the free Sarkosyl flows through. Any remaining

Sarkosyl is washed off the protein during the column wash, since there is no detectable Sarkosyl in the σ^{32} peak that elutes off the column.

3. One can assay for the presence of residual Sarkosyl as follows[6]:
   a. Add 5 μl of glacial acetic acid to a 100-μl sample and mix.
   b. Inject 50 μl of the mixture onto a C_{18} reversed-phase high-performance liquid chromatography (HPLC) column [e.g., Vydac (Hesperia, CA) 4.6 × 250 mm] equilibrated with 0.1% (v/v) trifluoroacetic acid (TFA).
   c. Wash with 0.1% (v/v) TFA for 5 min at 1 ml/min, elute with a linear gradient of 0.1% (v/v) TFA to 80% (v/v) acetonitrile–0.1% (v/v) TFA over 10 min, and wash with 80% (v/v) acetonitrile–0.1% (v/v) TFA for 3 min.
   d. Monitor the column with an ultraviolet (UV) detector at 215 nm.
   e. Calibrate the column by applying a 0.01% (w/v) Sarkosyl standard.
4. Contaminating proteins present in the inclusion body are in the POROS 50S column flow-through or they elute from the column differently than σ^{32}.
5. Soluble multimeric protein will still be present after dilution and/or dialysis, but this will tend to bind more tightly to the column and be removed.
6. Several variables increase the yield of monomeric, soluble refolded protein, including the following:
   a. Use freshly grown cells. Inclusion bodies seem to become less easy to solubilize in cells that have been frozen, especially if frozen for more than 1 day.
   b. Refold at low protein concentration, 0.1 mg/ml or less, to minimize aggregation.
   c. Additional variables that may need to be optimized for refolding of a specific protein include pH, salt concentration, temperature, and rate of removal of the denaturant. Likewise, the ion-exchange column used to isolate a monomeric refolded protein can be one

TABLE I
TRANSCRIPTION PROTEINS REFOLDED USING SARKOSYL

Protein	Ref.
Omega (ω) subunit	3
σ^{32}	4, 6
σ^{S}	5
MotA	7

of several different anion or cation exchangers, depending on the binding properties of the protein under study.

7. The procedure described above has been used successfully for refolding and purification of a number of transcription factors and polymerase subunits, shown in Table I.[3–7]

Acknowledgments

I thank Mark Knuth for introducing this use of Sarkosyl to the laboratory, and Mark, Dan Gentry, Lam Nguyen, Debra Jensen, Dayle Hager, and Tony Grabski for contributing to the development of the present methods for refolding proteins from Sarkosyl.

This work was supported by NIH Grants CA07175, CA23076, CA60896, and GM28575, and by NSF Grant CHE-8509625.

[7] M. Ouhammouch, K. Adelman, S. R. Harvey, G. Orsini, and E. N. Brody, *Proc. Natl. Acad. Sci. U.S.A.* **92,** 1 (1995).

[13] RNA Polymerase σ Factors of *Bacillus subtilis*: Purification and Characterization

By KATHLEEN M. TATTI and CHARLES P. MORAN, JR.

RNA polymerase from bacteria is composed of several protein subunits. The core RNA polymerase contains two α subunits, a β subunit, and a β' subunit. Most, if not all, bacteria contain multiple sigma (σ) subunits that associate with the core to form the holoenzyme. Each different σ confers on the holoenzyme a unique specificity for promoter utilization (reviewed in Gross *et al.*,[1] Haldenwang,[2] and Lonetto *et al.*[3]). The purification of the different holoenzyme forms is essential for *in vitro* studies of their interactions with their cognate promoters and with ancillary, regulatory factors. The gram-positive bacterium *Bacillus subtilis* has proven to be particularly useful for genetic and biochemical studies of RNA polymerase and its associated σ factors. This chapter describes procedures for the isolation of RNA polymerase from *B. subtilis*.

[1] C. A. Gross, M. Lonetto, and R. Losick, *in* "Transcriptional Regulation," pp. 129–176. Cold Spring Harbor Laboratory Press, Cold Spring Harbor, NY, 1992.
[2] W. G. Haldenwang, *Microbiol Rev.* **59,** 1 (1995).
[3] M. Lonetto, M. Gribskov, and C. A. Gross, *J. Bacteriol.* **174,** 3843 (1992).

Principle of Method

The isolation of RNA polymerase holoenzyme forms can be achieved by purifying the individual components of core polymerase and the specific σ factor; however, this approach relies on reconstituting an active holoenzyme form *in vitro*, which at times can be difficult. RNA polymerase holoenzymes also can be purified by procedures used to purify other DNA-binding proteins. These preparations are usually adequate for the study of their interactions with promoters and regulatory proteins.[4–7] To isolate a single form of RNA polymerase, several chromatographic steps are required. Before a purification scheme for a single holoenzyme is initiated, several important points should be considered. First, the growth phase when the cells are harvested and the medium used during the growth of the *Bacillus* strains will affect the yield of a specific holoenzyme. Second, the utilization of mutants that lack some forms of polymerase can facilitate the isolation of a single holoenzyme form. Third, different combinations of the procedures described in this chapter can be used to optimize purification of the specific holoenzyme, depending on the particular properties of the holoenzyme form. Finally, if several holoenzyme forms have been isolated, they can be separated by elution from a DNA–cellulose column using a linear salt gradient.

Many forms of RNA polymerase, each with different promotor specificities, are present during the exponential growth phase (Table I). Isolation of the various RNA polymerase holoenzymes from either exponentially growing or sporulating cells begins by removing the nucleic acids bound to the RNA polymerase holoenzymes. Two methods have been developed for this purpose: one uses the heparin–agarose column chromatography procedure[8,9] and the other uses the polyethylene glycol (PEG)–dextran phase-partitioning method.[10–12] The holoenzyme, $E\sigma^D$, involved in flagellar biosynthesis has been purified from vegetatively growing cells on a heparin–agarose column followed by DNA–cellulose chromatography.[13] Two peaks of polymerase activity were eluted from the heparin–agarose column: one

[4] S. W. Satola, J. M. Baldus, and C. P. Moran, Jr., *J. Bacteriol.* **174,** 1448 (1992).

[5] K. M. Tatti, H. L. Carter III, A. Moir, and C. P. Moran, Jr., *J. Bacteriol.* **171,** 5928 (1989).

[6] P. N. Rather, R. E. Hay, G. L. Ray, W. G. Haldenwang, and C. P. Moran, Jr., *J. Mol. Biol.* **192,** 557 (1986).

[7] C. H. Jones and C. P. Moran, Jr., *Proc. Natl. Acad. Sci. U.S.A.* **89,** 1958 (1992).

[8] B. L. Davidson, T. Leighton, and J. C. Rabinowitz, *J. Biol. Chem.* **254,** 9220 (1979).

[9] C. W. Cummings and W. G. Haldenwang, *J. Bacteriol.* **170,** 5863 (1988).

[10] W. G. Haldenwang, N. Lang, and R. Losick, *Cell* **23,** 615 (1981).

[11] W. G. Haldenwang and R. Losick, *Proc. Natl. Acad. Sci. U.S.A.* **77,** 7000 (1980).

[12] H. L. Carter III and C. P. Moran, Jr., *Proc. Natl. Acad. Sci. U.S.A.* **83,** 9438 (1986).

[13] J. L. Wiggs, M. Z. Gilman, and M. J. Chamberlin, *Proc. Natl. Acad. Sci. U.S.A.* **78,** 2762 (1979).

TABLE I
HOLOENZYME FORMS OF *Bacillus subtilis* RNA POLYMERASE AND THEIR PROPERTIES

Form	Subunit	Apparent molecular weight (SDS–PAGE)	Stage of development	Isolation period	[KCl][a] (M)	Refs.[b]
E	$\alpha_2\beta\beta'$	α = 45,000 β = 140,000 β' = 130,000	Vegetative	Mid-log	0.45	1
Eδ	δ	δ = 21,000	Vegetative	Mid-log	0.40	1
Eσ^A	σ^A	σ = 55,000	Vegetative	Mid-log	0.55	1
Eσ^B	σ^B	σ = 37,000	Vegetative	Late-log	0.60	1
Eσ^C	σ^C	σ = 32,000	Vegetative	Late-log	0.55	2
Eσ^D	σ^D	σ = 28,000	Vegetative	Log	0.55	3
Eσ^E	σ^E	σ = 29,000	Sporulation; stage II	T_2–T_3	0.8	1
Eσ^F	σ^F	σ = 29,000	Sporulation; stage II	T_2–T_3	0.6	4
Eσ^G	σ^G	σ = 30,000	Sporulation; stage III	T_3–T_4	0.6	4
Eσ^H	σ^H	σ = 30,000	Vegetative; stage 0	T_0–T_1	0.7	5
Eσ^K	σ^K	σ = 27,000	Sporulation; stage IV	T_4–T_5	0.8	6
Eσ^L	σ^L	σ = 50,000	Vegetative		Unknown	2

[a] KCl concentration that elutes the holoenzyme from a DNA–cellulose column.

[b] *Key to references:* (1) R. H. Doi, *in* "Molecular Biology of the Bacilli," pp. 71–110. Academic Press, New York, 1982; (2) W. C. Johnson, C. P. Moran, Jr., and R. Losick, *Nature* (*London*) **302,** 800 (1983); (3) J. L. Wiggs, M. Z. Gilman, and M. J. Chamberlain, *Proc. Natl. Acad. Sci. U.S.A.* **78,** 2762 (1979); (4) D. Sun, P. Stragier, and P. Setlow, *Genes Dev.* **3,** 141 (1989); (5) H. L. Carter and C. P. Moran, Jr., *Proc. Natl. Acad. Sci. U.S.A.* **83,** 9438 (1986); and (6) L. Kroos, B. Kunkel, and R. Losick, *Science* **243,** 526 (1989).

peak at the low salt concentration where Eσ^D elutes and another peak at the high salt concentration where the primary holoenzyme Eσ^A elutes. The phase-partitioning method followed by high-performance liquid chromatography using a Superose column also has been used to isolate Eσ^A,[14] while phase partitioning followed by Sephacryl S-300 and gradient elution from a DNA–cellulose column has been used to purify Eσ^B,[11] Eσ^H,[12] and Eσ^C.[15]

Several forms of RNA polymerase, each containing a different σ factor, can be isolated from sporulating *B. subtilis*. Once *B. subtilis* cells are deprived of nutrients to induce sporulation, a decrease in the concentration

[14] J. Weir, M. Predich, E. Dubnau, G. Nair, and I. Smith, *J. Bacteriol.* **173,** 521 (1991).

[15] W. C. Johnson, C. P. Moran, Jr., and R. Losick, *Nature* (*London*) **302,** 800 (1983).

of the primary holoenzyme form, $E\sigma^A$, occurs and coincides with the accumulation of the sporulation-specific holoenzymes.[9,16] Isolation of RNA polymerase from sporulating *Bacillus* cells is more complicated than from vegetative cells because (1) the cells are more difficult to disrupt; (2) the concentration of the secondary holoenzymes is low compared to the primary holoenzyme; (3) more polypeptides are associated with core enzyme in sporulating cells than in vegetative cells; and (4) more proteases are present. For these reasons, more precautions are taken during the isolation of sporulation-specific RNA polymerases than vegetative RNA polymerases.

Purification of $E\sigma^E$ RNA Polymerase

σ^E accumulates and is found associated with RNA polymerase from about the second to fourth hours of endospore development. Its purification is described here as an example of a procedure used to isolate a holoenzyme form from *B. subtilis*. Slight modifications of this procedure can be used to isolate the other forms of RNA polymerase from *B. subtilis*.

During the purification procedure, it is essential to maintain solutions at 4° and to adjust the pH of the buffers to pH 7.9–8.0 at 4°, the optimum pH for RNA polymerase. Phenylmethylsulfonyl fluoride (PMSF) is added to the buffers to inhibit serine proteases; however, it should be remembered that PMSF in aqueous solutions is active only from 3 to 25 hr at 4°, depending on the pH. In addition to the PMSF, a high concentration of KCl (1 *M*) is added to the harvest buffer to help inhibit proteolytic activity. Hemoglobin is added before the Sephacryl chromatography to protect the RNA polymerase against proteases and its color can be followed during the chromatography. To stabilize RNA polymerase, most buffers contain at least 5% (v/v) glycerol.

Solutions and Reagents

DS medium: Combine 8 g of Difco (Detroit, MI) nutrient broth, 0.25 g of $MgSO_4$, 1 g of KCl with distilled H_2O to 1 liter; adjust to pH 7.1 with NaOH. After autoclaving, add 1 ml of 0.1 *M* $Ca(NO_3)_2$, 1 ml of 0.01 *M* $MnCl_2$, and 1 ml of 0.001 *M* $FeSO_4$

PMSF stock: 6-mg/ml stock solution prepared fresh in 2-propanol and stored at −20°. (This solution lasts 9 months in 2-propanol, but only approximately 3 hr at 4° in aqueous solutions of pH 8.0)

[16] T. G. Linn, A. L. Greenleaf, R. G. Shorenstein, and R. Losick, *Proc. Natl. Acad. Sci. U.S.A.* **70,** 1865 (1973).

Dithiothreitol (DTT): 1 *M* solution prepared in distilled H_2O and stored at $-20°$ in aliquots

Harvest buffer: 0.05 *M* Tris–HCl (pH 8.0), 0.01 *M* EDTA, 10% (v/v) glycerol, 1 *M* KCl. Add 5% (v/v) PMSF at 6 mg/ml just before use, and adjust to pH 8.0 at 4° if necessary

Buffer I[17]: 0.01 *M* Tris–HCl (pH 8.4), 1 m*M* EDTA, 0.01 *M* $MgCl_2$, 0.3 m*M* DTT, 5% (v/v) PMSF stock. Adjust to pH 8.4 at 4°

Buffer A[17]: 0.01 *M* Tris–HCl (pH 8.0), 1 m*M* EDTA, 0.1 *M* KCl, 0.3 m*M* DTT, 10% (v/v) glycerol, 5% (v/v) PMSF stock (0.01 *M* $MgCl_2$).[18] Adjust to pH 8.0 at 4°

Buffer C[17]: 0.05 *M* Tris–HCl (pH 8.0), 1 m*M* EDTA, 0.3 m*M* DTT, 20% (v/v) glycerol, 5% (v/v) PMSF stock (0.1 *M* KCl).[19] Adjust to pH 8.0 at 4°

Storage buffer[17]: 0.01 *M* Tris–HCl (pH 8.0), 0.01 *M* $MgCl_2$, 0.01 m*M* EDTA, 0.1 *M* KCl, 0.3 m*M* DTT, 50% (v/v) glycerol, 5% (v/v) PMSF stock. Adjust to pH 8.0 at 4°

PEG (30%, w/w): 30 g of polyethylene glycol (PEG 6000; Sigma, St. Louis, MO) is dissolved in buffer I to a final weight of 100 g

Dextran (20%, w/w): 20 g of dextran T500 (Pharmacia, Piscataway, NJ) is dissolved in buffer I to a final weight of 100 g

Growth and Harvesting of Bacillus subtilis Cells

Bacillus subtilis wild-type strain SMY, a sporulating Marburg strain, is grown until T_3, 3 hr after the end of log-phase growth. Eighteen liters of DS medium is used to obtain a yield of approximately 60 g of cells, which provides a sufficient yield of $E\sigma^E$ RNA polymerase.

The cells are harvested by centrifugation at 7500 rpm for 10 min at 4° or by means of a harvester. The cells are washed once with harvest buffer followed by washing once with buffer I. The harvested cells can be stored at $-80°$, but they should be used within 2 days. The cell pellet is resuspended in 1–1.5 ml of buffer I per gram of sporulating cells. If vegetative cells are used, they are resuspended in 2–2.5 ml of buffer I per gram of cells. Resuspended cells are disrupted by passage through a French press cell, two times at 15,000 lb/in^2, and sonicated briefly to reduce the viscosity.

[17] All these buffers can be made as 10× stocks with the basic components and autoclaved, adding glycerol, PMSF, and DTT only to the 1× buffer before use.

[18] $MgCl_2$ (10 m*M*) is omitted in buffer A only if a DNA–cellulose column is the next step in the purification scheme.

[19] Buffer C is made with variable concentrations of KCl depending on the type of salt gradient utilized.

The cell debris is pelleted by centrifugation at 150,000–200,000 *g* for 90 min at 4°. Using a Beckman (Fullerton, CA) 70.1Ti rotor, this is 50,000 rpm, or 34,000 rpm in a Beckman 45Ti rotor. Accurately measure the volume of the supernatant, which for subsequent calculations is termed fraction 1. Pour off the supernatant, which should be straw colored, into a prechilled glass beaker and keep it at 4°.

Phase Partitioning

A phase-partitioning system of PEG–dextran is used to separate nucleic acids from protein. The lower dextran phase contains nucleic acids and those proteins that bind to nucleic acids, e.g., RNA polymerase, while the upper polyethylene glycol phase contains the remaining proteins. Gradually increasing the salt concentration to approximately 5 *M* releases the bound RNA polymerase into the upper phase of the polyethylene glycol–dextran suspension.

To calculate the volume of 30% PEG to use, multiply the volume of fraction 1 by 0.32 and add this volume to fraction 1 while stirring. To calculate the volume of 20% dextran to add to fraction 1, multiply the volume of fraction 1 by 0.115, and add this amount to the solution. Record the total volume (designated *a*), which includes the volume of fraction 1, the PEG added, and the dextran added. Stir the solution for 30 min at 4° and centrifuge the solution in a Sorvall (Norwalk, CT) SS34 rotor for 10 min at 10,000 rpm at 4°. Decant the upper phase and measure the volume of the upper phase (UPV). To calculate the lower phase volume (LPV), subtract the UPV from the total volume *a*.

To determine the volume of 30% PEG to add to the lower phase, multiply the LPV by 0.98 and add this amount to the lower phase. Multiply the LPV by 2.4 to determine the volume of buffer I and add this amount to the lower phase. Record the total volume (designated *b*). Add NaCl to 2 *M* and then stir for 30 min. Centrifuge for 10 min in a Sorvall SS34 rotor at 10,000 rpm. Pour off the upper phase and measure the volume (UPV). The lower phase volume (LPV) is equal to *b* (the total volume) minus the UPV at this stage.

Add 30% PEG ($0.98 \times$ LPV) to the lower phase and add buffer I ($2.4 \times$ LPV) to the lower phase. Record the total volume and add NaCl to 5 *M*. Allow the solution to stir for 45 min or longer until the NaCl solubilizes. It is important that all, or most, of the NaCl solubilizes to release the bound RNA polymerase to the upper phase. Centrifuge for 10 min in a Sorvall SS34 rotor at 10,000 rpm. Take the upper phase (designated *c*), which should contain the RNA polymerase. Add hemoglobin to upper phase *c* to 2 mg/ml. To remove the high salt, dialyze against buffer I containing 0.05 *M* KCl and 5% (v/v) glycerol for 0.5 hr at 4°. The volume of buffer should be at least 10 times the dialysate volume. Change the buffer once and save the second change of buffer to use after the ammonium sulfate precipitation.

Ammonium Sulfate Precipitation

Ammonium sulfate precipitation is used to remove the polyethylene glycol as well as to help concentrate the RNA polymerase before application to a sizing column. Measure the dialysate volume (DV) from the previous step. To determine the quantity of $(NH_4)_2SO_4$ to add, multiply the DV by 0.163 and slowly add this amount to the DV over a 15-min period while stirring. Centrifuge at 8000 rpm for 10 min at 4°. Remove the lower phase by pipetting from under the upper phase and measure the lower phase volume (LPV). Add $(NH_4)_2SO_4$ (0.26 × LPV) to the lower phase and slowly stir for 15 min at 4°. Centrifuge in an ultracentrifuge at 155,000 *g* for 30 min (50,000 rpm in a Beckman 70.1Ti rotor) at 4°. Pour off the supernatant solution. Resuspend the pellet in the least volume possible (about 5 ml) of the reserved buffer I containing 0.05 *M* KCl with 5% (v/v) glycerol from the previous dialysis. Dialyze against the reserved buffer for 2 hr or until clear at 4° to remove the ammonium sulfate.

Sizing Column

To isolate the holoenzyme forms of RNA polymerase and to remove small proteins, the sample is applied to either a Sephacryl S-300 or a Sepharose 6B column. Sephacryl S-300 resin is washed with buffer A (without Mg^{2+}), decanted to remove fines, degassed, and poured into a 44 × 2.5 cm column to give a packed bed volume of approximately 200 ml. The column is equilibrated with buffer A (without Mg^{2+}), and a sample of approximately 8 ml is loaded on the column and eluted with buffer A, using a flow rate of 40 ml/hr. Fractions of 200 drops (7.5 ml) are collected. The A_{280} of the samples can be monitored by using an ultraviolet (UV) monitor connected to a strip chart recorder. Aliquots of each fraction are removed and the total RNA polymerase activity is assayed for poly (dA-dT) · poly(dA-dT) transcribing activity (as described in the section Polydeoxyadenylic-thymidylic Acid Assay, below). The RNA polymerase elutes as a single peak in the light yellow samples (fractions 20–30 in Fig. 1) before the hemoglobin. These fractions are pooled and applied to a DNA-cellulose column. It is important that Mg^{2+} (e.g., $MgCl_2$) be omitted from buffers because its presence affects the binding of RNA polymerase to the DNA–cellulose.

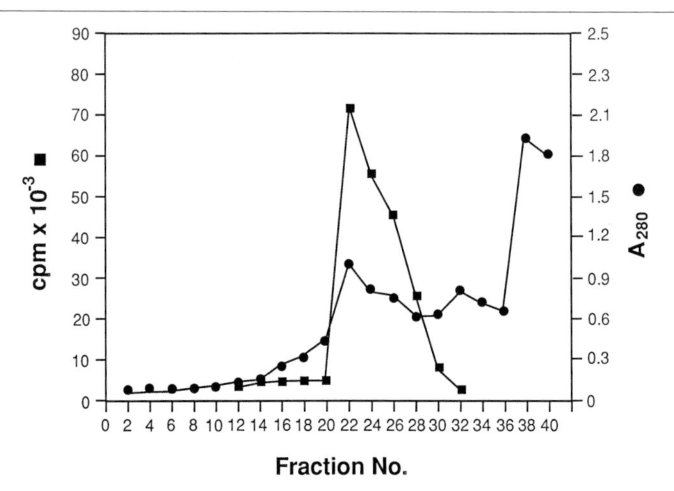

FIG. 1. Purification of Eσ^E RNA polymerase by Sephacryl S-300. The protein content of the fractions from the Sephacryl S-300 column is determined by monitoring the absorbance at A_{280}. The RNA polymerase activity of the fractions is determined by performing poly(dA-dT) assays (see text). The RNA polymerase activity (fractions 20–30) coincides with the first peak of protein. The second protein peak contains most of the added hemoglobin.

DNA–Cellulose Chromatography

The holoenzyme forms of *B. subtilis* RNA polymerase are separated on a DNA–cellulose column using a linear salt gradient. The concentrations of KCl that are used to elute the holoenzyme forms are indicated in Table I. The ancillary factor, δ, which is believed to prevent the binding of RNA polymerase holoenzyme at certain promoter sites *in vitro* and thus to inhibit transcription, elutes at a lower salt concentration from the DNA–cellulose column than Eσ^E. At T_3, the other holoenzymes present in the cell (Eσ^A, Eσ^B, Eσ^F, and perhaps some Eσ^G) elute at a KCl concentration of 0.55–0.6 *M*, while Eσ^E does not begin to elute until 0.8 *M* KCl.

The DNA–cellulose (5 g) is allowed to swell in buffer C containing 0.1 *M* KCl and washed in the same buffer three times. A 4-ml column is poured, washed with buffer C containing 1.0 *M* KCl, and then washed with buffer C containing 0.1 *M* KCl, using a flow rate of 4 ml/hr. The pooled samples from the Sephacryl S-300 column containing active RNA polymerase are loaded on the column, and the column is washed with 20 ml of the buffer C containing 0.1 *M* KCl. After loading the RNA polymerase, the column can be washed with 0.3 *M* KCl until no further A_{280}-absorbing material is eluted, which should eliminate many DNA-binding proteins associated with the Eδ form of RNA polymerase. The RNA polymerase is eluted with a linear 40-ml gradient of 0.3–0.9 *M* KCl–buffer C. The column is washed with 4 ml of buffer C containing 1.0 *M* KCl. Fractions of 0.9 ml are collected into tubes that contain $MgCl_2$ giving a final concentration of 10 m*M*. During the elution of the RNA polymerase, the fractions are monitored for their A_{280} activity and their conductivity. The fractions are dialyzed into storage buffer and stored at −20°. Each fraction is assayed for Eσ^E RNA polymerase activity by the *in vitro* transcription assay and for its sodium dodecyl sulfate-polyacrylamide gel electrophoresis (SDS–PAGE) pattern. As illustrated in the autoradiogram in Fig. 2, the Eσ^E activity starts eluting at approximately 0.8 *M* KCl from a DNA–cellulose column.

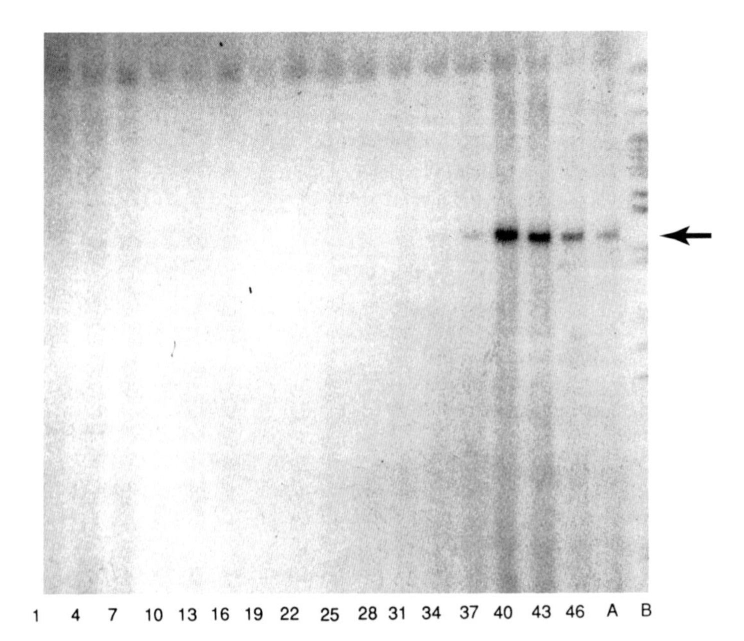

FIG. 2. *In vitro* transcription of pMB9 DNA by Eσ^E in fractions from a DNA–cellulose column. Fractions (1–46) eluted from a DNA–cellulose column with a linear salt gradient of 0.3–0.9 *M* KCl were assayed for their ability to transcribe pMB9 cleaved with *Hae*III. *In vitro*-synthesized RNA was subjected to electrophoresis on a 9% (w/v) polyacrylamide gel containing 7 *M* urea. A ^{32}P-labeled RNA "runoff" transcript of 100 bases, indicated by the arrow, was synthesized in fractions 37 to 46, demonstrating that Eσ^E RNA polymerase activity was present in these fractions. Lane A shows RNA polymerase activity in the 1.0 *M* KCl fraction from the DNA–cellulose column. Lane B is pBR322 DNA cleaved with *HPa*II, end labeled with [γ-^{32}P]ATP, and used as a molecular weight marker.

TABLE II

Purification of $E\sigma^E$ by Heparin–Agarose Chromatography

Preparation of *B. subtilis* cells

1. Twelve liters of *B. subtilis* SMY cells is grown in DS medium containing salts until T_3. Cells are harvested by centrifugation or by means of a harvester if one is available. Yield is approximately 50 g
2. Wash the cells in homogenization buffer[a] and store at −80° until ready to use, preferably only overnight
3. Thaw the pellets on ice and resuspend in 150 ml of homogenization buffer[a]
4. Pass the cells through a French press three times at 15,000 lb/in^2, keeping the cell at 4°
5. Sonicate using a Heat Systems (Farmingdale, NY) ultrasonic sonicator at a microtip setting of 7 for 15 sec, repeating this procedure twice or until the viscosity of the sample is reduced
6. Centrifuge in a 45Ti Beckman rotor for 90 min at 38,000 rpm at 4 °

Preparation of heparin–agarose column

7. Use 1 ml of heparin–agarose per gram of cells. Mix the heparin–agarose with the low-salt column buffer (0.05 *M* NaCl)[b] in a ratio of 1:1 and pour the column
8. After the column is packed, wash with 100 ml of 1 *M* NaCl column buffer[c] followed by 0.05 *M* NaCl column buffer.[b] Check the pH of the effluent, which must be pH 8.0, as the pH optimum for RNA polymerase is pH 7.9
9. Leave the column in this buffer at 4° until it is ready to use. Check the pH of the solution before using it

Heparin–agarose chromatography

10. Remove the heparin–agarose from the column and place in a shallow glass dish. Add the supernatant solution from the centrifugation to the heparin–agarose and place the dish on a rocking platform at 4° for 1 hr
11. Pour the heparin–agarose containing the bound RNA polymerase back into the column. Wash the column with 0.05 *M* NaCl buffer[b] until the effluent is colorless and start monitoring the absorption at A_{280}
12. When the absorbance at A_{280} levels off, start a 60-ml linear salt gradient containing 30 ml of 0.05 *M* NaCl buffer[b] and 30 ml of 1.0 *M* NaCl buffer.[c] Collect 2-ml fractions using a flow rate of 20 ml/hr. Monitor the absorbance at A_{280} until the A_{280} value levels off after the peak
13. Perform poly(dA-dT) assays on the fractions from the column (as described in the section Polydeoxyadenylic-thymidylic Acid Assay) to determine the total RNA polymerase activity
14. Pool the fractions from the second peak of poly(dA-dT) activity, which should occur at the high salt concentration and which should contain $E\sigma^E$
15. Perform *in vitro* transcription assays (as described in the section *In Vitro* Transcription Assay) on the fractions eluting at the higher salt concentration, using a linear template containing an $E\sigma^E$-dependent promoter if this is one of the last steps in the purification scheme
16. Dialyze the fractions containing RNA polymerase against storage buffer (see Solutions and Reagents) if this is the last column in the purification procedure. If additional purification is performed and a DNA–cellulose column or an HPLC column (see Table III) is the next column used, dialyze the active fractions against buffer C containing 0.05 *M* KCl (see Solutions and Reagents)

Characterization of RNA Polymerase

Polydeoxyadenylic-Thymidylic Acid

The total RNA polymerase activity of a sample is determined by its ability to transcribe a poly(dA-dT) · poly(dA-dT) (Sigma) template. Fractions from the Sephacryl S-300 column are assayed by this method. A reaction mixture is made containing 3.4 μl of ATP (30 mg/ml), 40 μl of poly(dA-dT) (0.5 mg/ml), 4.5 μl of DTT (0.1 *M*), 390.1 μl of H_2O, and 562 μl of 2× master mix. The 2× master mix is composed of 5 ml of 1 *M* Tris–HCl (pH 8.0), 0.125 ml of 0.1 *M* EDTA, 0.5 ml of 1 *M* $MgCl_2$, 0.5 ml of 0.1 *M* KPO_4, 62.5 mg of bovine serum albumin (BSA) (globulin free), and sterile H_2O to 35 ml, which should have a final pH of 7.9. To perform the assay, mix 25 μl of the reaction mixture with 2 μl of each fraction to be analyzed and 10 μCi of [α-^{32}P]UTP (800 Ci/mmol). Incubate the mixture at 37° for 10 min. Add 1 μl of unlabeled UTP (1.5 mg/ml) and incubate at 37° for 10 min. Stop the reaction by placing the mixture at 0° for 2 min.

To determine the ability of the RNA polymerase to transcribe the poly(dA-dT) template and, thus, to incorporate [α-^{32}P]UTP, 1 μl of the mixture is spotted 1 cm from the bottom of a strip of polyethyleneimine (PEI)–cellulose. After the sample has dried, the strip is placed in a small glass vial containing 100–200 μl of 2 *N* HCl. The spotted sample should be above the fluid level of HCl, and the strip of PEI–cellulose should be long enough to extend just above the top of the vial. Allow the fluid to ascend to the top of the cellulose. Remove the PEI–cellulose, cut the strip into three equal lengths, dry, and count. The unincorporated [^{32}P]UTP is at the top and the incorporated [^{32}P]UTP remains at the origin.

In Vitro Transcription Assay

One of the most common *in vitro* transcription assays used is the "run-off" assay. In this assay, a plasmid is cleaved by a restriction enzyme 50 to 200 base pairs downstream from a promotor recognized by a specific RNA

---

[a] Homogenization buffer: 10 m*M* Tris–Cl (pH 8.0), 10 m*M* $MgCl_2$, 1 m*M* EDTA, 7.5% (v/v) glycerol, 0.05 *M* NaCl. Immediately before use, add 0.3 m*M* DTT and 5% (v/v) PMSF (6 mg/ml).

[b] Low-salt column buffer: 10 m*M* Tris–Cl (pH 8.0), 10 m*M* $MgCl_2$, 1 m*M* EDTA, 7.5% (v/v) glycerol, 0.05 *M* NaCl. Immediately before use, add 0.3 m*M* DTT and 5% (v/v) PMSF (6 mg/ml).

[c] High-salt column buffer: 10 m*M* Tris–Cl (pH 8.0), 10 m*M* $MgCl_2$, 1 m*M* EDTA, 7.5% (v/v) glycerol, 1.0 *M* NaCl. Immediately before use, add 0.3 m*M* DTT and 5% (v/v) PMSF (6 mg/ml).

polymerase holoenzyme form. The RNA polymerase binds to the promoter, initiates transcription and transcribes the linear DNA fragment until it runs off the end of the template. The discrete ^{32}P-labeled RNA product can be visualized by autoradiography after electrophoresis on a polyacrylamide gel containing 7 *M* urea. This assay measures the transcriptional activity of a specific holoenzyme form.

A reaction mixture of 230 μl of 2× master mix (as described in the previous section), 80 μl of glycerol, 3.5 μl of 0.1 *M* DTT, 3.5 μl of 0.1 *M* EDTA (pH 8.0), and 183 μl of sterile H_2O is prepared. The DNA template used is either a pUC19 plasmid derivative containing the strong $E\sigma^E$ promoter G4,[6] cleaved by *Eco*RI 100 bp downstream from the start point of transcription, or pMB9, a plasmid containing a fortuitous $E\sigma^E$ promoter[10] that is cleaved by *Hae*III 100 bp downstream from the start point of transcription.

To determine the $E\sigma^E$ activity of fractions from the DNA–cellulose column, or the Sephacryl S-300 column, mix 1–2 μg of an appropriately cleaved DNA template, 25 μl of reaction mix, 5 μl of the fractions from the columns containing RNA polymerase, and distilled H_2O to a final volume of 40 μl. Incubate for 10 min at 37°. Add 1 μl of ATP, UTP, and GTP (1.5 mg/ml each) and 10 μCi of [α-^{32}P]CTP (800 Ci/mmol). Incubate for 1 min at 37° and add 1 μl of heparin (6-μg/μl stock in sterile H_2O). Incubate for 10 min at 37° and add 1 μl of CTP (1.5 mg/ml). Incubate for 5 min at 37°. Stop the reaction by adding 40 μl of stop mix containing 10 *M* urea, 0.02% (v/v) xylene cyanol, and 0.02% (w/v) bromphenol blue. Heat at 90° for 1 min and put on ice. Use 16 μl of each sample for electrophoresis on a 9% (w/v) acrylamide–7 *M* urea gel and visualize by autoradiography.

Sodium Dodecyl Sulfate-Polyacrylamide Gel Electrophoresis

A 10 or 12% (w/v) SDS-polyacrylamide gel is used to analyze the polypeptide composition of the fractions from the DNA–cellulose column. Aliquots of 200 to 250 μl are precipitated with an equal volume of 20% (w/v) trichloroacetic acid (TCA) and placed on ice for 10 min. Centrifuge for 15 min at 12,000 *g* and 4°. Remove the supernatant and wash the pellet with cold 1 *N* HCl. Centrifuge for 15 min at 12,000 *g* and 4°. Remove the supernatant and vacuum dry overnight. Resuspend the pellet in 10 μl of H_2O and 10 μl of 2× sample buffer[20] and subject to electrophoresis on a polyacrylamide gel containing SDS. The molecular weights of the core subunits and the different σ factors are listed in Table I.

[20] Sample buffer is made as a 2× stock containing 0.125 *M* Tris–HCl (pH 6.8), 20% (v/v) glycerol, 6% (w/v) SDS, 0.1% (w/v) bromphenol blue, 1.43 *M* 2-mercaptoethanol.

TABLE III
HPLC CHROMATOGRAPHY OF $E\sigma^E$ RNA POLYMERASE

Mono Q HR 5/5 chromatography[a]

1. Wash the column with 5 ml of low-salt (0.15 *M* NaCl) Mono Q buffer,[b] then with 10 ml of high-salt (1.0 *M* NaCl) Mono Q buffer[b]
2. Equilibrate with 5 ml of low-salt Mono Q buffer[b]
3. Filter the sample (prepared as described in Table II or in text) through 0.22 μm syringe or centrifuge at 4° for 10 min at 10,000 *g*
4. Apply the filtered sample at 4°, using a 50-ml superloop or any conveniently sized loop, at 1 ml/min
5. Apply a 20-ml linear salt gradient at 1 ml/min using 10 ml of 150 m*M* NaCl buffer[b] and 10 ml of 1 *M* NaCl buffer[b]
6. Wash with 10 ml of 1 *M* NaCl buffer.[b] Collect 0.9-ml fractions and monitor the A_{280} reading
7. $E\sigma^E$ elutes at a concentration of 0.4–0.55 *M* NaCl
8. Dialyze fractions against storage buffer (as described in Solutions and Reagents) and store at −20°

Superose 6 HR 10/30 chromatography[a]

1. Wash the Superose 6 column with 50 ml of buffer[c] at a flow rate of 1 ml/min
2. Centrifuge the sample to remove any particulate matter at 10,000 *g* for 10 min at 4°
3. Load the sample at 0.18 ml/min, using a 200-μl loop
4. Elute with 20 ml of the Superose 6 column buffer,[c] and immediately start collecting 0.9-ml fractions while monitoring the A_{280} reading. Collect samples until the A_{280} reading decreases
5. Twenty fractions are collected; the RNA polymerase elutes in a peak of activity, fractions 12–14, as determined by *in vitro* transcription assays (as described under *In Vitro* Transcription Assay)
6. Dialyze fractions 12–14 against storage buffer (prepared as described in Solutions and Reagents) and store at −20°

[a] All HPLC buffers must be filtered through a 0.22-μm pore size filter before being applied to the columns.

[b] Mono Q column buffer: 20 m*M* Tris–HCl (pH 8.0), 0.1 m*M* DTT, 0.1 m*M* EDTA, 10 m*M* $MgCl_2$, 5% (v/v) glycerol, 5% (v/v) PMSF, 0.150 *M* NaCl (for high-salt buffer use 1.0 *M* NaCl).

[c] Superose 6 column buffer: 50 m*M* Tris–Cl (pH 8.0), 10 m*M* $MgCl_2$, 0.1 m*M* EDTA, 0.1 m*M* DTT, 0.5 *M* NaCl, 5% (v/v) glycerol, 5% (v/v) PMSF.

Conclusion

The phase-partitioning method described provides a relatively pure preparation of $E\sigma^E$ with sufficient yield. If the enzyme preparations are contaminated with other holoenzymes, an additional DNA–cellulose column chromatography step with a shallow gradient (e.g., 0.7–0.9 *M* KCl) can be performed. It also may be advantageous to apply the sample to an HPLC Mono Q 5/5 (Pharmacia) column, eluting with a linear salt gradient.

Alternative purification schemes have been used to isolate Eσ^E and to resolve this holoenzyme from other species. Purification of Eσ^E can be performed by heparin–agarose chromatography (Table II) followed either by elution from a DNA–cellulose column with a linear salt gradient and/or by elution from HPLC columns (Table III).

Acknowledgments

We gratefully acknowledge the contributions of many colleagues in R. Losick's and our laboratories to the development of these procedures, especially C. Johnson, C. H. Jones, and H. L. Carter. In addition, we thank Kathryn Harrison for critically reading the manuscript. The work in our laboratory has been supported by PHS Grants AI20319 and GM39917.

Section IV

RNA Polymerase and Associated Factors from Eukaryotes

[14] Nuclear RNA Polymerases: Role of General Initiation Factors and Cofactors in Eukaryotic Transcription

By ROBERT G. ROEDER

The general objective of the experimental methods described in Sections II and IV of this volume and in Sections I and II of Volume 274 is to facilitate biochemical studies, ultimately with completely purified components, of the mechanisms involved both in transcription initiation, elongation, and termination by eukaryotic RNA polymerases and in the regulation of these processes by various gene- and cell-specific factors. Whereas prokaryotes contain a single major RNA polymerase that, with an appropriate sigma (σ) factor, can both accurately initiate transcription and respond to gene-specific regulatory factors, the situation in eukaryotic cells is considerably more complicated. First, there exist three distinct nuclear RNA polymerases that divide the responsibility for transcribing the major classes of genes in the nucleus.[1] Genes (class I) encoding the large ribosomal RNA precursor are transcribed by RNA polymerase I, genes (class II) encoding precursors to mRNA and U1–U5 small nuclear RNAs (snRNAs) are transcribed by RNA polymerase II, and genes (class III) encoding most small structural RNAs (including 5S RNA, tRNA, 7SK and 7SL RNAs, and U6 RNA) are transcribed by RNA polymerase III. The subunit structures of these enzymes are complex and distinct, although some polypeptides are common to two or to all three enzymes.[1,2] The RNA polymerases also show distinct sensitivities to α-amanitin, which has provided a convenient means to distinguish their functions in unfractionated systems. Second, despite the enormous complexity (12–16 subunits) of the three RNA polymerases, and the presence of core promoter elements common to many of the genes in each class, the "core" RNA polymerases possess no intrinsic capability to transcribe accurately (even at a low level) the corresponding target gene core promoters.[1,3,4] This requires for each RNA polymerase a distinct group of "general" accessory factors whose action is formally equivalent to that of a single polypeptide σ factor in conferring selectivity to the bacterial core RNA polymerase. Third, while the typical prokaryotic holoenzyme ($\alpha_2\beta\beta'\sigma$) responds directly to gene-specific activators bound to distal DNA

[1] A. Sentenac, *CRC Crit. Rev. Biochem.* **18,** 31 (1985).
[2] P. P. Sadhale and N. A. Woychik, *Mol. Cell. Biol.* **14,** 6164 (1994).
[3] L. Zawel and D. Reinberg, *Annu. Rev. Biochem.* **64,** 533 (1995).
[4] R. G. Roeder, *Trends Biochem. Sci.* **16,** 402 (1991).

regulatory elements, the situation is generally more complicated in eukaryotic cells. It may be that DNA-bound gene-specific activators for class I and class III genes generally function according to the bacterial paradigm and interact directly with core promoter-bound enzymes and accessory factors. In the case of the class II genes, however, the function of gene-specific activators requires a number of general and gene-specific cofactors that are not required for core promoter transcription by RNA polymerase II and the general factors.[3,4] The most recent developments in this area indicate that at least some of the cofactors may be present in a stable complex with the RNA polymerase and a subset of the general factors.[5] The existence of these various cofactors (both positive and negative) adds yet another layer of complexity to the regulation problem in eukaryotes.

The following sections review briefly some of the many factors that function in the three RNA polymerase systems, referring to studies in eukaryotes ranging from yeast to mammalian cells. One important point in this regard is that there is considerable (but not complete) evolutionary conservation of RNA polymerases, accessory factors, and at least some cofactors from yeast to humans. Hence, many of the fundamental mechanisms can be expected to be similar.

RNA Polymerase III Transcription

The class III genes fall into three subclasses on the basis of their promoter structures and transcription factor (TF) requirements[6,7]: (1) the tRNA, 7SL, and adenovirus VA RNA genes, as well as Alu sequences, have a simple internal promoter containing both an A-box and a B-box element, and require only the "common" factors TFIIIC and TFIIIB as accessory factors for RNA polymerase III; (2) the 5S RNA genes have more complex internal promoters containing an A-box, an I-box, and a C-box element and require the gene-specific factor TFIIIA in addition to the common factors TFIIIB and TFIIIC; (3) vertebrate U6 and 7SK RNA genes have typical class II upstream promoter elements, including a TATA box and a conserved proximal sequence element (PSE) that serve as core promoter elements and a distal sequence element (DSE) that serves as an enhancer.[8] Other class III genes may also use typical class II upstream

[5] A. J. Koleske and R. A. Young, *Trends Biochem. Sci.* **20,** 113 (1995).

[6] E. P. Geiduschek and G. A. Kassavetis, *in* "Transcriptional Regulation" (S. L. McKnight and K. R. Yamamoto, eds.), Vol. 1, p. 247. Cold Spring Harbor Laboratory Press, Cold Spring Harbor, NY, 1992.

[7] Z. Wang and R. G. Roeder, *Proc. Natl. Acad. Sci. U.S.A.* **92,** 7026 (1995).

[8] J.-B. Yoon, S. Murphy, L. Bai, Z. Wang, and R. G. Roeder, *Mol. Cell. Biol.* **15,** 2019 (1995).

promoter elements in conjunction with canonical class III internal promoter elements (7SL, selenocysteine tRNA, EBER, and yeast U6 genes).

In the simplest cases of promoter activation the core promoter elements of the tRNA and VA RNA genes are recognized by TFIIIC, and this is followed by the sequential binding of TFIIIB and Pol III.[6,7] At least in yeast, TFIIIB acts as a true initiation factor because it can remain stably bound (in a position-specific but sequence-independent manner) and facilitate Pol III recruitment and function following TFIIIC dissociation. In the case of the 5S RNA genes the promoter activation process is similar except that TFIIIC recruitment is dependent on prior binding of TFIIIA to the modified internal promoter. In this regard, the facilitated recruitment of common factors (TFIIIC, TFIIIB, and Pol III) by the gene-specific factor (TFIIIA) serves as a simple paradigm for the function of gene-specific activators in the more complex Pol II system. In the case of the mammalian U6 and 7SK genes, promoter activation requires a PSE-binding transcription factor (PTF), a modified form of TFIIIB, and a subcomponent (TFIIIC1) of TFIIIC;[7,8] and the functions of these components are enhanced by DSE-binding factors.

Apart from the gene-specific factors such as TFIIIA, the structures of the common factors are best understood in yeast.[9,10] Yeast TFIIIC contains six subunits, including two that recognize the A and B boxes, while yeast TFIIIB contains the TATA-binding protein (common to Pol I, II, and III factors), a TFIIB-related (BRF) subunit, and a third larger subunit. Mammalian TFIIIB appears similar (with conserved TBP and BRF subunits) to yeast TFIIIB,[7] whereas the mammalian TFIIIC can be split into a five-subunit component (TFIIIC2) that directly recognizes the promoter and a larger multisubunit component (TFIIIC1) that enhances the binding of TFIIIC2.[11] RNA polymerase III contains approximately 16 subunits, which are highly conserved from yeast to mammals.[1,2] At least in the simplest cases (5S and tRNA genes) active transcription (initiation and termination, the latter function provided by the RNA polymerase) can be observed in systems reconstituted with essentially homogenous components.

RNA Polymerase II Transcription

Class II genes typically contain both core promoter elements, which effect accurate transcription by RNA polymerase II in conjunction with a

[9] N. Chaussivert, C. Conesa, S. Shaaban, and A. Sentenac, *J. Biol. Chem.* **270,** 15353 (1995).

[10] G. A. Kassavetis, C. A. P. Joazeiro, M. Pisano, E. P. Geiduschek, T. Colbert, S. Hahn, and J. A. Blanco, *Cell* **71,** 1055 (1992).

[11] E. Sinn, Z. Wang, R. Kovelman, and R. G. Roeder, *Genes Dev.* **9,** 675 (1995).

group of general initiation factors, and distal regulatory elements, which effect the modulatory functions of cognate activators or repressors on core promoter ("basal") activity.[4] The best characterized core promoter elements are the TATA box and initiator (Inr) elements and eukaryotic promoters may contain one or the other or both of these elements. The naturally occurring general initiation factors implicated in TATA-dependent core promoter function include TFIIA, TFIIB, TFIID, TFIIE, TFIIF, and TFIIH.[4,12,13] Initiation commonly involves recognition of the TATA element by TFIID, the only one of these factors possessing an intrinsic site-specific DNA-binding capacity, followed by subsequent recruitment of the other general factors (and RNA polymerase II) into a functional preinitiation complex (PIC). Although a sequential pathway of factor recruitment can be demonstrated with purified components,[4,12] studies have suggested that some of the factors may enter the PIC in association with RNA polymerase II in the form of a holoenzyme complex.[5] Moreover, in more simplified systems accurate transcription by Pol II may be achieved with just the TATA-binding (TBP) subunit of the multisubunit TFIID and a subset (TFIIB, TFIIE, TFIIF, and TFIIH) of the other general factors. Depending on the specific promoter and the topological state of the template, accurate transcription initiation may also be achieved in the absence of TFIIE and TFIIH.[14] In the case of TATA-less Inr-containing promoters, core promoter function appears to require all the general factors, including both TBP and TBP-associated factors (TAF) subunits of TFIID as well as other factors; depending on the gene the latter may include factors (e.g., TFII-I or YYI) binding directly to Inr elements and/or cofactors including TAF–Inr interactions.[15] The TATA-less Inr-containing promoters generally show much weaker activities than TATA-containing promoters in cell-free systems, but represent an increasingly important class of promoters. Although the most emphasis to date has been on initiation factors, a number of factors (SII, SIII, TFIIF, and P-TEF) that affect elongation have also been defined.[16] As yet no factors essential for site-specific termination have been defined.

At the present time the above-mentioned general factors have been completely purified from yeast, human, and (in some cases) *Drosophila,*

[12] L. Zawel and D. Reinberg, *Curr. Opin. Cell Biol.* **4,** 488 (1992).

[13] E. Maldonado and D. Reinberg, *Curr. Opin. Cell Biol.* **7,** 352 (1995).

[14] J. D. Parvin, B. M. Shykind, R. E. Meyers, J. Kim, and P. A. Sharp, *J. Biol. Chem.* **269,** 18414 (1994).

[15] E. Martinez, C.-M. Chiang, H. Ge, and R. G. Roeder, *EMBO J.* **13,** 3115 (1994).

[16] T. Aso, W. S. Lane, J. W. Conaway, and R. C. Conaway, *Science* **269,** 1439 (1995).

and contain from 1 to more than 12 distinct subunits. These include 2–3 subunits for TFIIA, 1 for TFIIB, 10–14 for TFIID, 2 for TFIIE, 2 for TFIIF, approximately 9 for TFIIH, and 12 for Pol II.[13] cDNAs encoding virtually all of these polypeptides have been cloned and show considerable evolutionary conservation from yeast to human. From a practical standpoint, the cDNAs have provided a means to obtain large amounts of the factors following expression and reconstitution. Apart from RNA polymerase II, the only factor demonstrated to have enzymatic activities is TFIIH, which contains both a cyclin-regulated kinase (CAK kinase) that phosphorylates the heptapeptide repeat-containing C-terminal domain (CTD) of the large subunit of RNA polymerase II and ATPase/helicase activities involved in DNA melting and promoter clearance.[13] Also of major importance is TFIID, which contains both the TATA-binding polypeptide (TBP) common to Pol I, II, and III factors and a number of TBP-associated factors (type II TAFs) important both for Inr function (above) and for activator function (below).[17,18]

In the case of TATA-containing promoters, activator-dependent transcription requires all of the general components. Because TFIIA and the TAF subunits of TFIID are dispensable for basal transcription from the same TATA-containing promoters, they have been regarded as coactivators for those promoters (although they are essential basal factors for TATA-less promoters).[4,17,18] At the same time, and despite the enormous structural complexity of the GTFs and the presence of ample targets for various gene specific DNA-binding activators, the function of the latter still requires additional cofactors (coactivators) that do not directly bind DNA, at least in a site-specific way. Examples of the latter include the following: (1) positive cofactors (PCs) isolated either from the USA (upstream stimulatory activity) fraction (PC1, PC2, PC3, PC4) or from other chromatographic fractions (PC5) in mammalian cells.[19,20] At least in model activator/reporter assays these factors show little activator specificity and largely redundant functions, such that one (e.g., recombinant PC4) will suffice for high-level activation in reconstituted systems; (2) activator-specific coactivators that interact with, and in some cases may be isolated in association with, specific activators. Examples include OCA-B and CBP in mammalian cells,[21,22] as

[17] N. Hernandez, *Genes Dev.* **7,** 1291 (1993).

[18] J. A. Goodrich and R. Tjian, *Curr. Opin. Cell Biol.* **6,** 403 (1994).

[19] H. Ge and R. G. Roeder, *Cell* **78,** 513 (1994).

[20] J.-P. Halle, G. Stelzer, A. Goppelt, and M. Meisterernst, *J. Biol. Chem.* **270,** 21307 (1995).

[21] Y. Luo and R. G. Roeder, *Mol. Cell. Biol.* **15,** 4115 (1995).

[22] J. R. Lundblad, R. P. S. Kwok, M. E. Laurance, M. L. Harter, and R. H. Goodman, *Nature (London)* **374,** 85 (1995).

well as Gal-11 and the GCN4 · ADA2 · ADA3 complex in yeast[23,24]; (3) a complex of factors tightly associated with the CTD in a yeast RNA polymerase II holoenzyme complex.[5] These CTD-associated factors include the SRBs, as well as other genetically defined cofactors. Studies also point to the presence of potentially related coactivator-containing holoenzyme complexes in mammalian cells. As evidenced by the presence of Gal-11 in the yeast holoenzyme, independently isolated (biochemically or genetically) coactivators may derive from and function within the holoenzyme complex. However, from a practical standpoint it may be possible to study the function and specificity of such coactivators in systems reconstituted with isolated components. Although the mechanisms of action of the various coactivators are just being elucidated, studies of coactivators PC4 and Gal-11 indicate that they may function as adaptors between interacting activators and the basal transcription machinery.[19,23] Coupled with other demonstrations of direct activator interactions with several different components (TBP, TAFs, TFIIA, TFIIB, TFIIH) of the general transcriptional machinery,[25] it appears that concerted interactions of activators and coactivators with various general transcription factors (GTFs) enhance transcription by facilitating both the formation (e.g., via factor recruitment or stabilized binding) and function of preinitiation complexes. In addition to the above-described positive cofactors, a number of negative cofactors from mammalian (NC1, NC2/DR1, PC3/DR2)[4,19] and yeast (Mot I)[26] cells have been shown to enhance markedly the level of induction by activators via selective repressive effects on basal level transcription. Several of these factors appear to act via direct TBP interactions that prevent PIC formation, but can be reversed in a TFIIA-dependent manner. Chromatin-mediated repression mechanisms can also serve to enhance the level of induction by activators, and cofactors that may be involved in antirepression via chromatin remodeling have been described.[27]

RNA Polymerase I Transcription

Transcription by RNA polymerase I is unique among nuclear RNA polymerases in that it mediates transcription of a single homogeneous set of genes, albeit one that accounts for a large fraction of total RNA synthesis in growing cells. The promoters of these genes contain (in mammalian

[23] A. Barberis, J. Pearlberg, N. Simovich, S. Farrell, P. Reinagel, C. Bamdad, G. Sigal, and M. Ptashne, *Cell* **81,** 359 (1995).

[24] J. Horiuchi, N. Silverman, G. A. Marcus, and L. Guarente, *Mol. Cell. Biol.* **15,** 1203 (1995).

[25] R. Hori and M. Carey, *Curr. Opin. Genet. Dev.* **4,** 236 (1994).

[26] D. T. Auble, K. E. Hansen, C. G. F. Mueller, W. S. Lane, J. Thorner, and S. Hahn, *Genes Dev.* **8,** 1920 (1994).

[27] C. L. Peterson and J. W. Tamkun, *Trends Biochem. Sci.* **20,** 143 (1995).

cells) both a presumptive core promoter element (spanning the initiation site) and a more distal upstream control element.[28,29] The factors implicated in transcription of the rRNA genes include a species-specific DNA-binding factor (SLI in human, TIF-IB in mouse) containing TBP and associated type I TAFs, an enhancer-binding factor designated UBF, and two Pol I-associated factors designated TIF-IA and TIF-IC.[28,29] In a somewhat oversimplified view, it appears that SL1/TIF-IB recognizes the core promoter element, and that this interaction is stabhilized by UBF; reciprocally, UBF binding may also be stabilized by SL1/TIF-IB. RNA polymerase I and the remaining factors are presumably recruited to the preinitiation complex via interaction with UBF and/or SL1/TIF-IB. In contrast to the situation for Pol III (above), termination by Pol I involves a distinct factor (TTF-1/Rep1) that binds directly to the termination region.[30]

The class I transcription factors that have been purified and for which cDNA clones have been obtained include UBF (a single polypeptide with HMG boxes), SL1/TIF-IB (containing TBP and three type I TAFs), TTF-1/Reb1, and RNA polymerase I itself. Many of the corresponding polypeptides also show a strong evolutionary conservation.

Conclusions

It is clear that significant advances have made it possible to study fundamental transcription mechanisms in systems reconstituted with highly purified RNA polymerases and general transcription factors, many of which are available as functional recombinant proteins. Further, systems of variable purity can be used to study the mechanism of action of gene-specific activators, most notably with model promoters and activators. However, a major challenge will be to extend the analyses to natural promoters and activators, as well as to natural chromatin templates; and this will undoubtedly demand the identification, purification, and characterization of still other (co)factors—at least in the Pol II systems. In addition, the identification of multicomponent complexes of RNA polymerases, general factors, and cofactors necessitates that the physiological relevance of these complexes be further investigated and that the preinitiation complex assembly and activation pathways worked out with isolated components be evaluated accordingly. This undoubtedly will require the development of additional experimental protocols but the methods described in Sections II and IV of this volume and in Sections I and II of volume 274 offer a useful starting point for such studies.

[28] U. Rudloff, D. Eberhard, L. Tora, H. Stunnenberg, and I. Grummt, *EMBO J.* **13,** 2611 (1994).
[29] J. C. B. M. Zomerdijk, H. Beckmann, L. Comai, and R. Tjian, *Science* **266,** 2015 (1994).
[30] R. Evers and I. Grummt, *Proc. Natl. Acad. Sci. U.S.A.* **92,** 5827 (1995).

[15] Yeast RNA Polymerase II Holoenzyme

By YANG LI, STEFAN BJORKLUND, YOUNG-JOON KIM, and ROGER D. KORNBERG

Two forms of RNA polymerase II have been resolved from extracts of the yeast *Saccharomyces cerevisiae*, a 12-subunit "core" enzyme, and a much larger "holoenzyme" containing all core subunits and some 20 additional polypeptides.[1] The holoenzyme includes products of the *SRB* genes, previously shown to interact with RNA polymerase II in a larger complex.[2] The core and holoenzyme forms of the polymerase may be distinguished by three functional assays.[1] The holoenzyme is 10-fold more active than core enzyme in "basal" transcription reconstituted with pure general transcription factors (TFs) IIB, IIE, IIF, and IIH and TATA-binding protein. Holoenzyme is responsive to activator proteins in transcription of templates bearing activator-binding sites, whereas core enzyme is not; and holoenzyme is a 30- to 50-fold better substrate than is core enzyme for the C-terminal domain (CTD) kinase activity of TFIIH.

The set of polypeptides that associate with core RNA polymerase II to form the holoenzyme interact in a single large complex, previously termed a "mediator of transcriptional activation." These polypeptides are now known to include the products of the *GAL11*, *RGR1*, and *SIN4* genes, revealed by genetic studies as important for both negative and positive control of transcription. The complex is therefore more aptly termed a "mediator of transcriptional regulation."[3]

As many mediator polypeptides have yet to be identified, the full significance of RNA polymerase II holoenzyme is unknown. We present here our current protocol for holoenzyme purification, as an aid to further biochemical and molecular genetic analysis. This procedure entails a series of chromatographic steps, followed by gel filtration, which resolves an apparently homogeneous holoenzyme (Fig. 1) at the leading edge of a broad polymerase peak. Later-eluting fractions also contain high molecular weight polymerase complexes, which remain to be fully characterized.

[1] Y.-J. Kim, S. Bjorklund, Y. Li, M. H. Sayre, and R. D. Kornberg, *Cell* **77,** 599 (1994).
[2] C. M. Thompson, A. J. Koleske, D. M. Chao, and R. A. Young, *Cell* **73,** 1361 (1993).
[3] Y. Li, S. Bjorklund, Y. W. Jiang, Y.-J. Kim, W. S. Lane, D. J. Stillman, and R. D. Kornberg, *Proc. Natl. Acad. Sci. U.S.A.* **92,** 10864 (1995).

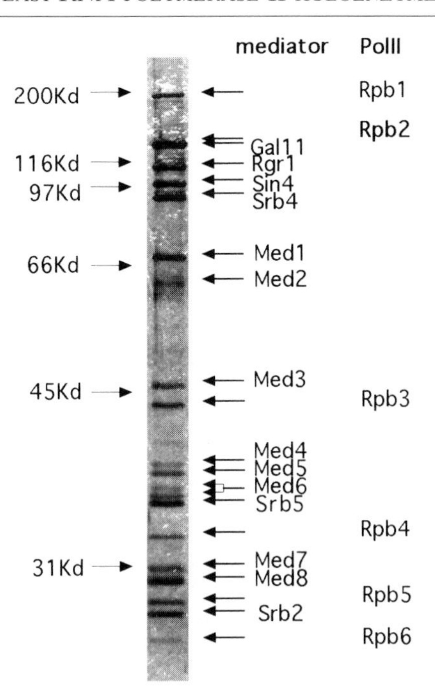

FIG. 1. SDS–PAGE of holo-RNA polymerase II. Polymerase fraction is from final gel-filtration column. RNA polymerase II holoenzyme and core enzyme subunits are indicated on the right. Rpb, RNA polymerase II subunits: Med, mediator of transcriptional regulation gene products. Molecular mass markers are on the left-hand side (kDa).

Buffers

Lysis buffer contains 0.15 *M* Tris–acetate (pH 7.8), 0.05 *M* potassium acetate, 20% (v/v) glycerol, 1 m*M* EDTA, 1 m*M* dithiothreitol (DTT), and protease inhibitors (1 m*M* phenylmethylsulfonyl fluoride, 2 m*M* pepstatin, 0.6 m*M* leupeptin, 2 m*M* benzamidine, prepared as 100× stock solution in 100% ethanol). Buffer A contains 25 m*M* HEPES–KOH (pH 7.6), 10% (v/v) glycerol, 1 m*M* EDTA, 1 m*M* dithiothreitol, protease inhibitors, and the molarity of potassium acetate indicated by a number after a hyphen. Buffer B contains 25 m*M* Tris–acetate (pH 7.8), 20% (v/v) glycerol, 1 m*M* EDTA, 1 m*M* DTT, 0.01% (v/v) Nonidet P-40 (NP-40: Sigma, St. Louis, MO), protease inhibitors, and potassium acetate indicated as for buffer A. Buffer C contains 0.01 *M* potassium phosphate (pH 7.7), 0.1 *M* potassium acetate, 20% (v/v) glycerol, 1 m*M* dithiothreitol, 0.01% (v/v) Nonidet P-40, 100 μM calcium chloride, and protease inhibitors. Buffer D is the same as buffer C but contains 0.2 *M* potassium phosphate. Buffer Q is the same as buffer B without 0.01% (v/v) Nonidet P-40.

Procedure

A culture (60 liters) of *S. cerevisiae* strain BJ926 (or strain of interest) in YPD [1% (w/v) yeast extract–2% (w/v) Bacto Peptone–2% (w/v) glucose] is grown at 30° to an A_{600} value of 3–4 [measured at a dilution of 1 : 10 in a Hewlett-Packard (Palo Alto, CA) 8451A diode array spectrophotometer]. The cells are harvested by centrifugation in a PR-6000 Centrifuge (DAMON/IEC Division) at 3000 rpm for 5 min at 4°. The cell pellet (about 500 g) is washed in 2 liters of cold distilled water and frozen in 3× lysis buffer (0.5 ml/g of paste). Portions (300 ml) of the cell suspension are placed in a stainless steel beat-beating apparatus (BioSpec Products, Bartlesville, OK) immersed in an ice–salt bath. Glass beads (0.5 mm) are added to fill the chamber, and the cells are disrupted by 20 cycles consisting of 30 sec of agitation and 90 sec of cooling. The temperature in the chamber remains less than 4° throughout. The cell lysate and glass beads are centrifuged in a Beckman (Fullerton, CA) JA-10 rotor at 9000 rpm for 20 min. To the supernatant is added 1/9 vol of 5 *M* potassium acetate (pH 8.0), followed by stirring for 15 min. Polyethyleneimine [10% (w/v), pH 7.8; Sigma] is slowly added to a final concentration of 0.1%, followed by stirring for another 30 min and centrifugation in a Beckman 45Ti rotor at 42,000 rpm for 90 min. About 15 g of protein is recovered in the clear amber supernatant.

The combined supernatants (whole-cell extract; about 600 ml) are diluted with 1.8 liters of 0.5× buffer A-0 and applied at 1 liter/hr to a Bio-Rex 70 column (25 × 5 cm, 500 ml; Bio-Rad, Richmond, CA) equilibrated with buffer A-0.15 (30–40 mg of protein/ml resin). The column is washed with 500 ml of buffer A-0.15 and 1 liter of buffer A-0.3, and holo-RNA polymerase II is eluted with buffer A-0.6. This eluate (BR600; 150 ml, 1.5 g of protein) is dialyzed for 3 hr against buffer B-0, diluted with about 150 ml of buffer B-0 to adjust the conductivity to that of buffer B-0.1, and applied at 100 ml/hr to a 50-ml DEAE-Sephacel column (10 × 2.5 cm, about 25–30 mg of protein/ml resin) equilibrated with buffer B-0.1. The column is washed with 50 ml of buffer B-0.1 and 100 ml of buffer B-0.2, and holo-RNA polymerase II is eluted with 100 ml of buffer B-0.55 (DE550; 50 ml, 400 mg of protein).

The DE550 fraction is applied at 100 ml/hr to a 50-ml hydroxylapatite column (30 × 1.4 cm, about 8 mg of protein/ml resin; Bio-Rad) equilibrated with buffer C. The column is washed with 50 ml of buffer C and developed with a linear gradient (500 ml) of buffer C to buffer D at 75 ml/hr. Fractions (7 ml) are collected and tested (5 μl) by Western blot analysis to locate the peak of holo-RNA polymerase II (at a phosphate concentration of about 90 m*M*).

Hydroxylapatite fractions containing holo-RNA polymerase II are pooled (35 ml), diluted 1:1 with buffer Q-0, and applied at 0.5 ml/min to a Mono Q HR5/5 column (Pharmacia, Piscataway, NJ) equilibrated with buffer Q-0.1. The column is washed with 3 ml of buffer Q-0.1, 6 ml of buffer Q-0.4, and 9 ml of buffer Q-0.7, and holo-RNA polymerase II is eluted with a linear gradient (10 ml) of buffer Q-0.7 to buffer Q-1.2. The peak of holo-RNA polymerase II elutes at 850 m*M* potassium acetate. The pooled peak fractions (250 μl) are applied at 0.25 ml/min to a Bio-Sil SEC 400 gel filtration column (Bio-Rad) equilibrated with buffer Q-0.8. Protein fractions are analyzed by sodium dodecyl sulfate-polyacrylamide gel electrophoresis (SDS–PAGE) with silver staining, by immunoblotting, and by transcription assays.

Assays

Reconstituted transcription is performed as described[4] with the following modifications. Reactions contain 100 ng each of two templates: pS(GCN4)^{2}CG$^-$, with two copies of a consensus Gcn4-binding sequence upstream of the yeast *CYC1* promoter fused to a 0.4-kbp G-less sequence in pSP73,[5] and pJJ470, with a consensus Gal4-binding sequence upstream of the yeast *CYC1* promoter fused to a 0.32-kbp G-less sequence in pUC18 (courtesy of M. Woontner and J. Jaehing). Reactions are in 40 m*M*, *N*-2-hydroxyethylpiperazine-*N*′-2-ethanesulfonic acid (HEPES; pH 7.7), 130 m*M* potassium acetate, 0.8 m*M* ATP, 0.8 m*M* CTP, 5 m*M* UTP, [α-^{32}P]UTP (0.4 mCi/ml), 1 m*M* dithiothreitol, and 7.5 m*M* magnesium acetate. Transcription factors are yeast TATA-binding protein (TBP; 60 ng), TFIIB (30 ng), TFIIE (40 ng), Gal4-VP16 (30 ng), and Gcn4 (30 ng) isolated from *Escherichia coli* expression strains as described,[6–8] and yeast TFIIF (40 ng), TFIIH (30 ng), and core or holo-RNA polymerase II purified to near homogeneity from yeast as described.[9–11]

[4] M. H. Sayre, H. Tshochner, and R. D. Kornberg, *J. Biol. Chem.* **267,** 23376 (1992).

[5] P. M. Flanagan, R. Kelleher III, M. H. Sayre, H. Tschochner, and R. D. Kornberg, *Nature (London)* **350,** 436 (1991).

[6] P. M. Flanagan, R. J. Kelleher III, W. J. Feaver, N. Lue, J. W. LaPointe, and R. D. Kornberg, *J. Biol. Chem.* **265,** 11105 (1990).

[7] D. I. Chasman, J. Leatherwood, M. Ptashne, and R. D. Kornberg, *Mol. Cell. Biol.* **9,** 4746 (1989).

[8] K. T. O'Neil, J. D. Shuman, C. Ampe, and W. F. DeGrado, *Biochemistry* **30,** 9030 (1991).

[9] A. M. Edwards, S. A. Darst, W. J. Feaver, N. E. Thompson, R. R. Burgess, and R. D. Kornberg, *Proc. Natl. Acad. Sci. U.S.A.* **87,** 2122 (1990).

[10] N. L. Henry, M. H. Sayre, and R. D. Kornberg, *J. Biol. Chem.* **267,** 23388 (1992).

[11] W. J. Feaver, J. Q. Svejstrup, L. Bardwell, A. J. Bardwell, S. Buratowski, K. D. Gulyas, T. F. Donahue, E. C. Friedberg, and R. D. Kornberg, *Cell* **75,** 1379 (1993).

For CTD phosphorylation assays, core or holo-RNA polymerase II (60 and 150 ng, respectively), TFIIH (30 ng), and 0.3 mCi of [γ-^{32}P]ATP are incubated for 30 min at room temperature in 15 μl of 20 m*M* HEPES-KOH (pH 7.8), 7.5 m*M* magnesium acetate, 1 m*M* DTT, 1 m*M* ATP, and 120 m*M* potassium acetate, followed by addition of 5 μl of 5$\times$ SDS gel-loading buffer and analysis by SDS–PAGE. Incorporation of ^{32}P into the largest polymerase subunit, Rpb1, is quantified with the use of a PhosphorImager (Molecular Dynamics), calibrated by liquid scintillation counting.

[16] Purification of Yeast RNA Polymerase II Holoenzymes

By ANTHONY J. KOLESKE, DAVID M. CHAO, and RICHARD A. YOUNG

RNA polymerase II can be purified from yeast in a high molecular weight complex called an RNA polymerase II holoenzyme.[1,2] This form of RNA polymerase II appears to be responsible for transcription initiation *in vivo*.[3] The holoenzyme form of RNA polymerase II contains multiple general transcription factors and a large multisubunit regulatory subcomplex. This regulatory subcomplex, which can be dissociated from RNA polymerase II with antibodies directed against the enzyme, contains the general transcription factor TFIIF, the SRB (suppressor of RNA polymerase B) proteins, and additional regulatory proteins, and has been called the Mediator.[4] Transcription by the RNA polymerase II holoenzyme can be stimulated by activator proteins,[1] a property that can be reconstituted with purified RNA polymerase II, general transcription factors, and the Mediator subcomplex.[4]

There are three forms of yeast RNA polymerase II holoenzyme that have been detected thus far, each of which was purified on the basis of its association with SRB proteins. The largest form contains RNA polymerase II, the Mediator subcomplex, and the general transcription factors TFIIH and TFIIB.[1] A second form of holoenzyme can be purified that has all these components except TFIIB (D. Chao, unpublished observation, 1995) and the third form lacks TFIIB and TFIIH.[4] It seems likely that the isolation of the smaller forms of the holoenzyme is a consequence of holoenzyme

[1] A. J. Koleske and R. A. Young, *Nature* (*London*) **368,** 466 (1994).
[2] A. J. Koleske and R. A. Young, *Trends Biochem.* **20,** 113 (1995).
[3] C. M. Thompson and R. A. Young, *Proc. Natl. Acad. Sci. U.S.A.* **92,** 4587 (1995).
[4] Y. J. Kim, *et al., Cell* **77**(4), 599 (1994).

instability. Such instability may be due to strain background, growth conditions, and extract and purification procedures.

In this chapter, we describe methods for the purification and assay of the two largest forms of RNA polymerase II holoenzyme from the yeast *Saccharomyces cerevisiae*. Use of the purified holoenzyme for transcription studies *in vitro* should yield new and exciting clues regarding the mechanisms underlying transcriptional regulation.

Growth of Yeast

The distinguishing feature of an RNA polymerase II holoenzyme is its association with multiple SRB regulatory proteins. The SRB proteins are somewhat more abundant in yeast cells grown in minimal medium than they are in cells grown in rich medium and it is possible that the largest form of the holoenzyme is more stable in minimal medium.[2,5] The total yield of yeast grown in minimal medium is only about 4 g (wet weight)/liter, so considerable effort is necessary to prepare the 500 g of yeast that is typically employed in the purification described here. We have used three approaches to obtaining 500 g of cells: (1) prepare cells using the protocol described below; (2) grow cells in a fermentor in rich medium; and (3) purchase commercial yeast. Although it is considerably more laborious, we find that the first approach is the most dependable in generating high-quality yeast extracts.

To generate approximately 500 g of cells, we grow six 3-liter cultures at a time. About 70 g of yeast (wet pellet weight) is obtained from six 3-liter cultures, so this procedure is repeated until 500 g is obtained. Those unfamiliar with the growth of yeast may wish to consult a reference before getting started.[6] The protease-deficient yeast strain BJ926 can be used to minimize the protease content of whole-cell extracts.[7]

1. Two days before starting cultures, patch several single colonies of freshly streaked BJ926 onto a rich plate (YPD) using a sterile toothpick. Make dime-sized patches and incubate them at 30° overnight.

2. Inoculate each of two 300-ml cultures of YNB medium [0.15% (w/v) Difco (Detroit, MI) yeast nitrogen base, 0.5% (w/v) ammonium sulfate, 200 m*M* inositol, 2% (w/v) glucose] in 1-liter flasks with the yeast scraped from a single patch with a sterile stick. Grow these cultures at 30°, shaking at approximately 200 rpm, for 20–24 hr.

[5] C. M. Thompson, A. J. Koleske, D. M. Chao, and R. A. Young, *Cell* **73,** 1367 (1993).
[6] F. Sherman, *Methods Enzymol.* **194**(1), (1991).
[7] E. W. Jones, *Methods Enzymol.* **194**(1), 428 (1991).

3. Inoculate each of six 3-liter cultures of 1× YNB (in 6-liter flasks) with 100 ml of the 24-hr culture. Allow the yeast to grow at 30°, shaking at 200 rpm, to an OD_{600} of 4.0 to 4.5 (20–24 hr).

4. Yeast cells are collected by centrifugation in six 1-liter bottles at 4000 rpm in a Sorvall (Newtown, CT) RC3B or similar centrifuge for 10 min at 4°. The supernatant can be carefully removed from the pelleted cells and another liter of culture can be added to the bottle and spun down on top of the previous yeast pellet. The pellets should be washed by resuspending in 100–200 ml of ice-cold buffer [20 m*M* *N*-2-hydroxyethylpiperazine-*N*′-2-ethanesulfonic acid (HEPES)–KOH (pH 7.5), 10% (v/v) glycerol, 50 m*M* potassium acetate, 1 m*M* dithiothreitol (DTT), and 1 m*M* EDTA]. This yeast suspension can be combined into one or two bottles and collected by centrifugation as described above for storage at −70°. Following centrifugation, discard the supernatant, weigh the bottle containing the pellet, and store it at −70° until enough yeast has been collected to proceed to the extraction process.

Yeast Extract Preparation

The following two procedures for producing yeast extracts can be adapted to 0.5–5 kg of yeast or even more. Owing to the labor involved in the purification, we do not recommend beginning with less than 500 g of yeast (wet pellet weight).

Procedure 1

In this procedure, the yeast whole-cell extracts are generated using methods similar to those described by Sayre *et al.*[8] The following protease inhibitors are used where indicated: 1 m*M* phenylmethylsulfonyl fluoride (PMSF), 2 m*M* benzamidine, 2 μM pepstatin A, 0.6 μM leupeptin, chymostatin (2 μg/ml), antipain hydrochloride (5 μg/ml; Sigma, St. Louis, MO). All procedures are carried out at 4°, unless otherwise noted.

1. Remove the yeast pellet from storage to 4° for about 1 hr to allow the pellet to thaw partially.

2. Resuspend the pellet in 3× lysis buffer plus protease inhibitors, using 0.5 ml/g (wet weight) of yeast. The pellet can be resuspended using a rubber policeman or by vigorous swirling of the bottle.

3. Rinse acid-washed glass beads (400–600 μm; Sigma) with 1× lysis buffer [20% (v/v) glycerol, 50 m*M* potassium acetate, 150 m*M* Tris–acetate (pH 7.9), 1 m*M* EDTA, 2 m*M* DTT].

[8] M. H. Sayre, H. Tschochner, and R. D. Kornberg, *J. Biol. Chem.* **267**(32), 23376 (1992).

4. Add glass beads and the cell suspension to a Bead-beater (Biospec Products). Add the glass beads to the chamber of the Bead-beater until it is 50–60% full. Fill the remaining space with the yeast–lysis buffer suspension. Assemble the chamber and seal. Fill the reservoir with an ice–salt–water bath. Use about 40 g of NaCl for the bath.

5. Run the beater 20 times for 30 sec. Allow 90 sec between runs to permit cooling of the chamber. Add more ice during the run if necessary. A digital controller/timer (VWR Scientific) is useful for automating this step.

6. Disassemble the chamber. Let the beads settle to the bottom and remove the lysate while minimizing the amount of bead carryover. Although tedious, pipetting the lysate works well. Rinse the beads further with 30–60 ml of 3× lysis buffer and decant the supernatant. The chamber can be refilled with yeast–lysis buffer solution for another round of lysis. The beads should be discarded after two rounds of lysis.

7. Spin the lysate in a Sorvall GSA rotor at 10,000 rpm for 20 min at 4° to pellet cell debris.

8. Adjust the supernatant slowly to 0.5 *M* in potassium acetate using a stock solution of 5 *M* potassium acetate with stirring, and stir gently for 30 min.

9. Spin the extract at 41,000 rpm in a Beckman (Fullerton, CA) Ti 50.2 rotor for 90 min. The lipid at the top of the tube is removed with a cotton-tipped applicator or by aspiration. Carefully remove the supernatant by pipetting, taking care to avoid the opaque and brown material at the bottom of the tube. The yield should be about 340–370 ml of 20- to 25-mg/ml protein extract for 500 g of starting material. Extracts can be frozen at −70°.

Procedure 2

The freeze-thaw procedure described below is extensively modified from the procedure of Dunn and Wobbe.[9] This procedure requires large amounts of liquid nitrogen, and it is important to take appropriate safety precautions. Handle the liquid nitrogen in a well-ventilated area, use appropriate containers, and wear protective clothing, glvoes, and safety goggles. Because this procedure has the potential to be quite messy, generous use of bench paper is recommended. Note that the form of holoenzyme purified by this procedure from BJ926 cells grown in rich medium does not contain TFIIB. All steps in this procedure are at room temperature unless otherwise specified.

[9] B. Dunn and C. R. Wobbe, Cell disruption using liquid nitrogen. *In*: "Current Protocols in Molecular Biology" (F. M. Ausubel *et al.*, eds.), pp. 13.5–13.6. John Wiley & Sons, New York, 1994.

1. Mix 500 g of frozen yeast with 500 ml of buffer containing 100 m*M* potassium acetate, 40 m*M* K-HEPES (pH 7.6), 2 m*M* Na-EDTA, 2 m*M* DTT plus 2× protease inhibitors, and thaw the mixture in a 65° water bath. Do not allow the temperature of the suspension to exceed 4°.

2. Pour the thawed suspension into 4-liter containers of liquid nitrogen while stirring with a plastic pipette. Pour the suspension into the liquid nitrogen at a speed that yields popcorn-size pellets of frozen cells. Drain the excess liquid nitrogen from the frozen pellets.

3. Place the frozen pellets in the 1-liter stainless steel chamber of a commercial Waring blender (VWR Scientific) until it is half full. Replace the lid, and blend for 2–4 min until the mixture has the consistency of a fine powder. It may be necessary to stir the mixture occasionally with a pipette to prevent clumping. Repeat the blending until all of the cells have been processed.

4. Thaw the disrupted cells at 65° in a beaker with stirring until the extract has the consistency of a viscous milkshake. In 750-ml batches, blend the lysate for 10 sec; wait 10 sec, and blend again for 10 sec.

5. Repeat steps 2–4. The number of cycles required for maximal cell breakage depends on the strain and growth conditions. For example, commercial yeast (Red Star) rehydrated for 30 min in 2% (w/v) glucose requires only one cycle, while BJ926 grown in a fermenter requires four cycles.

6. Spin the extract in a Sorvall H6000A rotor at 5000 rpm for 1 hr at 4°. Decant the supernatant and pour it through cheesecloth. Add 1/8 vol of glycerol and 1/8 vol of 5 *M* potassium acetate and stir at 4° for 15 min.

7. Spin the extract for 30 min at 13,000 rpm in a Sorvall GSA rotor at 4°. Pour the supernatant through cheesecloth at 4°. At this point, the extract can be stored temporarily at −70° if necessary.

8. Spin the extract for 90 min at 4° at 40,000 rpm in a Beckman 50 Ti rotor or at 42,000 rpm in a Beckman 45 Ti rotor. Remove lipids by aspiration, and recover the supernatant by pipetting.

9. The typical yield for this extraction procedure is 400 ml, with a protein concentration of 25–35 mg/ml. Extracts can be frozen at −70°.

Chromatography

The steps outlined here are scaled for 400 ml of extract with a protein concentration of approximately 25 mg/ml. The fractions can be stored at −20 or −70° between any chromatographic step in this sequence.

1. Prepare a column with Bio-Rex 70 resin 100–200 mesh (Bio-Rad, Richmond, CA). It is important to prepare this resin according to manufacturer directions and to equilibrate it with buffer A (100). Buffer A contains

20% (v/v) glycerol, 20 m*M* HEPES–KOH (pH 7.5), 1 m*M* DTT, 1 m*M* EDTA, and protease inhibitors. Buffer A (100) is buffer A plus 100 m*M* potassium acetate. As the protein is eluted in steps from this column, the exact column dimensions are not important; however, wider columns (>4-cm diameter) facilitate a faster flow rate. We use a 5 × 17 cm column. Typically, the column is loaded at 20–25 mg of protein per milliliter of column bed. Total onput onto this column should not exceed 50 mg of extract protein per milliliter of column bed.

2. If using an extract obtained with procedure 1, dilute the whole-cell extract 1:5 in buffer A so that the potassium acetate concentration is reduced to 100 m*M*. Load the extract onto the Bio-Rex 70 (Bio-Rad) column at a flow rate of 10 ml/min. Wash the column with buffer A (100) until no further protein can be eluted from the column, usually two to three column volumes. The protein concentration in the column eluate should be monitored by mini-Bradford assays, using a minimum of each fraction. The Bio-Rad protein assay is convenient for this; 10 μl of each fraction is added to 325 μl of Bio-Rad protein assay solution.

3. Elute the column with a step wash of buffer A (300) (buffer A plus 300 m*M* potassium acetate) until no further protein elutes from the column (approximately two to three column volumes). Although the Bio-Rex 70 flowthrough and buffer A (300) fractions are not used further in this protocol, they may provide a source of other factors and can be saved at $-70°$.

4. Elute the column with a step wash of buffer A (600) until no further protein elutes from the column. The RNA polymerase II holoenzyme should elute in this step. The holoenzyme is not yet sufficiently purified to be assayed for activity at this stage; it can be assayed for activity after chromatography on hydroxylapatite (see step 8). The presence of the holoenzyme can be monitored by Western blot using antibodies to SRB2, SRB4, SRB5, and SRB6. RNA polymerase II can be assayed in a similar fashion using a monoclonal antibody directed against the largest subunit of RNA polymerase II (Promega, Madison, WI). The column can be further eluted with a step wash of buffer A (1000). Eighty percent of the holoenzyme eluting from the column should be in the buffer A (600) fraction.

5. Dilute the Bio-Rex 70 (600) fraction 1:6 with buffer B [20% (v/v) glycerol, 20 m*M* Tris–acetate (pH 7.9), 1 m*M* DTT, 1 m*M* EDTA, 0.01% (v/v) Nonidet P-40 (NP-40) and protease inhibitors] to reduce the salt concentration to 100 m*M*. Load the solution onto DEAE-Sephacel (Pharmacia, Piscataway, NJ) resin at a flow rate of 4 ml/min. Again, column shape is not particularly important here, as stepwise elutions are used; we use a 2.5 × 8.5 cm column. The column should contain about 1 ml of bed volume per 5 mg of onput protein.

6. Wash the column extensively with buffer B (100) (buffer B plus 100 mM potassium acetate) and then elute with step washes of buffer B (400) and buffer B (650). The holoenzyme will elute from this column in the 400 mM potassium acetate step and can be monitored by Western blot analysis. TFIIE, which is not a component of yeast holoenzymes and is required for assays of holoenzyme activity, can be prepared from the 650 mM eluate.[10]

7. The next step involves chromatography on a BioGel HTP hydroxylapatite column. Prepare the resin by incubating in buffer C [20% (v/v) glycerol, 10 mM potassium phosphate (pH 7.7), 100 mM potassium acetate, 1 mM DTT, 0.01% (v/v) NP-40, and protease inhibitors]. It will take approximately 10 ml to swell 1 g of BioGel HTP.

8. Load the DEAE-Sephacel buffer B (400) fraction directly onto a 1.5 × 6.5 cm BioGel HTP hydroxylapatite column at a flow rate of 1 ml/min. The fraction does not need to be dialyzed or diluted prior to loading onto the column. Wash the column with 20 ml of buffer C (EDTA). Buffer C is 20% (v/v) glycerol, 10 mM potassium phosphate (pH 7.7), 100 mM potassium acetate, 1 mM DTT, 0.01% (v/v) NP-40, and protease inhibitors. Buffer C (EDTA) contains 0.25 mM EDTA. Elute with a 120-ml linear gradient of buffer C (EDTA) to buffer D [buffer D is identical to buffer C (EDTA) except that it contains 300 mM potassium phosphate, pH 7.7]. Assay fractions for transcriptional activity (see below) and for the presence of holoenzyme protein, using Western blot analysis. The holoenzyme should elute from this column in a peak corresponding to 68 to 112 mM potassium phosphate.

9. Dialyze the holoenzyme peak from the BioGel HTP (Bio-Rad) against buffer E (100). Buffer E is the same as buffer B except the EDTA is 0.25 mM; buffer E (100) is buffer E plus 100 mM potassium acetate. Centrifuge the dialyzed material in a Sorvall SS34 rotor at 10,000 rpm for 20 min at 4°. Load the supernatant (approximately 11 mg protein in 20 ml) onto a Mono Q HR 5/5 FPLC column (Pharmacia) at a flow rate of 0.5 ml/min. Elute the column with a 15-ml linear gradient from buffer E (100) to buffer E (2000). Assay fractions for transcriptional activity and the presence of holoenzyme protein. The holoenzyme elutes from this column at 0.95 M potassium acetate.

10. Pool the peak fractions containing holoenzyme activity and dilute them 1:6 with buffer F. Buffer F is the same as buffer A except the EDTA concentration is 0.25 mM. Load this material (approximately 1.1 mg of protein in 10 ml) onto a Mono S HR 5/5 FPLC column (Pharmacia), and elute with a 10-ml gradient from buffer F (100) to buffer F (1000) at a flow

[10] M. H. Sayre, H. Tschochner, and R. D. Kornberg, *J. Biol. Chem.* **267**(32), 23383 (1992).

rate of 0.5 ml/min. Assay fractions for transcriptional activity and the presence of holoenzyme protein. The holoenzyme should elute from this column at 450 mM potassium acetate.

11. Dilute the pooled peak fractions (approximately 0.6 mg in 8 ml) 1:4 in buffer E and immediately load onto a 1.5 × 1.5 cm DEAE-Sephacel column, wash with buffer E, and elute with a 12-ml gradient from buffer E (100) to buffer E (1000) at a flow rate of 0.3 ml/min. Assay fractions for transcriptional activity and the presence of holoenzyme protein. The holoenzyme elutes from this column at 400 mM potassium acetate.

This holoenzyme preparation should be approximately 90% pure. The total yield of the holoenzyme should be about 0.5 mg and the purification should be approximately 10,000-fold. Holoenzyme preparations have been stored in this buffer at −70° for up to 2 years without appreciable loss of activity.

Assay for Holoenzyme Activity

All forms of RNA polymerase II holoenzyme described thus far lack the general transcription factors TBP and TFIIE, and these factors must be added to obtain transcription *in vitro*. Protocols for the purification of recombinant TBP and yeast TFIIE have been published.[10,11] TFIIB is also required for transcription by some holoenzyme preparations. Holoenzyme and factor preparations should be dialyzed against a low-salt buffer, such as buffer F, to reduce the salt content prior to assay. Alternatively, the salt concentration can be determined by conductivity measurements, and the salt concentration in the transcription assay buffer can be adjusted accordingly. The following assay is a modification of that described in Ref. 12.

1. Make up 2× transcription buffer. 1× transcription buffer contains 50 mM HEPES-KOH (pH 7.3), 100 mM potassium glutamate, 15 mM magnesium acetate, 5 mM EGTA, 100 ng of GAL4 G-template, 3 mM DTT, enzyme-grade acetylated bovine serum albumin (BSA, 50 μg/ml; Promega), and 10% (v/v) glycerol. Transcription buffer is made up at 2× concentration to allow for the addition of further components (holoenzyme, TBP, TFIIE, activator, etc.). If TBP, TFIIE, or holoenzyme are in solutions containing glycerol, the glycerol content in the reaction must be adjusted accordingly.

2. Assemble the transcription reactions. The transcription reactions will have a 25-μl final volume, but for preincubation, they are assembled in a

[11] S. Buratowski, S. Hahn, L. Guarente, and P. A. Sharp, *Cell* **56,** 549 (1989).
[12] N. F. Lue and R. D. Kornberg, *Proc. Natl. Acad. Sci. U.S.A.* **84**(24), 8839 (1987).

volume of 22 μl at 24° and contain 20–40 ng of TBP, 20–40 ng of TFIIE, and 1 μg of holoenzyme. The remaining 3 μl of volume will contain the nucleotides added to initiate the reactions following the preincubation. Activators such as GAL4-VP16 are added at a concentration of 10 μ*M* before the preincubation step.

3. Preincubate the reactions at 24° for 1 hr.

4. Initiate the transcription reactions by addition of 3 μl of 3.33 m*M* ATP, 3.33 m*M* CTP, 15 μ*M* UTP, and [α-^{32}P]UTP (1.65 mCi/ml; 3000 Ci/mmol). The reactions are incubated at 24° for 40 min.

5. Terminate the transcription reaction by adding 100 μl of 10 m*M* Tris 8.0, 5 m*M* EDTA, 0.3 *M* NaCl, and RNase T_1 (2000 units/ml; Mannheim, Indianapolis, IN).

6. Prepare the transcripts for analysis by extracting the reactions twice with a 1:1 mixture of Tris-buffered phenol and chloroform. Precipitate RNA by addition of 2.5 vol of ethanol and incubation on dry ice–ethanol for 20 min, and pellet by high-speed centrifugation in a microcentrifuge for 15 min. Wash the pellet with 70% (v/v) ethanol. Allow to dry for several minutes and resuspend it in 10 μl of sample buffer [0.1× TBE, 80% (v/v) formamide, 0.01% (v/v) xylene cyanol, and 0.01% (w/v) bromphenol blue].

7. Analyze the transcripts by subjecting them to electrophoresis on a 6% (w/v) acrylamide [19:1 (v/v) acrylamide:bisacrylamide ratio]–7 *M* urea gel in 1× TBE buffer at 30 V/cm. Heat the resuspended pellets at 80° for 10 min, cool them briefly on ice, and spin to collect any condensation prior to loading the samples. The gel should be prerun at least 15–30 min before loading samples. Following electrophoresis, dry the gel and expose to autoradiography. The transcripts from the GAL4 G-less template are 350–375 nucleotides long.

Acknowledgments

We thank Sha-Mei Liao, who initiated biochemical analysis of transcription in the Young laboratory. We also thank Steve Buratowski, John Feaver, Michael Sayre, and Alan Sachs for reagents and advice on chromatography and Phil Johnson and John Harper for growing the yeast used in some of these experiments. D.M.C. is a predoctoral fellow of the Howard Hughes Medical Institute.

[17] Phosphorylation of Mammalian RNA Polymerase II

By MICHAEL E. DAHMUS

The largest subunit of RNA polymerase (RNAP) II contains at its C terminus an unusual domain composed of multiple repeats of the consensus sequence Try-Ser-Pro-Thr-Ser-Pro-Ser. This domain, designated the C-terminal domain (CTD), is hyperphosphorylated in RNAP IIO and unphosphorylated in RNAP IIA. The largest subunit of RNAPs IIA and IIO is designated IIa and IIo, respectively. Although RNAPs IIA and IIO appear to have distinct functions in the transcription cycle, the function of the CTD and the role of phosphorylation in mediating CTD activity remains unclear. Procedures are described below for the *in vitro* phosphorylation of RNAP IIA with CTD kinases. The ability to generate RNAP IIO from RNAP IIA using CTD kinases of known specificities should facilitate the analysis of CTD function and the role of specific CTD kinases.

The *in vitro* phosphorylation of mammalian RNAP IIA with casein kinase II (CKII) in the presence of [γ-^{32}P]ATP has been extensively used in the analysis of RNAP II and in the characterization of CTD kinases and CTD phosphatases. The CTD of mammalian RNAP II contains the sequence Asp-Asp-**Ser**-Asp-Glu-Glu-Asn at its very C terminus. The most C-terminal serine (in boldface, above) is flanked by acidic residues and is a consensus recognition site for CKII. The preparation of ^{32}P-labeled RNAP IIA by phosphorylation with CKII is also discussed. Because it is frequently preferable first to label RNAP II with ^{32}P at the CKII site before phosphorylation with a CTD kinase, the CKII reaction is discussed first.

Phosphorylation of RNA Polymerase II with Casein Kinase II

Assay Procedure

Buffer A contains 25 m*M* Tris–HCl (pH 7.9), 5 m*M* $MgCl_2$, 0.5 m*M* dithiothreitol (DTT), and 20% (v/v) glycerol.

The standard reaction mixture contains 5 μg of RNAP IIA (specific activity, 200–400 units/mg), 5 units of CKII, 2 μM [γ-^{32}P]ATP (2000 Ci/mmol), 20 m*M* Tris–HCl (pH 7.9), 20 m*M* KCl, 7 m*M* $MgCl_2$, 0.5 m*M* DTT, 50 μM EDTA, 0.025% (v/v) Tween 80, and 14% (v/v) glycerol in a final reaction volume of 70 μl. The reaction is incubated at 37° for 10 min. One unit of CKII activity is defined as the amount of enzyme that catalyzes

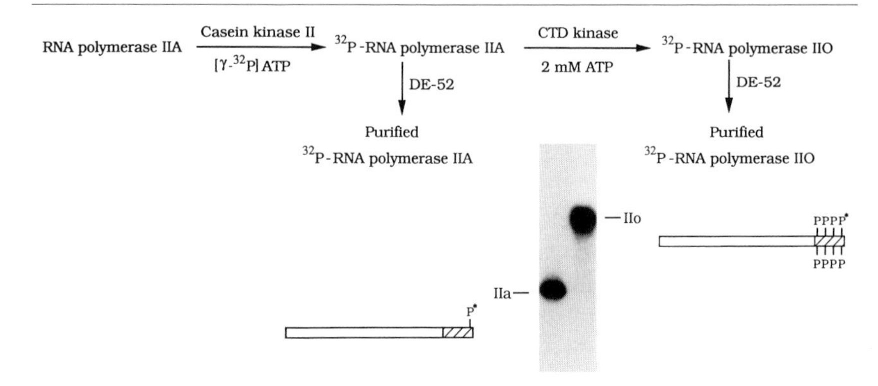

FIG. 1. The preparation of ^{32}P-labeled RNAPs IIA and IIO. The conditions for the CKII and CTD kinase reactions, as well as the procedure for the DE-52 purification of RNAP II, are given in text. The autoradiogram is of a 5% (w/v) polyacrylamide-SDS gel used to resolve the subunits of calf thymus RNAPs IIA and IIO. (From Chesnut *et al.*[2])

the transfer of 1 pmol of phosphate from ATP to phosvitin in 1 min under standard reaction conditions.[1]

^{32}P-Labeled RNAP IIA is purified from CKII and free nucleotide by chromatography on DE-52. DE-52 is equilibrated with buffer A containing 150 m*M* KCl, and a 150-μl column is packed in a microcolumn (Isolabs, Akron, OH) or a 1-ml disposable pipette tip containing a siliconized glass wool plug. The salt concentration of the CKII reaction is adjusted to 150 m*M* KCl by the addition of an aliquot of 2 *M* KCl and the sample is then loaded onto the DE-52 column. The flowthrough is collected and reloaded onto the column. This cycle is repeated for a total of four times to ensure efficient binding of RNAP II. The column is then washed with 1–2 ml of buffer A containing 150 m*M* KCl and 0.025% (v/v) Tween 80. Two-drop fractions are collected and monitored with a Geiger counter for the elution of free nucleotide. When the counts fall to about 10% of the maximum level, 1-drop fractions are collected and labeled RNAP IIA is eluted with buffer A containing 500 m*M* KCl and 0.025% (v/v) Tween 80. The peak of RNAP II is located by precipitation of 0.5-μl aliquots of each fraction with 5% (w/v) trichloroacetic acid (TCA) in the presence of carrier bovine serum albumin (BSA). Precipitates are collected on nitrocellulose filters and counted in a liquid scintillation counter. Peak fractions are dialyzed in microcollodion bags (Sartorius, Edgewood, NY) against buffer A containing 75 m*M* KCl and 0.025% (v/v) Tween 80. ^{32}P-Labeled RNAP II can be stored at −80°. Figure 1[2] is a schematic of the reaction and purification

[1] M. E. Dahmus, *J. Biol. Chem.* **256,** 3319 (1981).

[2] J. D. Chesnut, J. H. Stephens, and M. E. Dahmus, *J. Biol. Chem.* **267,** 10500 (1992).

protocol and also contains an autoradiogram of ^{32}P-labeled RNAP IIA resolved by electrophoresis on a 5% (w/v) polyacrylamide-sodium dodecyl sulfate (SDS) gel.

Precautions

Because the CKII site is at the very C terminus of the CTD, the ability to label RNAP IIA efficiently is dependent on an intact C terminus. The CTD is susceptible to limited proteolysis during the purification of RNAP II. Accordingly, the quality of RNAP IIA is related to the efficiency of inhibiting proteolytic activity during the purification process. RNA polymerase II purified by the Hodo and Blatti method[3] contains a frayed C terminus. This is apparent from the presence of multiple closely spaced bands in the region of subunit IIa on SDS–PAGE. Only the uppermost band labels with CKII, indicating that the other polypeptides have lost the CKII site. RNA polymerase IIA purified by the method of Kim and Dahmus[4] results in a single subunit IIa band in SDS–PAGE and labels efficiently with CKII.

It is also important to assess the purity of ^{32}P-labeled RNAP II by SDS–PAGE and autoradiography. The full spectrum of RNAP II subunits as well as contaminating proteins is best displayed by electrophoresis on a 5–17.5% (w/v) polyacrylamide-SDS gel.[5] If either RNAP II or CKII is not pure, background labeling can be extensive. The 20.5-kDa subunit of RNAP II is also phosphorylated by CKII.[6] In addition, the 26-kDa subunit of CKII is autophosphorylated.[1]

The phosphorylation of RNAP II with CKII does not result in a shift in the electrophoretic mobility of subunit IIa (Fig. 1). Furthermore, it does not appear to alter the activity of RNAP II. Labeling RNAP IIA by the protocol described above results in recovery of greater than 80% of the original polymerase activity. Activity can be monitored by either promoter-independent assays[4] or in a promoter-dependent assay.[2,7]

Phosphorylation of RNA Polymerase II with C-Terminal Domain Kinases

General Considerations

Although a variety of CTD kinases have been identified on the basis of *in vitro* assays, only a limited number of these enzymes have been used

[3] H. G. Hodo and S. P. Blatti, *Biochemistry* **16,** 2334 (1977).
[4] W. Y. Kim and M. E. Dahmus, *J. Biol. Chem.* **263,** 18880 (1988).
[5] U. K. Laemmli, *Nature (London)* **227,** 680 (1970).
[6] M. E. Dahmus, *J. Biol. Chem.* **256,** 3332 (1981).
[7] M. E. Kang and M. E. Dahmus, *J. Biol. Chem.* **268,** 25033 (1993).

to prepare RNAP IIO *in vitro*. One difficulty is that certain CTD kinases actively phosphorylate RNAP II assembled into preinitiation complexes but do not readily phosphorylate free RNAP II.[8–11] RNAP IIO has been prepared by the phosphorylation of RNAP IIA with partially purified serine/threonine CTD kinases from HeLa cells[2,7] and with recombinant c-Abl tyrosine kinase (S. S. Lee and M. E. Dahmus, unpublished results, 1995).

RNA polymerase IIO is operationally defined on the basis of the electrophoretic mobility of the largest subunit in SDS–PAGE.[12] Phosphorylation of the CTD at multiple sites results in a marked decrease in the electrophoretic mobility of the largest RNAP II subunit. The magnitude of the mobility shift is related in a complex way to the level of phosphate incorporation.[13,14] A subunit that contains on the order of one phosphate per repeat has an electrophoretic mobility comparable to that of *in vivo*-phosphorylated RNAP II. The incorporation of additional phosphate does not alter the electrophoretic mobility of subunit IIo. Therefore, the precise number and position of phosphorylated residues within the CTD are unknown. Furthermore, the indication that different CTD kinases preferentially phosphorylate residues at specific positions within the CTD suggests that various conformations of the CTD might be generated by phosphorylation with specific CTD kinases. The most striking difference is the phosphorylation of the CTD on serine/threonine or tyrosine.

Two distinct assays can be used to monitor CTD kinase activity. The first assay is based on the direct transfer of ^{32}P from [γ-^{32}P]ATP to the CTD of RNAP II. The second assay is based on the mobility shift in SDS–PAGE of CKII-labeled subunit IIa to the position of subunit IIo.

Direct Transfer Assay

The standard reaction contains CTD kinase, 0.05 μg of RNAP IIA, 10 μM [γ-^{32}P]ATP (4 Ci/mmol), 20 mM Tris–HCl (pH 7.9), 8 mM $MgCl_2$, 25 μM EDTA, 0.01% (v/v) Triton X-100, 0.5 mM DTT, and 10% (v/v) glycerol in a final reaction volume of 25 μl. Reactions are incubated at 30° for 10 min, stopped by the addition of 5 μl concentrated Laemmli sample buffer, and resolved by SDS–PAGE. The amount of phosphate incorpo-

[8] P. J. Laybourn and M. E. Dahmus, *J. Biol. Chem.* **265,** 13165 (1990).

[9] H. Serizawa, R. C. Conaway, and J. W. Conaway, *Proc. Natl. Acad. Sci. U.S.A.* **89,** 7476 (1992).

[10] H. Lu, L. Zawel, L. Fisher, J. M. Egly, and D. Reinberg, *Nature (London)* **358,** 641 (1992).

[11] R. Shiekhattar, F. Mermelstein, R. P. Fisher, R. Drapkin, B. Dynlacht, H. C. Wessling, D. O. Morgan, and D. Reinberg, *Nature (London)* **374,** 283 (1995).

[12] D. L. Cadena and M. E. Dahmus, *J. Biol. Chem.* **262,** 12468 (1987).

[13] J. M. Payne and M. E. Dahmus, *J. Biol. Chem.* **268,** 80 (1993).

[14] J. Zhang and J. L. Corden, *J. Biol. Chem.* **266,** 2297 (1991).

rated can be quantitated by scanning the dried gel in a PhosphorImager (Fuji Medical Systems, Inc., Stamford, CT) or Betascope blot analyzer (Betagen Corp., Waltham, MA).

Although 5–17.5% (w/v) polyacrylamide-SDS gels are frequently used

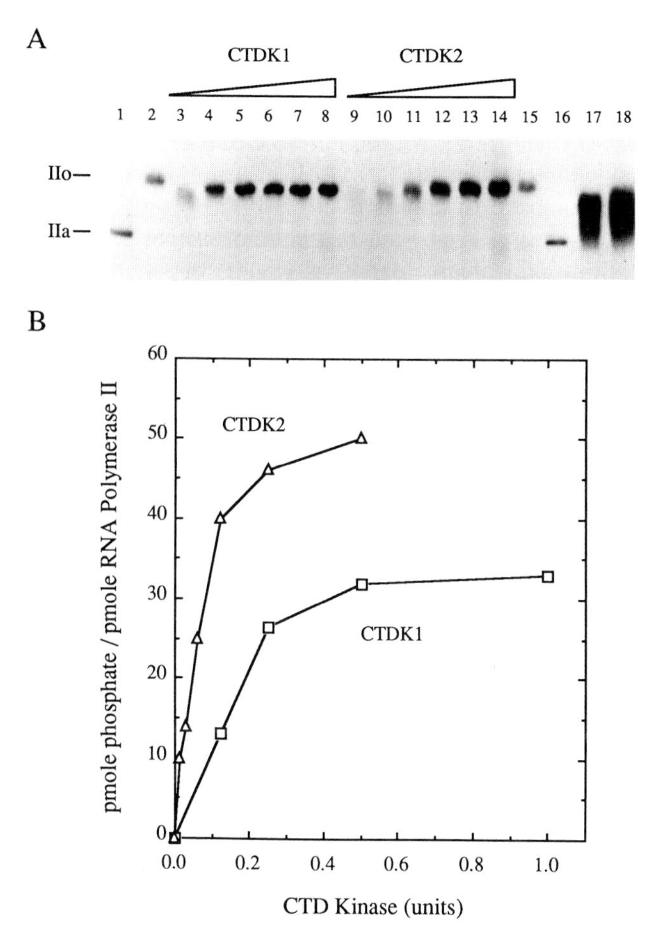

FIG. 2. Stoichiometry of phosphorylation of calf thymus RNAP II by HeLa cell CTDK1 and CTDK2. Reactions contained 0.05 μg of RNAP II and increasing amounts of either CTDK1 or CTDK2. The precise reaction conditions are described in test. Panel A is an autoradiogram of the dried gel. Marker lanes 1 and 16 contain CKII-labeled RNAP IIA and lanes 2 and 15 contain CKII-labeled RNAP IIO. Lanes 17 and 18 are assays to confirm the units of CTDK1 and CTDK2, respectively. One unit of CTD kinase activity is defined as the amount of enzyme that catalyzes the transfer of 1 pmol of phosphate from ATP to RNAP subunit IIa per minute under the conditions described. Panel B shows the quantitation of ^{32}P incorporated into the largest subunit catalyzed by either CTDK1 (□) or CTDK2 (△). (From Payne and Dahmus.[13])

in the analysis of RNAP II, a 5% (w/v) gel affords better resolution of subunits IIo and IIa. Figure 2A is an autoradiogram of a 5% (w/v) polyacrylamide-SDS gel that shows the incorporation of increasing amounts of ^{32}P into the largest RNAP II subunit as a function of increasing amounts of CTDK1 or CTDK2.[13] Marker lanes show the positions of subunits IIa and IIo.

RNA polymerase IIA labeled with ^{32}P by CKII provides a convenient marker in SDS–PAGE for subunit IIa. It is more difficult, however, to obtain markers for subunit IIo. The most reliable marker is that of native RNAP IIO as visualized in an immunoblot of a crude cell extract probed with antibody directed against RNAP II.[15] The difficulty with this approach is that it is not easy to compare directly the position of bands on an autoradiogram with the position of immunoreactive bands on an immunoblot. A more convenient marker is RNAP II, labeled with ^{32}P at the CKII site, that has been converted *in vitro* to RNAP IIO by incubation with a known CTD kinase (see following section). Accordingly, both ^{32}P-labeled subunits IIa and IIo can be included on the same gel and visualized directly on the autoradiogram. Before such a IIo standard is used, it is important to confirm that the *in vitro*-phosphorylated subunit IIo has an electrophoretic mobility comparable to that of *in vivo*-phosphorylated subunit IIo.

Advantages of the direct transfer assay are that it is sensitive and quantitative and hence provides a basis for defining units of CTD kinase activity and for measuring the stoichiometry of phosphorylation. One disadvantage is that large amounts of [γ-^{32}P]ATP are required because of the relatively high K_m of certain CTD kinases for ATP.[8] Furthermore, the purified CTD kinase must not contain potential substrates in the molecular weight range of 150,000 to 250,000, which would interfere with the quantitation of phosphate incorporation into the RNAP II largest subunit. Finally, for a variety of experiments examining the activity of RNAPs IIA and IIO, it is convenient to have ^{32}P-labeled RNAP II that is of uniform radiolabel-specific activity irrespective of the level of CTD phosphorylation. Clearly, the direct transfer assay results in ^{32}P-labeled RNAP II with a radiolabel specific activity that is directly proportional to the level of CTD phosphorylation.

Mobility Shift Assay

The interconversion of RNAPs IIA and IIO is generally determined by the mobility shift of the largest RNAP II subunit in SDS–PAGE. To increase the sensitivity of such assays, RNAP II can be labeled at the CKII

[15] W. Y. Kim and M. E. Dahmus, *J. Biol. Chem.* **261,** 14219 (1986).

site by phosphorylation in the presence of [γ-^{32}P]ATP as described above. Casein kinase II-labeled RNAP II is a convenient substrate for CTD kinase assays in that phosphorylation reactions can be carried out in the presence of high concentrations of unlabeled ATP.[16] Furthermore, because the shift in electrophoretic mobility of prelabeled subunit IIa to the position of subunit IIo is used to follow the progress of the reaction, contaminating protein kinase(s) and substrates do not interfere with the assay. Therefore, this method provides a sensitive and specific assay for CTD kinases. A potential problem with the mobility shift assay is the presence of protein phosphatases that strip the labeled phosphate from the CKII site. This is generally only a problem in crude extracts and can be circumvented by the inclusion of okadaic acid in the assay. A second difficulty with this assay is that it is not quantitative and hence cannot be used to determine units of CTD kinase activity.

The electrophoretic mobility shift of subunit IIa to IIo has also been used to monitor the phosphorylation of RNAP II as it progresses through the transcription cycle. *In vitro* transcription can be initiated by the addition of ^{32}P-labeled RNAP IIA and the state of phosphorylation determined at various points in the transcription cycle by SDS–PAGE.[17] Casein kinase II-labeled RNAP IIO also serves as an ideal substrate for CTD phosphatase assays.[18] C-Terminal domain phosphatase activity is defined as that activity that shifts the mobility of subunit IIo to the position of subunit IIa. C-Terminal domain phosphatase from HeLa cells selectively dephosphorylates serine and threonine residues within the consensus repeat but does not remove phosphate from the CKII site.

The procedure below describes the preparation of ^{32}P-labeled RNAP IIO in which only the phosphate at the CKII site is labeled. The CKII and CTD kinase reactions are carried out sequentially and involve an initial reaction with CKII in the presence of [γ-^{32}P]ATP followed by the addition of a large excess of unlabeled ATP and CTD kinase. The reaction is shown schematically in Fig. 1.

The conditions of the CKII reaction are identical to those described above. At the end of the CKII reaction, unlabeled ATP is added to a final concentration of 2 m*M*. The amount of CTD kinase required for the conversion of RNAP IIA to IIO must be determined in a pilot experiment in which increasing amounts of CTD kinase are incubated with CKII-labeled RNAP IIA. A sufficient amount of CTD kinase is added to quantitatively convert RNAP IIA to IIO. C-Terminal domain kinase reactions are incubated for 30 min at 30°.

[16] J. M. Payne, P. J. Laybourn, and M. E. Dahmus, *J. Biol. Chem.* **264,** 19621 (1989).
[17] M. E. Dahmus, *Prog. Nucleic Acid Res. Mol. Biol.* **48,** 143 (1994).
[18] R. S. Chambers and M. E. Dahmus, *J. Biol. Chem.* **269,** 26243 (1994).

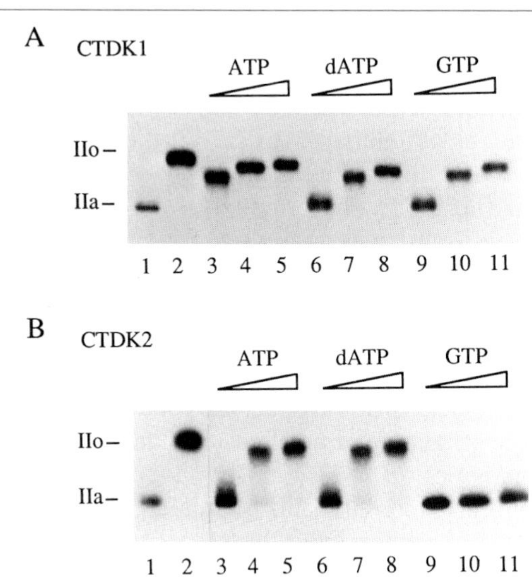

FIG. 3. Effect of nucleotide concentration on the mobility shift assay. CTDK1 (0.2 units; A) or CTDK2 (0.05 units; B) was incubated with ^{32}P-labeled RNAP IIA (2000 cpm) in the presence of increasing amounts of ATP, dATP, or GTP for 10 min at 30°. Reactions contained 1, 10, or 30 μM ATP (lanes 3–5), dATP (Lanes 6–8), or GTP (lanes 9–11), respectively. Lanes 1 and 2 are marker lanes containing ^{32}P-labeled RNAP IIA and IIO, respectively. (From Payne and Dahmus.[13])

^{32}P-Labeled RNAP IIO is purified from free nucleotide and protein kinases by chromatography on DE-52 as described above for ^{32}P-labeled RNAP IIA. The purity and elution properties of the CTD kinase on DE-52 must be taken into account in the purification of ^{32}P-labeled RNAP IIO. If the CTD kinase employed is not resolved from RNAP IIO by chromatography on DE-52, either heparin–Sepharose or P11 can be used. An additional concern is that the CTD kinase may be contaminated with factors that either bind to or coelute with RNAP II and influence subsequent assays. Peak fractions are dialyzed in microcollodion bags (Sartorius) against buffer A containing 75 mM KCl and 0.025% (v/v) Tween 80. ^{32}P-Labeled RNAP IIO can be stored at −80°.

The sequential phosphorylation of RNAP II with CKII and CTD kinase does not appreciably affect the activity of RNAP II in promoter-independent assays.[2,4] It is more difficult to follow the recovery of RNAP II activity by promoter-dependent assays because the transcriptional activity of RNAP IIO is low in the absence of CTD phosphatase (J. H. Stephens and M. E. Dahmus, unpublished observations, 1992). RNA polymerase IIO does not efficiently assemble into preinitiation complexes and, therefore, is transcriptionally inactive in the absence of enzymes that can generate RNAP IIA.[2,7,19]

An example of the mobility shift assay carried out on an analytical scale can be found in Fig. 3. Reactions conditions were identical to those described above for the direct transfer assay except that the reaction contained 2000 cpm of ^{32}P-labeled RNAP IIA and various amounts of ATP, dATP, or GTP.[13] The final reaction volume was 25 μl.

Summary

This chapter has described procedures for the phosphorylation of mammalian RNAP II with CKII and CTD kinases. The naturally occurring CKII site at the C terminus of the largest mammalian RNAP II subunit provides a convenient means to tag RNAP II with ^{32}P. The positioning of ^{32}P at the CKII site, in conjunction with the change in electrophoretic mobility that accompanies the multisite phosphorylation of the CTD, has provided a convenient assay for CTD kinase and CTD phosphatase. Furthermore, the utilization of CKII-labeled RNAP II in *in vitro* transcription reactions has made it possible to establish the state of CTD phosphorylation as a function of the progression of RNAP II through the transcription cycle. The ability to convert RNAP IIA to RNAP IIO by phosphorylation with different CTD kinases should facilitate both an analysis of the function of specific CTD kinases and a critical examination of the idea that phosphorylation of the CTD with CTD kinases of different specificities differentially affect RNAP II activity.

Acknowledgments

I gratefully acknowledge my colleagues Ross Chambers, Mona Kang, and Alan Lehman for their careful review of this manuscript. Research in our laboratory was supported by a grant (GM33300) from the National Institutes of Health (M.E.D.).

[19] H. Lu, O. Flores, R. Weinmann, and D. Reinberg, *Proc. Natl. Acad. Sci. U.S.A.* **88,** 10004 (1991).

[18] Purification of RNA Polymerase II General Transcription Factors from Rat Liver

By RONALD C. CONAWAY, DANIEL REINES, KARLA PFEIL GARRETT, WADE POWELL, and JOAN WELIKY CONAWAY

Eukaryotic messenger RNA synthesis is a complex biochemical process requiring the concerted action of multiple "general" transcription factors (TFs) that control the activity of RNA polymerase II at both the initiation[1] and elongation[2,3] stages of transcription. Because the general transcription factors are present at low levels in mammalian cells, their purification is a formidable undertaking. For this reason we explored the feasibility of using rat liver as a source for purification of the general factors. Rat liver has proven to be an ideal model system for biochemical studies of transcription initiation and elongation by RNA polymerase II (Figs. 1 and 2). In our hands the yield of general transcription factors per gram of rat liver is roughly equivalent to their yield per gram of cultured HeLa cells. Moreover, we have been able to develop convenient and reproducible methods for preparation of rat liver extracts from as much as 1 kg of liver per day. Because it is both technically difficult and expensive to obtain such quantities of cultured cells on a daily basis, rat liver provides a significant logistic advantage for purification of the general transcription factors.

Materials

Reagents

We use male Sprague-Dawley rats (200–300 g) from SASCO, Harlan Sprague-Dawley, and Simonson. Unlabeled ultrapure ribonucleoside 5′-triphosphates and RNAguard are from Pharmacia-LKB Biotechnology (Piscataway, NJ). [α-^{32}P]CTP (>400 Ci/mmol) is from Amersham (Arlington Heights, IL). Phenylmethylsulfonyl fluoride (PMSF) is from Sigma (St. Louis, MO) and is dissolved in dimethyl sulfoxide (DMSO) to 1 *M* and added to buffers just prior to use. Leupeptin and antipain are from Sigma and are dissolved in water to 25 mg/ml. Bovine serum albumin (BSA)

[1] R. C. Conaway and J. W. Conaway, *Annu. Rev. Biochem.* **62,** 161 (1993).
[2] C. M. Kane, *in* "Transcription: Mechanisms and Regulation" (R. C. Conaway and J. W. Conaway, eds.), p. 279. Raven, New York, 1994.
[3] T. K. Kerppola and C. M. Kane, *FASEB J.* **5,** 2833 (1991).

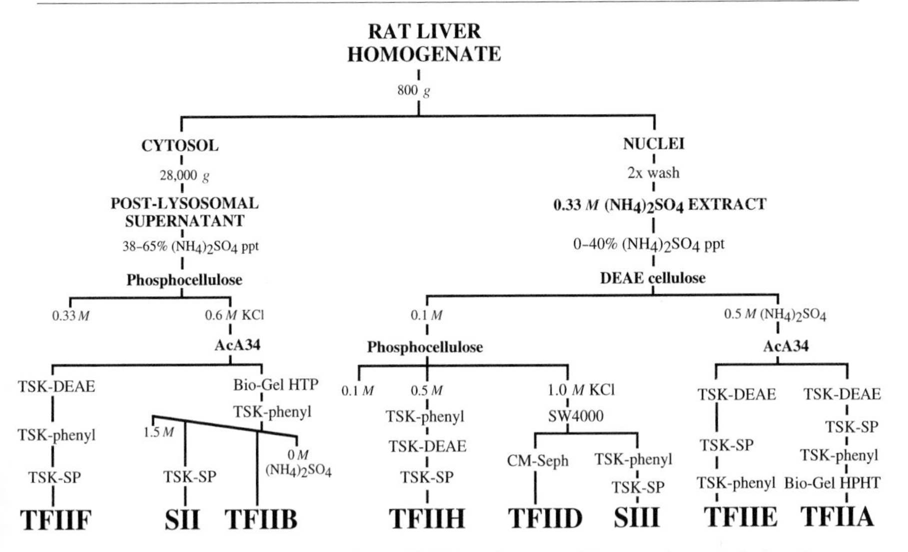

FIG. 1. Resolution and purification of RNA polymerase II general transcription factors from rat liver. TFIIF (rat $\beta\gamma$)[14]; TFIIB (rat α)[8]; TFIIH (rat δ)[13]; TFIID (rat τ)[16]; SII[18]; TFIIE (rat ε)[10]; TFIIA[15]; SIII (Elongin).[17]

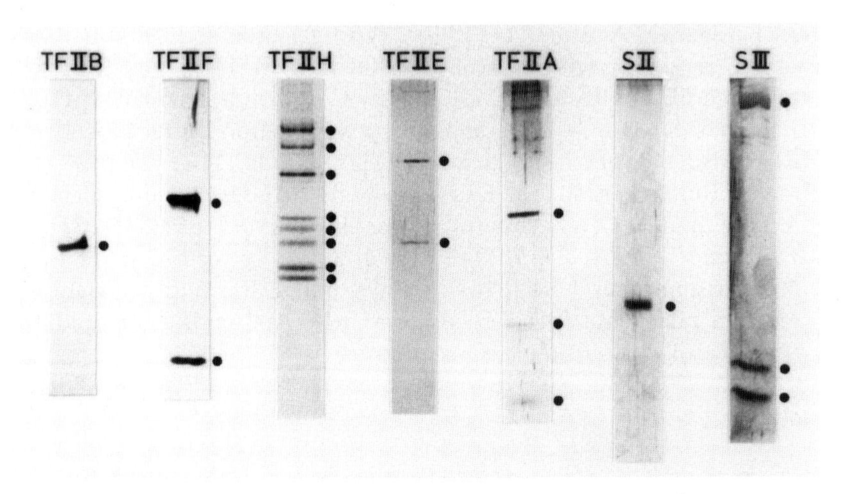

FIG. 2. Sodium dodecyl sulfate-polyacrylamide gel analysis of RNA polyacrylamide gel analysis of RNA polymerase II general transcription factors from rat liver. Subunits of purified general transcription factors are indicated by black circles.

(Pentex, fraction V) is from ICN Immunogiologicals (Costa Mesa, CA). Polyvinyl alcohol (PVA, type II), *Torula* yeast RNA, and heparin are from Sigma. Schwarz/Mann ultrapure ammonium sulfate and sucrose are from ICN Biomedicals, Inc. Glycerol (spectranalyzed grade) is from Fisher (Pittsburgh, PA). Acetylated BSA and RNasin are from Promega (Madison, WI). Formalin-fixed *Staphylococcus aureus* is from Bethesda Research Laboratories/Life Technologies (Gaithersburg, MD).

Chromatography Supplies

Phosphocellulose (P11) and DEAE-cellulose (DE-52) are from Whatman (Clifton, NJ). Phosphocellulose is precycled according to manufacturer instructions and washed once before use with three packed column volumes of buffer A (see below) containing 0.1 *M* KCl and BSA (0.2 mg/ml). Ultrogel AcA 34 is from IBF Biotechnics (Columbia, MD). Phenyl-Sepharose 6 FF (low sub), CM-Sephadex (C-25), and DEAE-Sephadex (A-25) are from Pharmacia. BioGel HTP and HPHT are from Bio-Rad (Richmond, CA). Collodion bags are from Sartorius (Gottingen, Germany). All high-performance liquid chromatography (HPLC) is performed using a Beckman System Gold chromatograph.

Buffers

TMSD is 10 m*M* Tris–HCl (pH 7.5), 1.5 m*M* $MgCl_2$, 0.25 *M* sucrose, 0.5 m*M* dithiothreitol (DTT), and 0.5 m*M* PMSF. Buffer A is 20 m*M* *N*-2-hydroxyethylpiperazine-*N'*-2-ethanesulfonic acid (HEPES)–NaOH (pH 7.9), 1 m*M* EDTA, 1 m*M* DTT, 10% (v/v) glycerol, and 0.5 m*M* PMSF. Buffer B is 50 m*M* Tris–HCl (pH 7.9), 0.1 m*M* EDTA, 1 m*M* DTT, 10% (v/v) glycerol, and 0.5 m*M* PMSF. Buffer C is 40 m*M* Tris–HCl (pH 7.9), 0.5 m*M* EDTA, 1 m*M* DTT, and 10% (v/v) glycerol. Buffer D is 40 m*M* HEPES–NaOH (pH 7.9), 0.5 m*M* EDTA, 1 m*M* DTT, and 10% (v/v) glycerol. Buffer E is 40 m*M* HEPES–NaOH (pH 7.9), 0.1 m*M* EDTA, 1 m*M* DTT, and 10% (v/v) glycerol. Buffer G is 40 m*M* HEPES–NaOH (pH 7.0), 0.5 m*M* EDTA, 1 m*M* DTT, and 10% (v/v) glycerol. Buffer I is 10 m*M* potassium phosphate (pH 7.5) and 1 m*M* DTT. Buffer P is 10 m*M* potassium phosphate (pH 7.6), 0.1 m*M* EDTA, 1 m*M* DTT, and 10% (v/v) glycerol.

Methods

Runoff Transcription Assay of General Initiation Factors

The general initiation factors are purified as described below. Reaction mixtures (60 μl) containing 20 m*M* Tris–HCl (pH 7.9), 20 m*M* HEPES–

NaOH (pH 7.9), 60 mM KCl, 0.1 mM EDTA, 1 mM DTT, 3% (v/v) glycerol, BSA (0.5 mg/ml), 2% (w/v) PVA, 6 units of RNasin, 200 ng of *Nde*I-digested pDN-AdML plasmid DNA[4] [which contains adenovirus ML (AdML) core promoter sequences from −50 to +10 and directs synthesis of a 254-nucleotide runoff transcript], and approximately 10 ng of TFIIA, 5 ng of TFIIB, 60 ng of TFIID, 20 ng of TFIIE, 10 ng of TFIIF, 40 ng of TFIIH, and 0.003 unit of RNA polymerase IIA,[5] are preincubated for 30 min at 28°. Transcription is initiated by addition of 50 μM ATP, 50 μM UTP, 10 μM CTP, 10 μCi of [α-^{32}P]CTP, and 7 mM $MgCl_2$. After 3 min, heparin (10 μg/ml) or 0.5% (w/v) Sarkosyl is added to prevent further initiations, and 50 μM GTP is added to allow synthesis of full-length runoff transcripts. After incubation for 30 min at 28°, reaction mixtures are diluted with an equal volume of 0.2 M Tris–HCl (pH 7.5), 25 mM EDTA, 0.3 M NaCl, 2% (w/v) sodium dodecyl sulfate (SDS), digested with proteinase K at 1 mg/ml for 10 min at room temperature, and ethanol precipitated with 20 μg of torula yeast RNA as carrier. Runoff transcripts are suspended in 9 M urea, 0.025% (w/v) bromphenol blue, and 0.025% (w/v) xylene cyanol FF and analyzed by electrophoresis through 6% (w/v) polyacrylamide, 7 M urea gels.

Assay of SII

SII is an RNA polymerase II-binding protein that activates a nascent RNA cleavage activity in elongation complexes and permits readthrough of intrinsic arrest sites by RNA polymerase II.[6] Readthrough can be assayed in standard runoff transcription assays using templates containing transcription arrest sites. The cleavage activity is more difficult to detect unless elongation complexes are washed free of ribonucleoside triphosphates to prevent renewed elongation of truncated RNA chains. Preparation of washed elongation complexes allows both assays to be conducted side by side by withholding or restoring ribonucleoside triphosphates with SII.

Assembly of Arrested RNA Polymerase II Elongation Complexes. The plasmid pADTerm-2 contains a *Taq*I fragment from the human histone H3.3 gene, which bears a transcription arrest site for RNA polymerase II (site Ia).[7] The fragment was inserted into the unique *Acc*I site of pDN-AdML. When cleaved with *Nde*I this plasmid yields a runoff transcript of

[4] R. C. Conaway and J. W. Conaway, *J. Biol. Chem.* **263,** 2962 (1988).

[5] H. Serizawa, R. C. Conaway, and J. W. Conaway, *Proc. Natl. Acad. Sci. U.S.A.* **89,** 7476 (1992).

[6] D. Reines, *in* "Transcription: Mechanisms and Regulation" (R. C. Conaway and J. W. Conaway, eds.), p. 263. Raven, New York, 1994.

[7] D. Reines, M. J. Chamberlin, and C. M. Kane, *J. Biol. Chem.* **264,** 10799 (1989).

530 nucleotides and a transcript of 205 nucleotides when arrest occurs at site Ia. Transcription typically employs partially purified RNA polymerase II and general transcription factors isolated from rat liver[7] by a modification of the methods of Conaway *et al.*[8] One hundred nanograms of DNA template is incubated with rat liver RNA polymerase II (0.5 μg; DEAE-Sephadex fraction) and fraction D (TFIIH and TFIID; 2 μg, CM-Sephadex fraction) in 20 μl of 20 m*M* HEPES–NaOH (pH 7.9), 20 m*M* Tris–HCl (pH 7.9), 2% (w/v) PVA, acetylated BSA (0.4 mg/ml), 12 units of RNAguard, 0.15 *M* KCl, 2 m*M* DTT, and 7% (v/v) glycerol for 30 min at 28°. Thirty-three microliters of a solution containing TFIIF/E (1 μg; BioGel HTP fraction) and TFIIB (3 ng; TSK phenyl-5PW fraction) in the same buffer without KCl is then added, and the incubation is continued for another 20 min. $MgCl_2$, ATP, UTP, and [α-^{32}P]CTP are added in 6 μl to final concentrations of 7 m*M*, 20 μM, 20 μM, and approximately 0.6 μM, respectively. In the absence of GTP, ternary complexes containing a 14-nucleotide transcript are synthesized because the first G residue appears in the transcript at position +15. After 20 min, heparin (10 μg/ml) is added to ternary complexes followed by more of each ribonucleoside triphosphate (800 μM), and incubation is continued for 15 min at 28°.

Immunoprecipitation and Washing of Elongation Complexes for SII-Mediated Cleavage and Readthrough Assays. Elongation complexes are immunoprecipitated with a monoclonal antibody (D44) against RNA developed by Eliat *et al.*[9] Elongation complexes are made 10 μg/ml in protein A-purified D44 IgG and incubated at 4° for 5 min. Ten microliters of formalin-fixed *S. aureus* washed in reaction buffer containing 20 m*M* Tris–HCl (pH 7.9), 3 m*M* HEPES–NaOH (pH 7.9), 62 m*M* KCl, 2.2% (w/v) PVA, 3% (v/v) glycerol, 2 m*M* DTT, 0.5 m*M* EDTA, and acetylated BSA (0.3 mg/ml) is added to each reaction equivalent and incubated at 4° for 5 min. Complexes are collected by centrifugation at 16,000 *g* for 2 min at 4° in a microcentrifuge. Complexes are washed in 1.2 vol of reaction buffer by gentle resuspension in reaction buffer and centrifugation. The final resuspension is in 55 μl of reaction buffer. Three cycles of resuspension and centrifugation are sufficient to reduce nucleotide concentrations to submicromolar levels. SII activity can be followed by assaying the nascent RNA cleavage reaction where SII is mixed with elongation complexes and 7 m*M* $MgCl_2$ at 28°. The extent of cleavage is a function of SII concentration and incubation time. The propensity of elongation complexes to hydrolyze their RNA chains varies between different elongation complexes, with

[8] J. W. Conaway, M. W. Bond, and R. C. Conaway, *J. Biol. Chem.* **262,** 8293 (1987).
[9] D. Eilat, M. Hochberg, R. Fischel, and R. Laskov, *Proc. Natl. Acad. Sci. U.S.A.* **79,** 3818 (1982).

arrested (SII-dependent) complexes cleaving more rapidly at a given concentration of SII, or requiring lower SII concentrations to achieve half-maximal cleavage, than complexes stalled at other template locations (SII-independent) by nucleotide starvation. Hence, kinetic experiments are required to assess the activity of an SII preparation. Readthrough activity of an SII preparation can be determined using arrested complexes that have not been isolated using the above-described anti-RNA immunoselection or by restoring ribonucleoside triphosphates and $MgCl_2$ to such washed complexes.

Assay of SIII

Transcription reaction conditions are the same as those used for assays of the general initiation factors except that (1) 50 ng of recombinant yeast TATA-binding protein (TBP; AcA 44 fraction)[10] replaces TFIID; (2) transcription is initiated by addition of 50 μM ATP, 1 μM UTP, 10 μM CTP, 50 μM GTP, and 10 μCi of [α-^{32}P]CTP; and (3) heparin or Sarkosyl are not added to reaction mixtures. Note that SIII increases the rate of elongation of runoff transcripts, so the optimal length of reaction incubations must be determined by kinetic measurements.

Protein Determination

Protein concentrations are measured using the protein dye assay (Bio-Rad) according to manufacturer instructions. Bovine serum albumin is the standard.

Resolution and Purification of RNA Polymerase II General Transcription Factors

Preparation of Rat Liver Homogenate

Fifty to 60 male Sprague-Dawley rats weighing 200–300 g each are fasted overnight (with water *ad libitum*) to reduce glycogen stores, which can complicate subcellular fractionation. Rats are anesthetized with ether or carbon dioxide and decapitated using a small rodent guillotine (Harvard Apparatus, South Natick, MA). Livers are rapidly excised and immersed in ice-cold TMSD. We note that failure to excise livers within 1–3 min after decapitation, or freezing livers at this stage, will result in significantly lower yields of general transcription factors.

[10] J. W. Conaway, J. P. Hanley, K. P. Garrett, and R. C. Conaway, *J. Biol. Chem.* **266,** 7804 (1991).

All further operations are carried out at 4° in a cold room. Livers (~500 g total) are minced into small pieces using scalpels, suspended in TMSD to a final volume of 1400 ml, and homogenized by two passes through a Ziegler–Pettit continuous-flow homogenizer.[11]

Preparation of Cytosolic and Nuclear Ammonium Sulfate Fractions

Subcellular fractionation of rat liver[8] is carried out by a modification of the methods of Fleischer and Kervina.[12] The homogenate is distributed to six 250-ml conical bottom polypropylene bottles (Corning, Corning, NY) and centrifuged for 10 min at 2000 rpm (800 *g*) in a J-6 centrifuge. The supernatants (cytosol) are pooled and centrifuged for 90 min at 13,500 rpm (28,000 *g*) in a JA-14 rotor. Solid $(NH_4)_2SO_4$ is then added slowly with stirring to the resulting postlysosomal supernatant to 38% saturation [0.213 g of $(NH_4)_2SO_4$ per milliliter]. Thirty minutes after the ammonium sulfate has dissolved, the pH is adjusted by addition of NaOH [1 μl of 1 *N* NaOH per gram of $(NH_4)_2SO_4$ added], and the suspension is centrifuged for 45 min at 9500 rpm (16,000 *g*) in a JA-10 rotor. The resulting pellets (cytosolic 0–38% ammonium sulfate fraction), which contain RNA polymerase II, can be quick-frozen in liquid nitrogen and stored at −80° without significant loss of activity. Solid $(NH_4)_2SO_4$ is then added slowly with stirring to the supernatant to 65% saturation [0.153 g of $(NH_4)_2SO_4$ per milliliter]. Thirty minutes after the ammonium sulfate has dissolved, the pH is adjusted by addition of NaOH [1 μl of 1 *N* NaOH per gram of $(NH_4)_2SO_4$ added], and the suspension is centrifuged for 45 min at 9500 rpm (16,000 *g*) in a JA-10 rotor. The resulting pellets (cytosolic 40–65% ammonium sulfate fraction), which contain TFIIB, TFIIF, and SII, can be quick-frozen in liquid nitrogen and stored at −80° without significant loss of activity.

The pellets (nuclei) of the first centrifugation step are washed twice more by resuspension in TMSD and centrifugation at 2000 rpm (800 *g*) in the J-6 centrifuge. Crude nuclei are then resuspended in TMSD to a final volume of 2000 ml and extracted with 0.32 *M* $(NH_4)_2SO_4$ by dropwise addition of 175 ml of saturated $(NH_4)_2SO_4$ with gentle stirring.[13] Thirty minutes after addition of ammonium sulfate, the extract is centrifuged for 90 min at 9500 rpm (16,000 *g*) in a JA-10 rotor. Solid $(NH_4)_2SO_4$ is then added slowly with stirring to the resulting postnuclear supernatant to 40% saturation [0.186 g of $(NH_4)_2SO_4$ per milliliter]. Thirty minutes after the ammonium sulfate has dissolved, the pH is adjusted by addition of NaOH [1 μl of 1 *N* NaOH per gram of $(NH_4)_2SO_4$ added], and the suspension is

[11] D. M. Ziegler and F. H. Pettit, *Biochemistry* **5,** 2932 (1966).
[12] S. Fleischer and M. Kervina, *Methods Enzymol.* **XXXI,** 6 (1974).
[13] R. C. Conaway and J. W. Conaway, *Proc. Natl. Acad. Sci. U.S.A.* **86,** 7356 (1989).

centrifuged for 45 min at 9500 rpm (16,000 *g*) in a JA-10 rotor. The resulting pellets (nuclear 0–40% ammonium sulfate fraction), which contain TFIIA, TFIID, TFIIE, TFIIH, and SIII, can be quick-frozen in liquid nitrogen and stored at $-80°$ for at least 2 years without significant loss of activity.

Purification of TFIIB, TFIIF, and SII

TFIIB,[8] TFIIF,[14] and SII can be purified to homogeneity from the cytosolic 38–65% ammonium sulfate fraction from as few as 50 rats. Here we describe purification of TFIIB, TFIIF, and SII from 1 kg of rat liver (~100 rats).

The 38–65% ammonium sulfate fraction from ~100 rats is dissolved in 200 ml of buffer A containing leupeptin and antipain at 10 μg/ml each. The solution is dialyzed 2–4 hr against buffer A, diluted with buffer A to a conductivity equivalent to that of buffer A containing 0.2 *M* KCl, and centrifuged for 10 min at 7000 rpm (7500 *g*) in a JA-14 rotor. The resulting supernatant is mixed with phosphocellulose (~30 mg of protein per milliliter packed column bed volume) equilibrated in buffer A containing 0.2 *M* KCl for 1 hr with occasional stirring in a column with the following dimensions: diameter $<$ height $<$ $2\times$ diameter. The slurry is then filtered at one to two packed column volumes per hour and washed at the same flow rate with buffer A containing 0.33 *M* KCl until the eluate reaches a conductivity equivalent to that of buffer A containing 0.33 *M* KCl and contains < 50 μg of protein per milliliter. TFIIB, TFIIF, and SII are eluted stepwise from phosphocellulose at the same flow rate with buffer A containing 0.6 *M* KCl. One-fifth column volume fractions are collected. Note that active fractions can be quick-frozen at this and all subsequent steps if necessary and stored at $-80°$ for at least 2 years without significant loss of activity.

Active fractions from phosphocellulose are dialyzed briefly against buffer A to reduce the concentration of KCl to 0.1–0.3 *M*. Solid $(NH_4)_2SO_4$ is then added slowly with stirring to this fraction to 65% saturation [0.4 g of $(NH_4)_2SO_4$ per milliliter]. Thirty minutes after the ammonium sulfate has dissolved, the pH is adjusted by addition of NaOH [1 μl of 1 *N* NaOH per gram of $(NH_4)_2SO_4$ added], and the suspension is centrifuged for 45 min at 13,500 rpm (28,000 *g*) in a JA-14 rotor. The ammonium sulfate precipitate is then dissolved in 15 ml of buffer A containing leupeptin (10 μg/ml) and antipain (10 μg/ml). The resulting solution is dialyzed in collodion bags against buffer A to a conductivity approximately equivalent to that of 1 *M* $(NH_4)_2SO_4$ and then centrifuged for 15 min at 10,000 rpm (12,000 *g*) in a JA-20 rotor. The resulting supernatant is applied at 35 ml/

[14] J. W. Conaway and R. C. Conaway, *J. Biol. Chem.* **264,** 2357 (1989).

hr to an Ultrogel AcA 34 column (2.6 × 100 cm) equilibrated in buffer A containing 0.4 *M* KCl but lacking PMSF. The column is eluted at the same flow rate, and 10-ml fractions are collected.

Purification of TFIIB and SII. Active TFIIB and SII fractions, which elute from AcA 34 with an apparent native molecular mass of ~40 kDa, are applied directly at three packed column volumes per hr to a BioGel HTP column (~1 mg of protein per milliliter packed column bed volume) that is equilibrated in buffer P and that has the following dimensions: diameter < height < 2× diameter. The column is washed at the same flow rate with buffer P containing 150 m*M* potassium phosphate (pH 7.6) until the eluate contains <50 μg of protein per milliliter. TFIIB and SII are eluted stepwise at the same flow rate with buffer P containing 400 m*M* potassium phosphate (pH 7.6). One-fifth column volume fractions are collected. Active fractions are diluted with an equal volume of buffer A containing 3.0 *M* $(NH_4)_2SO_4$ and centrifuged for 20 min at 35,000 rpm (100,000 *g*) in a 50 Ti rotor. The supernatant is applied at 1 ml/min to a TSK phenyl-5PW column (7.5 × 75 mm) equilibrated in buffer A containing 1.5 *M* $(NH_4)_2SO_4$ but lacking PMSF. TFIIB and SII are eluted at the same flow rate with a 30-ml linear gradient from buffer A containing 1.5 *M* $(NH_4)_2SO_4$ to buffer A. One-milliliter fractions are collected. TFIIB elutes from TSK phenyl-5PW at ~0.5 *M* $(NH_4)_2SO_4$. Active TFIIB fractions are then rechromatographed on TSK phenyl-5PW under the same conditions. At this stage TFIIB is >95% pure.

Active SII fractions, which elute from TSK phenyl-5PW at ~1.2 *M* $(NH_4)_2SO_4$, are dialyzed against buffer A containing 0.1 *M* KCl but lacking PMSF and applied at 1 ml/min to a TSK SP-5PW column (7.5 × 75 mm) equilibrated in the same buffer. SII is eluted at the same flow with a 30-ml linear gradient from 0.1 to 0.5 *M* KCl in buffer A lacking PMSF. One-milliliter fractions are collected. SII elutes from TSK SP-5PW at ~0.17 *M* KCl. At this stage SII is >95% pure.

Purification of TFIIF. Active TFIIF fractions, which elute from AcA 34 near the void volume, are dialyzed against buffer C to a conductivity equivalent to that of buffer C containing 90 m*M* KCl and centrifuged for 20 min at 30,000 rpm (100,000 *g*) in a 45 Ti rotor. The resulting supernatant is applied at 5 ml/min to a TSK DEAE-5PW column (21.5 × 150 mm) equilibrated in buffer C containing 90 m*M* KCl. TFIIF is eluted at the same flow rate with a 500-ml linear gradient from 90 to 320 m*M* KCl in buffer C. Ten-milliliter fractions are collected. The active fractions, which elute from TSK DEAE-5PW at ~0.2 *M* KCl, are dialyzed against buffer D to a conductivity equivalent to that of buffer D containing 0.1 *M* KCl and centrifuged for 20 min at 30,000 rpm (100,000 *g*) in a 45 Ti rotor. The resulting supernatant is applied at 1 ml/min to a TSK SP-5PW column

(7.5 × 75 mm) equilibrated in buffer D containing 0.1 *M* KCl. TFIIF is eluted at the same flow rate with a 30-ml linear gradient from 0.1 to 0.5 *M* KCl in buffer D. One-milliliter fractions are collected. The active fractions, which elute from TSK SP-5PW at ~0.2 *M* KCl, are diluted with an equal volume of buffer E containing 3.0 *M* $(NH_4)_2SO_4$ and centrifuged for 20 min at 35,000 rpm (100,000 *g*) in a 50 Ti rotor. The resulting supernatant is applied at 1 ml/min to a TSK phenyl-5PW column (7.5 × 75 mm) equilibrated in buffer E containing 1.5 *M* $(NH_4)_2SO_4$. TFIIF is eluted at the same flow rate with a 30-ml linear gradient from buffer E containing 1.5 *M* $(NH_4)_2SO_4$ to buffer E. One-milliliter fractions are collected. At this stage TFIIF is >95% pure.

Purification of TFIIA, TFIID, TFIIE, TFIIH, and SIII

TFIIA,[15] TFIID,[16] TFIIE,[10] TFIIH,[13,17] and SIII[18] as well as RNA polymerase IIA[5] can be purified from the nuclear extract. Here we describe purification of TFIIA, TFIID, TFIIE, TFIIH, and SIII from 1 kg of rat liver (~100 rats).

The nuclear 0–40% ammonium sulfate fraction from ~100 rats is dissolved in ~200 ml of buffer B containing leupeptin and antipain at 10 μg/ml each. The solution is then diluted with buffer B to a conductivity equivalent to that of buffer B containing 0.1 *M* $(NH_4)_2SO_4$ and centrifuged for 10 min at 7000 rpm (7500 *g*) in a JA-14 rotor. The resulting supernatant is mixed with DEAE cellulose (~10 mg of protein per milliliter packed column bed volume) for 1 hr with occasional stirring in a column that is equilibrated in buffer B containing 0.1 *M* $(NH_4)_2SO_4$ and that has the following dimensions: diameter < height < 2× diameter. The slurry is then filtered at one to two packed column volumes per hour and washed at the same flow rate with buffer B containing 0.1 *M* $(NH_4)_2SO_4$ until the eluate contains < 50 μg of protein per milliliter. The eluate contains TFIID, TFIIH, and SIII. TFIIA, TFIIE, and RNA polymerase IIA are eluted stepwise from DEAE-cellulose at the same flow rate with buffer B containing 0.5 *M* $(NH_4)_2SO_4$. One-fifth column volume fractions are collected.

For further purification of TFIID, TFIIH, and SIII, the 0.1 *M* $(NH_4)_2SO_4$ eluate from DEAE-cellulose is dialyzed briefly against buffer A to a conductivity equivalent to that of buffer A containing 0.15 *M* KCl and centrifuged for 10 min at 7000 rpm (7500 *g*) in a JA-14 rotor. The resulting supernatant

[15] T. Aso, H. Serizawa, R. C. Conway, and J. W. Conaway, *EMBO J.* **13,** 435 (1994).

[16] J. W. Conaway, D. Reines, and R. C. Conaway, *J. Biol. Chem.* **265,** 7552 (1990).

[17] J. W. Conaway, J. N. Bradsher, and R. C. Conaway, *J. Biol. Chem.* **267,** 10142 (1992).

[18] J. N. Bradsher, K. W. Jackson, R. C. Conaway, and J. W. Conaway, *J. Biol. Chem.* **268,** 25587 (1993).

is mixed with phosphocellulose (~15 mg of protein per milliliter packed column bed volume) for 1 hr with occasional stirring in a column that is equilibrated in buffer A containing 0.15 *M* KCl and that has the following dimensions: diameter < height < 2× diameter. The slurry is then filtered at one to two packed column volumes per hour and washed at the same flow rate with buffer A containing 0.15 *M* KCl until the eluate contains < 50 μg of protein per milliliter. TFIIH is then eluted stepwise from phosphocellulose at the same flow rate with buffer A containing 0.5 *M* KCl. One-fifth column volume fractions are collected. TFIID and SIII are eluted stepwise at the same flow rate with buffer A containing 1.0 *M* KCl. One-fifth column volume fractions are collected.

Purification of TFIIH. Active TFIIH fractions, which elute from phosphocellulose with 0.5 *M* KCl, are diluted with an equal volume of buffer A containing 2.0 *M* $(NH_4)_2SO_4$ and centrifuged for 10 min at 7000 rpm (7500 *g*) in a JA-14 rotor. The supernatant is applied at one to two packed column volumes per hour to a phenyl-Sepharose 6FF column (~15 mg of protein per milliliter packed column bed volume) that is equilibrated in buffer A containing 1.0 *M* $(NH_4)_2SO_4$ and that has the following dimensions: height = 2–3× diameter. TFIIH is eluted at the same flow rate with a 10× column volume linear gradient from buffer A containing 1.0 *M* $(NH_4)_2SO_4$ to buffer A. One-third column volume fractions are collected. Active fractions, which elute from phenyl-Sepharose at ~0.1 *M* $(NH_4)_2SO_4$, are dialyzed against buffer C to a conductivity equivalent to that of buffer C containing 50 m*M* KCl and centrifuged for 20 min at 30,000 rpm (100,000 *g*) in a 45 Ti rotor. The resulting supernatant is applied at 5 ml/min to a TSK DEAE-5PW column (21.5 × 150 mm) equilibrated in buffer C containing 50 m*M* KCl. TFIIH is eluted at the same flow rate with a 500-ml linear gradient from 50 to 300 m*M* KCl in buffer C. Ten-milliliter fractions are collected. The active fractions, which elute from TSK DEAE-5PW at ~0.2 *M* KCl, are dialyzed against buffer D to a conductivity equivalent to that of buffer D containing 0.1 *M* KCl and centrifuged for 20 min at 30,000 rpm (100,000 *g*) in a 45 Ti rotor. The resulting supernatant is applied at 1 ml/min to a TSK SP-5PW column (7.5 × 75 mm) equilibrated in buffer D containing 0.1 *M* KCl. TFIIH is eluted at the same flow rate with a 30-ml linear gradient from 0.1 to 0.4 *M* KCl in buffer D. One-milliliter fractions are collected. At this stage TFIIH, which elutes from TSK SP-5PW at ~0.23 *M* KCl, is >90% pure.

Purification of TFIID and SIII

Active TFIID and SIII fractions, which elute from phosphocellulose with 1.0 *M* KCl, are dialyzed against buffer A containing 0.5 *M* $(NH_4)_2SO_4$

for 2–3 hr to reduce the concentration of KCl to 0.1–0.3 *M*. Solid $(NH_4)_2SO_4$ is then added slowly with stirring to this fraction to ~60% saturation [0.3 g of $(NH_4)_2SO_4$ per milliliter]. Thirty minutes after the ammonium sulfate has dissolved, the pH is adjusted by addition of NaOH [1 μl of 1 *N* NaOH per gram of $(NH_4)_2SO_4$ added], and the suspension is centrifuged for 45 min at 13,500 rpm (28,000 *g*) in a JA-14 rotor. The ammonium sulfate precipitate is then dissolved in 4–5 ml of buffer G. The resulting solution is dialyzed briefly against buffer G to a conductivity equivalent to that of buffer G containing 0.5 *M* $(NH_4)_2SO_4$ and then centrifuged for 20 min at 35,000 rpm (100,000 *g*) in a 50 Ti rotor. The resulting supernatant is applied at 4 ml/min to a TSK SW4000 column (21.5 × 600 mm) equilibrated in buffer G containing 0.4 *M* KCl. The column is eluted at the same flow rate, and 5-ml fractions are collected. Active TFIID fractions, which elute from TSK SW4000 with an apparent native molecular mass of ~1300 kDa, are dialyzed against buffer A to a conductivity equivalent to that of buffer A containing 35 m*M* $(NH_4)_2SO_4$ and centrifuged for 15 min at 10,000 rpm (12,000 *g*) in a JA-20 rotor. The resulting supernatant is applied at one or two packed column volumes to a CM-Sephadex (C-25) column (~4 mg of protein per milliliter packed column bed volume) that is equilibrated in buffer A containing 35 m*M* $(NH_4)_2SO_4$ and that has the following dimensions: height = 2–3× diameter. The column is washed at the same flow rate until the eluate contains <50 μg of protein per milliliter. TFIID is eluted stepwise at the same flow rate with buffer A containing 140 m*M* $(NH_4)_2SO_4$. One-fifth column volume fractions are collected.

Purification of SIII. Active SIII fractions, which elute from TSK SW4000 with an apparent native molecular mass of ~140 kDa, are diluted with an equal volume of buffer E containing 2.0 *M* $(NH_4)_2SO_4$ and centrifuged for 20 min at 30,000 rpm (100,000 *g*) in a 45 Ti rotor. The resulting supernatant is applied at 5 ml/min to a TSK phenyl-5PW column (21.5 × 150 mm) equilibrated in buffer E containing 1.0 *M* $(NH_4)_2SO_4$. SIII is eluted at the same flow rate with a 500-ml linear gradient from buffer E containing 1.0 *M* $(NH_4)_2SO_4$ to buffer E. Ten-milliliter fractions are collected. The active fractions, which elute from TSK phenyl-5PW at ~0.4 *M* $(NH_4)_2SO_4$, are dialyzed against buffer D containing 50 m*M* KCl to a conductivity equivalent to that of buffer D containing 100 m*M* KCl and centrifuged for 20 min at 30,000 rpm (100,000 *g*) in a 45 Ti rotor. The resulting supernatant is applied at 1 ml/min to a TSK SP-5PW column (7.5 × 75 mm) equilibrated in buffer D containing 100 m*M* KCl. SIII is eluted at the same flow rate with a 50-ml linear gradient from 100 to 800 m*M* KCl in buffer D. One-milliliter fractions are collected. SIII elutes at ~350 m*M* KCl. At this stage SIII is >95% pure.

Purification of TFIIE and TFIIA

For further purification of TFIIE and TFIIA, the 0.5 *M* $(NH_4)_2SO_4$ fraction from DEAE cellulose is precipitated with ammonium sulfate. Solid $(NH_4)_2SO_4$ is added slowly with stirring to this fraction to ~60% saturation [0.35 g of $(NH_4)_2SO_4$ per milliliter]. Thirty minutes after the ammonium sulfate has dissolved, the pH is adjusted by addition of NaOH [1 μl of 1 *N* NaOH per gram of $(NH_4)_2SO_4$ added], and the suspension is centrifuged for 45 min at 13,500 rpm (28,000 *g*) in a JA-14 rotor. The ammonium sulfate precipitate is then dissolved in 10 ml of buffer A containing leupeptin (10 μg/ml) and antipain (10 μg/ml). The resulting solution is dialyzed in collodion bags against buffer A to a conductivity approximately equivalent to that of 1 *M* $(NH_4)_2SO_4$ and then centrifuged for 15 min at 10,000 rpm (12,000 *g*) in a JA-20 rotor. The resulting supernatant is applied at 35 ml/hr to an Ultrogel AcA 34 column (2.6 × 100 cm) equilibrated in buffer A containing 0.4 *M* KCl but lacking PMSF. The column is eluted at the same flow rate, and 10-ml fractions are collected.

Purification of TFIIE. Active TFIIE fractions, which elute from AcA 34 near the void volume, are dialyzed against buffer C to a conductivity equivalent to that of buffer C containing 70 m*M* KCl and then centrifuged for 20 min at 30,000 rpm (100,000 *g*) in a 45 Ti rotor. The resulting supernatant is applied at 5 ml/min to a TSK DEAE-5PW column (21.5 × 150 mm) equilibrated with buffer C containing 70 *M* KCl. TFIIE is eluted at the same flow rate with a 500-ml linear gradient from 70 to 375 m*M* KCl in buffer C. Ten-milliliter fractions are collected. The active fractions, which elute from TSK DEAE-5PW at ~0.2 *M* KCl, are dialyzed against buffer D to a conductivity equivalent to that of buffer D containing 40 m*M* KCl and then centrifuged for 20 min at 30,000 rpm (100,000 *g*) in a 45 Ti rotor. The resulting supernatant is applied at 1 ml/min to a TSK SP-5PW column (7.5 × 75 mm) equilibrated in buffer D containing 40 m*M* KCl. TFIIE is eluted at the same flow rate with a 40-ml linear gradient from 40 to 400 m*M* KCl in buffer D. One-milliliter fractions are collected. The active fractions, which elute from TSK SP-5PW at ~250 m*M* KCl, are diluted with an equal volume of buffer D containing 3.0 *M* $(NH_4)_2SO_4$ and then centrifuged for 20 min at 35,000 rpm (100,000 *g*) in a 50 Ti rotor. The resulting supernatant is applied at 1 ml/min to a TSK phenyl-5PW column (7.5 × 75 mm) equilibrated in buffer D containing 1.5 *M* $(NH_4)_2SO_4$. TFIIE is eluted at the same flow rate with a 50-ml linear gradient from buffer D containing 1.5 *M* $(NH_4)_2SO_4$ to buffer D. One-milliliter fractions are collected. TFIIE elutes at ~0.3 *M* $(NH_4)_2SO_4$. At this stage TFIIE is >90% pure.

Purification of TFIIA. Active TFIIA fractions, which elute from AcA 34 with an apparent native molecular mass of ~160 kDa, are dialyzed against buffer C to a conductivity equivalent to that of buffer C containing 70 m*M* KCl and then centrifuged for 20 min at 30,000 rpm (100,000 *g*) in a 45 Ti rotor. The resulting supernatant is applied at 5 ml/min to a TSK DEAE-5PW column (21.5 × 150 mm) equilibrated in buffer C containing 70 m*M* KCl. TFIIA is eluted at the same flow rate with a 500-ml linear gradient from 70 to 375 m*M* KCl in buffer C. Ten-milliliter fractions are collected. The active fractions, which elute from TSK DEAE-5PW at ~250 m*M* KCl, are dialyzed against buffer D to a conductivity equivalent to that of buffer D containing 50 m*M* KCl and then centrifuged for 20 min at 30,000 rpm (100,000 *g*) in a 45 Ti rotor. The resulting supernatant is applied at 1 ml/min to a TSK SP-5PW column (7.5 × 75 mm) equilibrated in buffer D containing 50 m*M* KCl. TFIIA does not bind to TSK SP-5PW under these conditions. The active fractions are diluted with an equal volume of buffer D containing 2.0 *M* $(NH_4)_2SO_4$ and then centrifuged for 20 min at 35,000 rpm (100,000 *g*) in a 50 Ti rotor. The resulting supernatant is applied at 1 ml/min to a TSK phenyl-5PW column (7.5 × 75 mm) equilibrated in buffer D containing 1.0 *M* $(NH_4)_2SO_4$. TFIIA is eluted at the same flow rate with a 30-ml linear gradient from buffer D containing 1.0 *M* $(NH_4)_2SO_4$ to buffer D. One-milliliter fractions are collected. The active fractions, which elute from TSK phenyl-5PW at ~0.2 *M* $(NH_4)_2SO_4$, are dialyzed in collodion bags against buffer I to a conductivity equivalent to that of buffer I and then centrifuged for 20 min at 35,000 rpm (100,000 *g*) in a 50 Ti rotor. The resulting supernatant is applied at 0.5 ml/min to a BioGel HPHT column (7.8 × 100 mm) equilibrated in buffer I. TFIIA is eluted at the same flow rate with a 27-ml linear gradient from 10 to 600 m*M* potassium phosphate in buffer I. One-milliliter fractions are collected. TFIIA elutes at ~0.2 *M* potassium phosphate.

Acknowledgments

Work in the authors' laboratories is supported by NIH Grant GM41628 and funds provided to the Oklahoma Medical Research Foundation by the H. A. and Mary K. Chapman Charitable Trust (J.W.C. and R.C.C) and by NIH Grant GM46331 and American Cancer Society Grant JFRA-394 (D.R).

[19] Reconstitution of TATA-Binding Protein-Associated Factor/TATA-Binding Protein Complexes for *in Vitro* Transcription

By JIN-LONG CHEN and ROBERT TJIAN

The transcription factor IID (TFIID) is a multiprotein complex comprised of the TATA-binding protein (TBP) and at least eight TBP-associated factors (TAFs).[1,2] The TBP subunit of TFIID was characterized first and found to be responsible for binding to the TATA box and directing basal levels of transcription.[3] However, later studies revealed profound functional differences between TBP and TFIID.[3] Most important, TBP alone is not sufficient to support transcriptional activation by upstream enhancer-binding factors.[4] Instead, activation of transcription requires a complement of TAFs acting as coactivators to mediate activation.[5] In addition, although TBP alone is sufficient to direct basal transcription, TAFs have been shown also to be essential to modulate basal transcription in a promoter-specific manner.[6,7] In some extreme cases, TBP alone barely supports basal transcription, whereas TFIID directs highly active initiation of transcription at these same promoters. Taken together, the TBP–TAF subunits of the TFIID complex play a central role in the control of RNA polymerase (Pol) II transcription. Therefore, a number of laboratories have set out to clone, express, and characterize biochemically all of the major subunits of TFIID. To dissect the function of TAFs, we have developed a method for assembling a variety of partial TAF/TBP complexes as well as holo-TFIID *in vitro*.[5]

In this chapter we describe our strategies for assembling functionally active TAF/TBP complexes.[5] First, methods we have developed to build TAF/TBP complexes *in vitro* are discussed. A general description of procedures used to prepare the building blocks, TBP and TAFs, needed for the assembly process follows. Finally, a cookbook-style protocol for the assembly of two different partial TAG/TBP complexes as well as recombinant holo-TFIID is included.

[1] R. Tjian and T. Maniatis, *Cell* **77,** 5 (1994).
[2] J. A. Goodrich and R. Tjian, *Curr. Opin. Cell Biol.* **6,** 403 (1994).
[3] N. Hernandez, *Genes Dev.* **7,** 1291 (1993).
[4] B. F. Puge and R. Tjian, *Cell* **61,** 1187 (1990).
[5] J.-L. Chen, L. D. Attardi, C. P. Verrijzer, K. Yokomori, and R. Tjian, *Cell* **79,** 93 (1994).
[6] C. P. Verrijzer, J.-L. Chen, K. Yokomori, and R. Tjian, *Cell* **81,** 1115 (1995).
[7] J. Kaufmann and S. Smale, *Genes Dev.* **8,** 821 (1994).

Assembly of TAF/TBP Complexes

Strategy and General Considerations

Given the fact that attempts to supplement transcription reactions with TAFs as free subunits failed to reconstitute most activities associated with TFIID, the preassembly of a a stable TAF/TBP complex appears to be a prerequisite for TFIID function. We have therefore used the results from our studies on subunit interactions within the TFIID complex as a blueprint to design a stepwise assembly protocol *in vitro*. The strategy for assembly of the TFIID complex is outlined in Fig. 1. In our current procedure, an epitope-tagged core subunit is first immobilized on protein

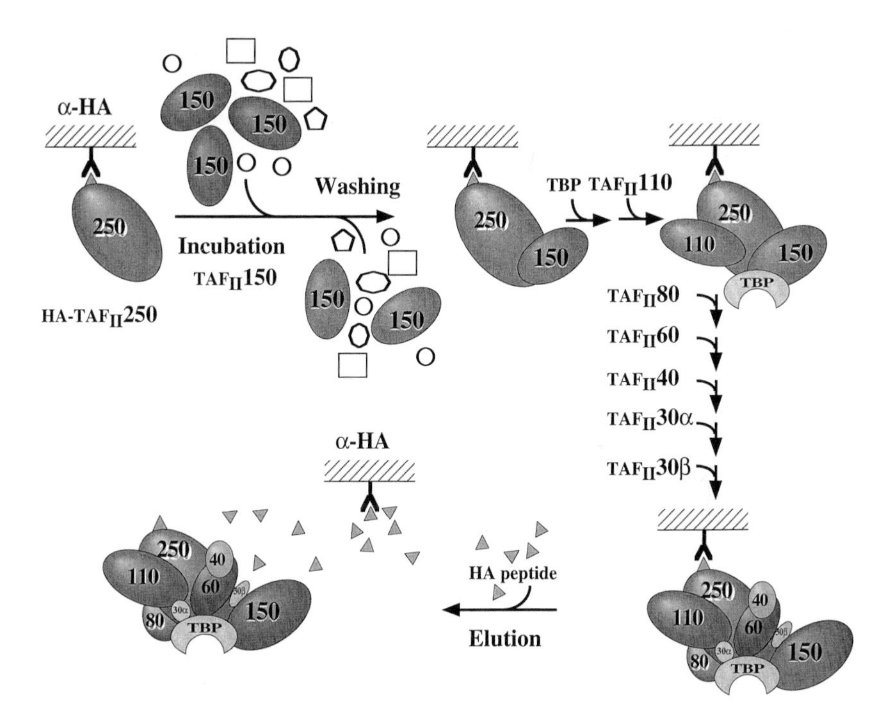

FIG. 1. The strategy of *in vitro* assembly of TBP/TAF complexes. As the first step, an epitope-tagged core subunit was immobilized on the beads. In this case, TAF$_{II}$250 fused to an N-terminal HA-epitope peptide was immobilized on protein A-Sepharose beads conjugated covalently with monoclonal antibodies directed against the HA-epitope. At each assembly step, the immobilized subunit or intermediate complexes were incubated with an excess of the next subunit. After incubation, excess subunits and impurities were removed by extensive washes. This procedure was repeated for each assembly step. Finally, recombinant holo-TFIID or partial complexes of interest were eluted under native conditions with the HA peptide and used in *in vitro* transcription reactions.

A–Sepharose beads covalently conjugated with either anti-HA (12CA5) or anti-FLAG (Kodak, Rochester, NY) monoclonal antibodies.[8,9] This strategy was chosen because it enables the assembled complexes to be conveniently eluted with peptides under native conditions. In addition to immunnoaffinity resins, other affinity resins, such as Ni-agarose for His-tagged proteins, can also be used.[10] More important, in this design, only the core subunit (first subunit) can possess the epitope tag that is recognized by the antibody attached to the beads. However, the other TAF subunits can bear different tags for the purpose of rapid purification prior to the assembly process.

Ha-tagged hTAF$_{II}$250 is often chosen as the "core" subunit and foundation for building complexes, because it has been shown to interact with TBP and many of the other TAFs. In this case, the FLAG epitope tag can be used to purify and to concentrate the other TFIID subunits. To ensure the formation of stoichiometric complexes, excess amounts (at least 10-fold) of each successive subunit are added to limiting amounts of the immobilized core subunit or intermediate complexes. After an incubation period, excess and therefore unincorporated subunits as well as any impurities are removed by extensive washing. This assembly process results in a substantial level of purification such that the resulting complexes usually are near homogenous even though relatively crude materials are used initially. In fact, one of the major advantages of this stepwise assembly is that it allows us to avoid the difficult and tedious task of purifying each of the TAFs and TBP individually. Not surprisingly, the concentration and quality of TAFs and TBP used as the starting material are usually more critical than the absolute purity of the materials in determining the success rate of reconstituting active TFIID complexes. In the following section, the preparation of each TFIID subunit is described.

Expression and Purification of Recombinant TAFs and TBP

The aim of this section is not to document the procedure for purifying TAFs and TBP to homogeneity, but instead to help prepare the building blocks for the assembly of TFIID complexes in the most expedient manner possible. To obtain functionally active proteins, all the recombinant subunits used in the assembly process are expressed in a baculovirus expression system, although in some cases *Escherichia*

[8] Q. Zhou, P. M. Lieberman, T. G. Boyer, and A. J. Berk, *Genes Dev.* **6,** 1964 (1992).

[9] C.-M. Chiang and R. G. Roeder, *Peptide Res.* **6,** 62 (1993).

[10] D. R. O'Reilly, L. K. Miller, and V. A. Luckow, *in* "Baculovirus Expression Vectors: A Laboratory Manual." W. H. Freeman, New York, 1992.

coli-expressed proteins can also be used. The general methodologies of baculovirus expression are described elsewhere.[10] Most of the expression constructs consist of full-length cDNAs clones into the pVL1392 (or pVL1393) baculovirus expression vector (PharMingen, San Diego, CA).[5] In addition, modified versions of these vectors are used for expression of N-terminal HA- or FLAG-tagged subunits.[5,6] All recombinant viruses are plaque purified and amplified. As mentioned earlier, we have found that good expression of each protein is one of the most decisive factors in the assembly process. Optimal conditions for the expression of each protein are determined empirically. In general, Sf9 (*Spodoptera frugiperda*, fall armyworm ovary) cells are infected with an MOI (multiplicity of infection) between 3 and 10 and then harvested between 36 and 60 hr postinfection; either whole-cell or nuclear extracts are then prepared from them. The standard buffer for protein preparations, purification, and assembly is HEMG [25 m*M* *N*-2-hydroxyethylpiperzine-*N'*-2-ethanesulfone acid (HEPES; pH 7.6), 0.1 m*M* EDTA, 12.5 m*M* $MgCl_2$, 10% (v/v) glycerol] containing 0.1% (v/v) Nonidet P-40 (N), 1.0 m*M* dithiothreital (D), a cocktail of proteinase inhibitors (I) [0.2 m*M* AEBSF (CalBiochem, La Jolla, CA), 1 m*M* sodium metabisulfite, pepstatin (0.7 mg/ml), leupeptin (2 mg/ml)], and varying concentrations of KCl. For convenience, 0.1-HEMG-NDI represents HEMG buffer plus Nonide P-40, dithiothreitol, proteinase inhibitors and 0.1 *M* KCl, while 0.4-HEMG-DI is a similar buffer with 0.4 *M* KCl, but no Nonidet P-40. The preparation of each protein is briefly described below.

Drosophila TATA-Binding Protein: Nuclear extracts from Sf9 cells infected with a recombinant baculovirus expressing dTBP are prepared, and purified by step-gradient chromatography on an SP-Sepharose column (Pharmacia, Piscataway, NJ) (0.1, 0.2, 0.4 *M* KCl).[11] dTBP elutes with 0.4 *M* KCl. To select for active molecules, dTBP is further purified on a DNA affinity column consisting of the adenovirus major late TATA box sequence $(GCTATAAAAGG)_3$ coupled to streptavidin–agarose (Pierce, Rockford, IL).[6] In the case of $TAF_{II}250$, we often use human $TAF_{II}250$ instead of the *Drosophila* protein, because $hTAF_{II}250$ is more highly expressed than $dTAF_{II}250$. Whole-cell extracts (0.4 *M* KCl) are prepared by sonicating infected Sf9 cells on 0.4-HEMG-NDI, followed by centrifugation. Usually the HA-$hTAF_{II}250$ sonicate can be applied directly to the HA immunoaffinity beads without further purification. However, if the results of the assembly process are not satisfactory, owing to a high level of impurities or low efficiency of incorporation, HA-$hTAF_{II}250$ can be purified over a

[11] R. Weinzierl, B. D. Dynlacht, and R. Tjian, *Nature (London)* **362,** 511 (1993).

Q-Sepharose column (Pharmacia) before the start of the assembly reactions. $hTAF_{II}250$ elutes in the 0.2–0.4 *M* step. Similarly, $TAF_{II}30\alpha$ and $TAF_{II}30\beta$ are also well expressed in the baculovirus system.[12] Whole-cell extracts (0.1 *M* KCl) containing these TAFs can be directly used for reconstitution experiments. By contrast, Sf9 extract expressing $TAF_{II}150$ is usually prepurified on SP-Sepharose and eluted in the 0.2–0.4 *M* KCl step.[13] Owing to the intrinsic properties of $TAF_{II}150$, such as its low solubility and sensitivity to proteolytic degradation, the efficiency of incorporation of $dTAF_{II}150$ into complexes varies considerably from preparation to preparation. To overcome this problem, we have developed a modified protocol for assembling $TAF_{II}150$-containing complexes. FLAG-tagged $TAF_{II}150$ (F-$dTAF_{II}150$) from whole-cell sonicate (0.4 *M* KCl) is first immobilized and purified on an anti-FLAG resin as the first step in assembly process. The detailed protocol is described in the next section. The FLAG immunopurification technique can also be used for the rapid purification of the other building blocks, F-$dTAF_{II}80$ and F-$dTAF_{II}40$.[14] For example, $dTAF_{II}80$ is ordinarily expressed at rather low levels and can be difficult to purify by conventional chromatography. However, using the anti-FLAG M2 immunopurification procedure, 40 ml of a $dTAF_{II}80$ crude Sf9 extract can be easily purified, concentrated on 50 μl of anti-FLAG beads, and eluted in 50 μl. In this way, the concentration of $TAF_{II}80$ can reach up to 1 mg/ml. In the case of $TAF_{II}110$, whole-cell extracts are prepared by sonication in 0.1-HEMG-NDI. The sonicate is passed over Q- and SP-Sepharose columns. The flowthrough material is suitable for use in the assembly of TAF/TBP complexes without the need of further purification. It should be noted that $TAF_{II}110$ will bind significantly to both columns, if run in the absence of Nonidet P-40. $TAF_{II}60$ is purified by DEAE-celluose column chromatography (DE-52; Whatman, Clifton, NJ), using a stepwise elution (0.1, 0.3 *M* KCl). The resulting $TAF_{II}60$ is partially purified and concentrated in the 0.3 *M* fraction (10–20% pure).

Protocols

The first protocol represents a typical procedure used for the assembly of a partial complex. The second protocol contains a modified version involving two sequential immunopurification steps. This facilitates the incorporation efficiency of certain subunits, such as $TAF_{II}150$. Because the

[12] K. Yokomori, J.-L. Chen, A. Adman, S. Zhou, and R. Tjian, *Genes Dev.* **7,** 2587 (1993).
[13] C. P. Verijzer, K. Yokomori, J.-L. Chen, and R. Tjian, *Science* **264,** 933 (1994).
[14] C. J. Thut, J.-L. Chen, R. Klem, and R. Tjian, *Science* **267,** 100 (1995).

basic principle and procedures are described in the first two protocols, we simply list the order of the assembly process and indicate some of the key points for assembly of holo-TFIID in the last protocol.

I. Assembly of $TAF_{II}250/TAF_{II}60/TAF_{II}40/TBP$ Complex

The following procedure is taken from Ref. 14.

1. Add 50 μl of a 30% (v/v) slurry of anti-HA beads (protein A–Sepharose beads covalently conjugated with anti-hemagglutinin antibodies) to a 1.5-ml Eppendorf tube. *Note:* For the preparation of anti-HA beads, the procedures are described in detail elsewhere.[8,15] The beads are stored in 0.1-HEMG containing 0.02% (v/v) NaN_3 at 4°.
2. Centrifuge at 4000 rpm, 4° for 30 sec to 1 min in a microcentrifuge.
3. Remove the supernatant by aspiration with a 23-gauge needle (cut the sharp tip off). A similar procedure is described in detail in *Antibodies: A Laboratory Manual.*[16]
4. Wash the beads once with 1 ml of 0.1-HEMG-NI.
   a. Add the buffer, resuspend the beads by inverting the tube, and occasionally flick it with your fingers. Avoid foaming!
   b. Spin at 4000 rpm, 4°, 30 sec–1 min.
   c. Remove the supernatant by aspiration as described in step 2.
5. Add 90 μl of an appropriately diluted Sf9 cell extract (see below) expressing $hTAF_{II}250$. To dilute extracts, add 2 vol of 0.1-HEMG-NDI to 1 vol of the $hTAF_{II}250$ containing extract (0.4 *M* KCl) slowly, and remove any precipitates by centrifugation at full speed in a microcentrifuge at 4° for 20 min. Ultracentrifugation (100,000 *g*) is recommended, if nonspecific background proteins are detected in the resulting complex.
6. Incubate for 2 hr at 4° with nutation (orbital rocker; nutator, Clay Adams).
7. After incubation, remove the supernatant as described in step 2 and 3.
8. Wash the beads twice, each time with 0.5 ml of 0.1-HEMG-NDI, as described in step 4. This washing step takes approximately 5 min in total.

[15] E. Harlow and D. Lane, *in* "Antibodies: A Laboratory Manual," p. 522. Cold Spring Harbor Laboratory Press, Cold Spring Harbor, NY, 1988.

[16] E. Harlow and D. Lane, *in* "Antibodies: A Laboratory Manual," p. 468. Cold Spring Harbor Laboratory Press, Cold Spring Harbor, NY, 1988.

9. Add 50 μl of diluted dTBP (see below). To dilute TBP, add 50 μl of 0.1-HEMG-NDI containing insulin 0.2 mg/ml to 3 μl of pure dTBP (~2 mg/ml), and then preclear the sample by centrifugation.
10. Incubate for 1.5 hr at 4° with nutation.
11. Remove the supernatant as described.
12. Wash twice with 0.5 ml of 0.1-HEMG-NDI for about 5 min.
13. Add 200 μl of the partially purified $dTAF_{II}60$ preparation. [Prior to addition, spin the extract in a microcentrifuge for 20 min (to preclear).]
14. Incubate for 2.5 hr at 4° with nutation.
15. Remove the supernatant.
16. Wash with 0.5 ml of 0.1-HEMG-NDI twice for 5 min.
17. Add 50 μl of diluted F-$dTAF_{II}40$ (see below). [Dilute 3 vol of immunopurified $dTAF_{II}40$ (~1 mg/ml) with 2 vol of 0.1-HEMG-NDI, and preclear by centrifugation.]
18. Incubate for 2 hr at 4° with nutation.
19. Remove the supernatant.
20. After the last component has been incorporated, wash the beads thoroughly four times with 1 ml of 0.1-HEMG-NDI. The final washing procedure usually takes more than 30 min.

Elution of Complex from Beads

21. Add 70 μl of the elution buffer [0.1-HEMG-NDI plus HA peptide (YPYDVPDYA; 1 mg/ml) and 200 μg of insulin per milliliter].
22. Incubate at 4° for 2 hr with nutation. Alternatively, if the efficiency of the elution is low, then the complex can be eluted at room temperature. Under this circumstance, the last wash immediately prior to the elution should be also carried out at room temperature.
23. Collect the eluate by certrifugation at 4000 rpm for 30 sec to 1 min in a microcentrifuge at 4°. Remove as much of the supernatant (eluate) as possible. It is not necessary to avoid the beads.
24. To remove any residual beads left in the supernatant, pass the supernatant through an empty spin column (1-ml Bio-spin; Bio-Rad, Richmond, CA). For this purpose, a spin column is placed in a 1.5-ml Eppendorf tube, and spun at 1000 rpm for several seconds in a microcentrifuge with the lid open.
25. The eluted complex is divided into 10 to 20-μl aliquots, frozen in liquid nitrogen, and stored at −80°.

II. Assembly of $TAF_{II}250/TAF_{II}150/TAF_{II}110/TBP$ Complex (Two-Way Immunoassembly Protocol)

Assembly of Stable $TAF_{II}250/TAF_{II}150$ Dimer on Anti-FLAG Beads (First Affinity Resin)

1. Add 200 μl of crude Sf9 extracts containing F-d$TAF_{II}150$ (0.4 *M* KCl) to 60 μl (120 μl of 50% slurry) of M2 anti-FLAG resins (Kodak). The beads are prewashed with 0.2-HEMG-NI (no dithiothreitol).
2. Incubate for 2 hr at 4° with nutation.
3. Remove the supernatant.
4. Add 1.2 ml of diluted HA-h$TAF_{II}250$ extract. [Dilute 1 vol of the HA-h$TAF_{II}250$ extract (0.4 *M* KCl) with 2 vol of 0.1-HEMG-NDI and spin. It is critical to add the buffer to the extract drop by drop while mixing to ensure that the salt concentration is decreased gradually to avoid precipitation.]
5. Incubate for 2.5 hr at 4° with nutation.
6. Remove the supernatant.
7. Wash twice with 1 ml of 0.16-HEMG-NDI.

Elution of F-$TAF_{II}150$/HA-$TAF_{II}250$ from Anti-FLAG Beads

8. Elute with 150 μl of elution buffer [0.16-HEMG-NDI plus FLAG peptide (DYKDDDDK; 0.2 mg/ml)] for 1.5 hr with nutation.
9. Collect the eluate as described in the previous protocol (steps 21–24).

Assembly of Higher-Order Complex on Anti-HA Beads (Second Affinity Resin)

10. Add the eluted dimeric intermediate to anti-HA beads as described in step 1 of protocol I.
11. Incubate for 2 hr at 4° with nutation.
12. Wash once with 0.5 ml of 0.16-HEMG-NDI, and then wash with 0.5 ml of 0.1-HEMG-NDI.
13. Add the third component-dTBP, incubate, and wash as described in Protocol I for the incorporation of dTBP.
14. Add 200 μl of the d$TAF_{II}110$ extract and incubate for 2.5 hr at 4°.
15. After the assembly process is complete, wash the beads extensively as described in protocol I.
16. Elute the complex with the HA peptide; the remaining steps are essentially identical to those in protocol I.

III. Assembly of Complete Nine-Subunit TFIID Complex

60 μl of anti-FLAG beads → + F-dTAF$_{II}$150 → + HA-hTAF$_{II}$250 → Elute with FLAG peptide →

Anti-HA beads → + dTBP → + dTAF$_{II}$110 → + 200 μl of F-TAF$_{II}$80 (0.1 M KCl) without dilution, for 3 hr → + dTAF$_{II}$60 →

+ F-dTAF$_{II}$40 → + dTAF$_{II}$30α 200-μl extract, 0.1 M KCl, 2 hr → + dTAF$_{II}$30β 100-μl extract, 0.1 M KCl, 2 hr → Wash extensively → Elute with HA peptide →

Assays for Assembled TAF/TBP Complexes

To determine the stoichiometry and purity of the assembled complexes, 5–10 μl of eluted material can be analyzed by sodium dodecyl sulfate-polyacrylamide gel electrophoresis (SDS–PAGE), followed by silver staining or Western blotting. The integrity of the eluted complexes can be verified by reimmunoprecipitation using antibodies directed against different subunits.[11] A relatively large amount (>60 μl) of the eluted complexes is required for one reimmunoprecipitation. In general, we examined the quality of each newly assembled complex by silver staining only, because we already knew that the procedures for a specific complex are highly reproducible. To assay transcriptional activities of the assembled complexes, a relatively simple TFIID-dependent *in vitro* transcription system can be utilized.[5,17]

Concluding Remarks

In this chapter, we have described various methods developed for reconstituting active TAF/TBP complexes. Using this powerful strategy, our studies have revealed that different classes of activators function through different mechanisms involving distinct sets of TAFs to regulate transcription.[5] Moreover, the functional analysis of partial complexes has shown that TAFs are also involved in core promoter recognition in basal transcription.[6] Reconstitution experiments have suggested that the multisubunit nature of the TFIID complex indeed contributes to transcriptional synergism and core promoter switching, two processes involved in important biological regulatory events.[18,19] Interestingly, there are additional TBP-containing

[17] J. D. Dignam, P. L. Martin, B. S. Shastry, and R. G. Roeder, *Methods Enzymol.* **101,** 582 (1983).

[18] F. Sauer, S. K. Hansen, and R. Tjian, *Science* **270,** 1783 (1995).

[19] S. K. Hansen and R. Tjian, *Cell* **82,** 565 (1995).

complexes that consist of distinct TAFs important for Pol I and Pol II transcription.[2,3] Using a similar assembly design, we have succeeded in building transcriptionally active SL1 complex from recombinant TBP and the three Pol I TAFs.[20,21] In addition to TBP-containing complexes, several other transcription factors are composed of multiple subunits.[22] For example, TFIIH has been shown to consist of at least 7 polypeptides, while RNA Pol II contains more than 10 subunits. Even though a great deal of progress has been made in the molecular cloning of those subunits, the reconstitution of functional multiprotein complexes has been a substantial obstacle. It is hoped that the development of an assembly procedure for the multisubunit TFIID complex described here will provide a paradigm for similar studies with other complexes.

Acknowledgments

We thank Edith Wang and Tim Hoey for helpful discussion. This work was supported in part by a grant from the National Institutes of Health to R.T.

[20] J. C. B. M. Zomerdijk, H. Beckmann, L. Comai, and R. Tjian, *Science* **266,** 2015 (1994).
[21] H. Beckmann, J.-L. Chen, T. O'Brien, and R. Tjian, *Science* **270,** 1506 (1995).
[22] S. Buratowski, *Cell* **77,** 1 (1994).

[20] Purification, Characterization, and Role of CCAAT-Binding Factor in Transcription

By SANKAR N. MAITY and BENOIT DE CROMBRUGGHE

Different CCAAT-Binding Proteins

Many eukaryotic class II promoters contain a pentanucleotide CCAAT motif that is often located around −80 upstream of the start of transcription.[1] To date several proteins have been isolated that bind to CCAAT motifs. The CCAAT-binding factor (CBF; also designated NF-Y and CP1) is one of these and is capable of forming a specific complex with a number of promoter DNAs containing a CCAAT motif (Table I)[2–12] Point muta-

[1] P. Bucher, *J. Mol. Biol.* **212,** 563 (1990).
[2] A. Hatamochi, P. T. Golumbek, E. V. Schaftingen, and B. de Crombrugghe, *J. Biol. Chem.* **263,** 5940 (1988).
[3] S. N. Maity, P. T. Golumbek, G. Karsenty, and B. de Crombrugghe, *Science* **241,** 582 (1988).
[4] A. Dorn, J. Bollekens, A. Staub, C. Benoist, and D. Mathis, *Cell* **50,** 863 (1987).
[5] J. Wuarin, C. Mueller, and U. Schibler, *J. Mol. Biol.* **214,** 865 (1990).

TABLE I
CCAAT-BINDING FACTOR BINDING SITE IN VARIOUS EUKARYOTIC PROMOTERS

Promoter	Sequence	Ref.
Mouse α2(1) collagen	−75 CTCCACCAATGGGAG −89	2
Mouse α1(1) collagen	−91 CCCAGCCAATCAGAA −105	3
Mouse MHC class II Eα	−47 TTTAACCAATCAGAA −61	4
Mouse albumin	−93 AGGAACCAATTGAAA −79	5
Human α-globin	−76 GCCAGCCAATGAGCG −62	6,7
Rat aldolase B	−136 ACGCGCCAATCAGAG −122	8
Human thrombospondin 1	−57 TCCGGCCAATGGGCG −71	9
Human argininosuccinate lyase	−84 CCGCGCCAATAGGAG −98	10
GRP78/BiP (glucose-regulated protein)	−103 GTTCACCAATCGGAG −89	11
Chicken β-actin	−96 GGCAGCCAATCAGAG −82	12
RSV LTR[b]	−121 TTCCACCAATCGGCA −135	2
	−57 TTCGTCCAATCCATG −85	
Advenovirus major late	−71 TATAACCAATCACCT −85	6

[a] Direct binding of CBF/NFY/CP1 to the CCAAT motifs that are shown have been demonstrated.

[b] RSV LTR, Rous sarcoma virus long terminal repeat.

tions in each base of the CCAAT motif of these DNAs resulted in either loss or decrease in formation of the DNA–CBF complex, indicating that DNA binding of CBF requires a high degree of conservation of the CCAAT sequence. Moreover, mutations in the CCAAT motif that inhibit formation of the CBF–DNA complex, also resulted in reduced transcription activity.[3–5,9–12] These and other results indicated that CBF is a transcription factor that can activate various eukaryotic genes.

Other CCAAT-binding polypeptides include CTF/NF1 and C/EBP, but the DNA sequences in the binding sites for these two proteins often do

[6] L. A. Chodosh, A. S. Baldwin, R. W. Carthew, and P. A. Sharp, *Cell* **53,** 11 (1988).

[7] C. G. Kim and M. Sheffery, *J. Biol. Chem.* **265,** 13362 (1990).

[8] M. Raymondjean, A.-L. Pichard, C. Gregori, F. Ginot, and A. Kahn, *Nucleic Acids Res.* **19,** 6145 (1991).

[9] P. Framson and P. Bornstein, *J. Biol. Chem.* **268,** 4989 (1993).

[10] T. Matsubasa, M. Takiguchi, I. Matsuda, and M. Mori, *J. Biochem.* **116,** 1044 (1994).

[11] B. Roy and A. S. Lee, *Mol. Cell. Biol.* **15,** 2263 (1995).

[12] W. W. Quitschke, Z.-Y. Lin, L. DePontozilli, and B. M. Paterson, *J. Biol. Chem.* **264,** 9539 (1989).

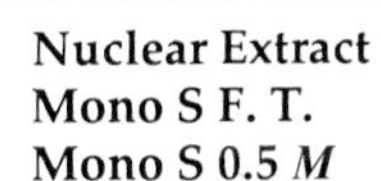

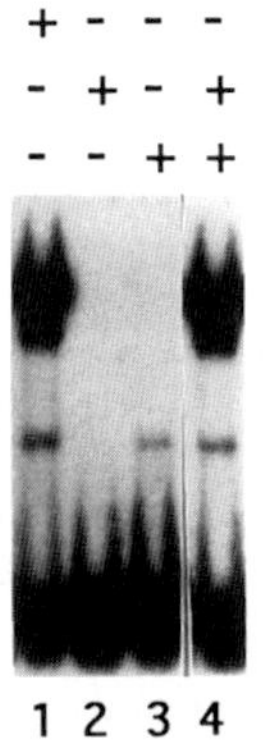

FIG. 1. Separation of CBF into two chromatographic fractions. Nuclear extracts from rat liver (10 mg of protein) were applied on a 1-ml Mono S column equilibrated with buffer 4 plus 75 m*M* NaCl. The column was first washed with the same buffer and bound protein was eluted with buffer 4 plus 500 m*M* NaCl. One microliter of each fraction was used in gel-retardation assays to monitor CBF activity. The major DNA–protein complex formed with nuclear extracts (lane 1) is the CBF–DNA complex. No CBF–DNA complex was formed either with flowthrough (lane 2) or with the 500 m*M* NaCl eluate (lane 3) fractions of Mono S. When the gel-retardation assay contains both the flowthrough and the 500 m*M* NaCl eluate fractions, the CBF–DNA complex was formed (lane 4).

not contain a complete CCAAT motif.[13,14] Hence, CBF can be distinguished from these two proteins on the basis of the sequence requirements in the DNA-binding sites.

DNA-Binding Activity of CCAAT-Binding Factor Separated into Two Chromatographic Fractions

A nuclear protein that binds to a DNA fragment of the $\alpha 2(1)$ collagen promoter containing the CCAAT sequence on the lower strand has been initially identified both in mouse NIH 3T3 and in rat liver nuclear extracts.[2] To characterize this protein, the nuclear extracts are fractionated by ion-exchange chromatography. Interestingly, the DNA-binding activity of CBF separates into two fractions: a flowthrough fraction and a salt eluate fraction of Mono S (Pharmacia, Piscataway, NJ). Each of these two fractions is unable to bind to DNA by itself, but when the two fractions are added together, the binding activity is reconstituted (Fig. 1). This result indicates that CBF is a heteromeric protein consisting of two chromatographic components designated A and B.[2]

[13] N. Mermod, E. A. O'Neill, T. J. Kelly, and R. Tjian, *Cell* **58,** 741 (1989).
[14] C. R. Vinson, P. B. Sigler, and S. L. McKnight, *Science* **246,** 911 (1989).

Purification of CCAAT-Binding Factor from Rat Liver Nuclear Extracts

To better understand the role of CBF in transcription, CBF is purified to homogeneity from rat liver nuclear extracts.[15,16]

Purification Scheme

When nuclear extracts are applied to a DNA affinity Sepharose column to which a double-stranded oligonucleotide corresponding to the sequence of the CCAAT motif of the mouse $\alpha 2(1)$ collagen promoter is covalently linked, CBF activity is tightly bound to the column and is eluted with 1 *M* NaCl buffer. A high concentration of EDTA (5 m*M*) is included in buffer solutions to prevent nucleases from degrading the DNA of the affinity resin. Thus a purification scheme has been developed to purify first the entire CBF molecule from nuclear extracts, using a DNA affinity resin, followed by FPLC (fast protein liquid chromatography) ion-exchange columns to purify each CBF component (Fig. 2).

Materials and Reagents

Sprague-Dawley rats: Purchase from Charles River, (Wilmington, MA)
Waring blender
LSC Homogenizer CH41 (Yamato, Orangeburg, NY)
DNA affinity resin: Affinity column consists of a double-stranded oligonucleotide [corresponding to the sequence between −105 and −65 of the $\alpha 2(1)$ collagen promoter] covalently linked through an eight-nucleotide long 5′ overhead of one DNA strand to CNBr-activated Sepharose 4B.[2] The coupling of DNA to the activated Sepharose beads is done according to protocols described by Kadonaga and Tjian.[17]
FPLC columns: Mono Q and Mono S ion-exchange columns are run on an FPLC system P500. The columns and the FPLC system are purchased from Pharmacia Biotech

Assay of CCAAT-Binding Factor Activity

Monitor CBF activity in each step of the purification by gel-retardation assay.

[15] S. N. Maity, T. Vuorio, and B. de Crombrugghe, *Proc. Natl. Acad. Sci. U.S.A.* **87,** 5378 (1990).
[16] T. Vuorio, S. N. Maity, and B. de Crombrugghe, *J. Biol. Chem.* **265,** 22480 (1990).
[17] J. T. Kadonaga and R. Tjian, *Proc. Natl. Acad. Sci. U.S.A.* **83,** 5889 (1986).

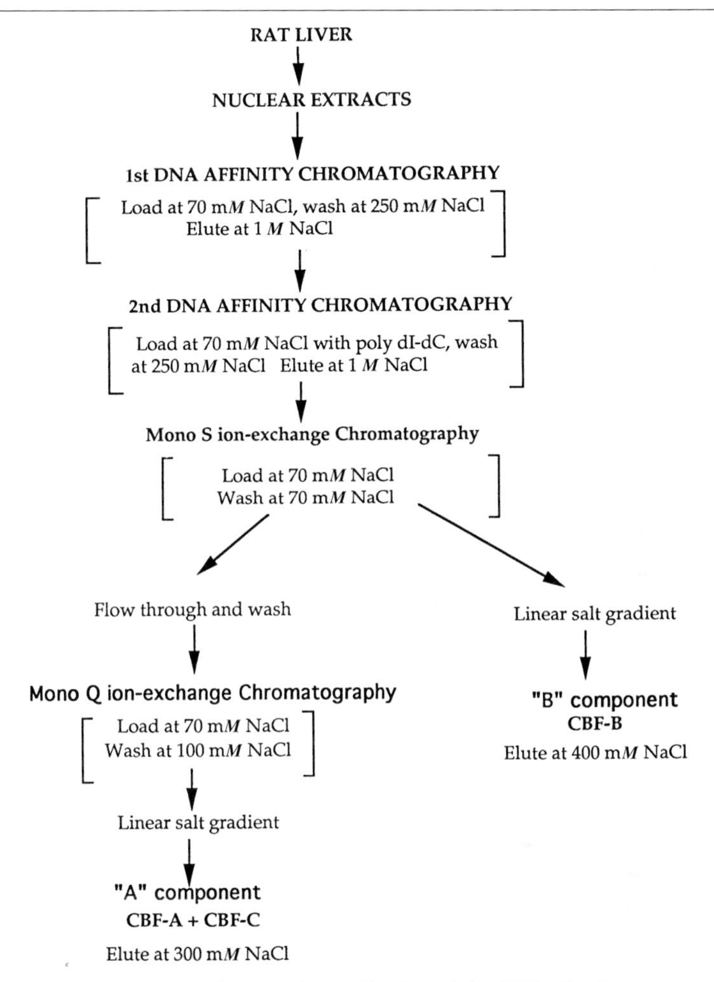

FIG. 2. Flow chart for the purification of the CBF subunits.

1. Label 1 pmol of the double-stranded oligonucleotide used to prepare the DNA affinity resin, with [α-^{32}P]dCTP and Klenow DNA polymerase, using standard protocol.

2. Add 1–2 μl of protein sample to a final total volume of 10 μl of a mixture containing 25 m*M* *N*-2-hydroxyethylpiperazine-*N'*-2-ethanesulfonic acid (HEPES, pH 7.9), 75 m*M* NaCl, 5 m*M* EDTA, 10% (v/v) glycerol, 0.05% (v/v) Nonidet P-40, 0.5 m*M* phenylmethylsulfonyl fluoride, 0.5 m*M* dithiothreitol (DTT), bovine serum albumin (0.25 mg/ml), 4 μg of poly[d(I-C)/d(I-C)], and 10 fmol of labeled oligonucleotide.

When highly purified preparations of CBF are used, reduce the amount of poly[d(I-C)/d(I-C)] to 0.5 μg. The DNA-binding activity of each CBF component is assayed after complementation with the other chromatographic CBF component.

3. Incubate the mixture for 10 min at room temperature. Add 0.5 μl of dye solution containing 0.1% (w/v) xylene cyanol and 0.1% (w/v) bromphenol blue.

4. Load the mixture on a 5% (w/v) polyacrylamide gel prepared in 0.5× TBE (44 m*M* Tris, 44 m*M* boric acid, and 1 m*M* EDTA) and fractionate by electrophoresis for 2 hr at 150 V in 0.5× TBE running buffer.

5. Dry the gel by vacuum dryer, and detect radiolabeled DNA bands in the gel by autoradiography. The electrophoretic mobility of a protein bound to DNA is slower than that of free DNA. The major protein–DNA complex in the autoradiogram is the CBF–DNA complex.

Preparation of Rat Liver Nuclear Extracts

1. Anesthetize Sprague-Dawley rats with carbon dioxide, remove livers by dissection, collect in ice-cold 0.9% (w/v) NaCl solution, and wash extensively in the same solution to remove blood.

2. Homogenize the washed tissue in a Waring blender four times (10 sec each) at low speed with buffer 1 (4 ml/g), containing 10 m*M* HEPES, (pH 7.9), 5 m*M* KCl, 1.5 m*M* $MgCl_2$, 0.1 m*M* EDTA, 0.25 *M* sucrose, 0.5% (v/v) Nonidet P-40, 0.5 m*M* phenylmethylsulfonyl fluoride, leupeptin (2 μg/ml), and pepstatin A (2 μg/ml). Further homogenize the extract in a motor-driven LSC homogenizer at 800 rpm, and then centrifuge at 2000 *g* for 10 min.

3. Discard the supernatant and resuspend the pellet containing the nuclear fraction in half of the initial volume of buffer 1 by gentle mixing. Recentrifuge as described above and discard the supernatant. Repeat this procedure by resuspending the pellet in one-quarter of the initial volume of buffer 1, centrifuge, and discard the supernatant.

4. Resuspend the pellet in 2 vol of buffer 2 (i.e., approximately 0.4 ml/g of liver) containing 50 m*M* HEPES, (pH 7.9), 1.5 m*M* $MgCl_2$, 0.1 m*M* EDTA, 0.5 m*M* dithiothreitol, 400 m*M* NaCl, 10% (v/v) glycerol, 0.5% (v/v) Nonidet P-40, 0.5 m*M* phenylmethylsulfonyl fluoride, leupeptin (2 μg/ml), and pepstatin A (2 μg/ml). Stir the mixture for 30 min at 4° and centrifuge at 80,000 *g* for 30 min. The supernatant is the nuclear extract.

Purification of CCAAT-Binding Factor Using DNA Affinity Resin

1. Dilute 1 vol of nuclear extract with 2 vol of buffer 3 containing 25 m*M* HEPES (pH 7.9), 10% (v/v) glycerol, 5 m*M* EDTA, 0.5% (v/v)

Nonidet P-40, 0.5 m*M* phenylmethylsulfonyl fluoride, 0.5 m*M* DTT, leupeptin (2 μg/ml), and pepstatin A (2 μg/ml). Centrifuge the diluted extract at 12,000 *g* for 30 min at 4°.

2. Load the supernatant onto the DNA affinity resin column (approximately 10 vol of nuclear extracts loaded on 1 vol of affinity resin). The resin should be preequilibrated with buffer 3 plus 70 m*M* NaCl.

3. Wash the resin sequentially with 10 column volumes of buffer 3 plus 70 m*M* NaCl and buffer 3 plus 250 m*M* NaCl.

4. Elute CBF from the resin with buffer 3 plus 1 *M* NaCl.

5. Dialyze the eluted CBF against 10 vol of buffer 3 plus 70 m*M* NaCl solution for 3 hr with a fresh buffer change each hour. Measure the conductivity of the dialyzed CBF, and if the salt concentration is greater than 100 m*M* NaCl, dilute the CBF solution with buffer 3 to 100 m*M* salt.

6. Centrifuge the CBF solution at 10,000 *g* for 30 min at 4° to remove any aggregate. Add poly[d(I-C)/d(I-C)] (5 μg/ml final concentration) to the supernatant, incubate for 10 min, and load the solution for the second time into the DNA affinity resin equilibrated with buffer 3 plus 70 m*M* NaCl (approximately 1 vol of resin for 30 vol of nuclear extracts).

7. Wash the resin with 10 vol of buffer 3 plus 250 m*M* NaCl solution, and elute CBF from the resin as before. The eluted sample contains a 250-fold higher specific DNA-binding activity per microgram of protein than that of crude liver nuclear extracts.

Purification of CCAAT-Binding Factor Components Using Fast-Protein Liquid Chromatography Columns

1. Dialyze the salt-eluted CBF sampler after the second DNA affinity chromatography against 10 vol of 75 m*M* NaCl in buffer 4 containing 25 m*M* HEPES (pH 7.9), 10% (v/v) glycerol, 1 m*M* EDTA, 0.05% (v/v) Nonidet P-40, 0.5 m*M* phenylmethylsulfonyl fluoride, 0.5 m*M* DTT, leupeptin (2 μg/ml), and pepstatin A (2 μg/ml), for 3 hr with fresh buffer change each hour. Measure conductivity of the dialyzed sample, and if the salt concentration is greater than 75 m*M* NaCl, dilute the sample with buffer 4 to 75 m*M* salt.

2. Equilibrate a 1-ml Mono S column with 10 column volumes of buffer 4 plus 75 m*M* NaCl. *Note*: All the operations with Mono S or Mono Q columns are run on an FPLC P500 system and the columns are maintained according to manufacturer protocols.

3. Apply the dialyzed sample on the equilibrated column. Collect the flowthrough material, which contains the "A" component of CBF.

4. Wash the column with 10 column volumes of buffer 4 plus 75 m*M* NaCl.

5. Elute the column with a salt gradient of 75 m*M* NaCl to 600 m*M* NaCl in buffer 4. Increase the salt concentration at a rate of 25 m*M*/ml/min and collect 1-ml fractions. Peak "B" activity elutes at 400 m*M* NaCl.

6. Equilibrate the Mono Q column with 10 column volumes of buffer 5 plus 75 m*M* NaCl. Buffer 5 contains all the component as in buffer 4 except that buffer 5 has 25 m*M* Tris–Cl, pH 7.9, instead of 25 m*M* HEPES, pH 7.9.

7. Load the flowthrough material of Mono S column on the equilibrated Mono Q column.

8. Wash the column with 10 column volumes of buffer 5 plus 75 m*M* NaCl.

9. Elute the column with a salt gradient of 75 m*M* NaCl to 500 m*M* NaCl in buffer 5. Run the gradient as above. Peak "A" activity elutes at 300 m*M* NaCl.

Polypeptides of "B" and "A" Components

The two purified CBF components are analyzed on a 10% (w/v) sodium dodecyl sulfate (SDS)-polyacrylamide gel and the gels are stained with either silver or Coomassie blue.

"B" Component

Two major polypeptides of 40 and 37 kDa are present in the "B" component. Both polypeptides possess "B" activity. DNA-binding activity of the polypeptides is determined by the following two methods.

Elution and Renaturation of Each Polypeptide from Gel according to Method of Hager and Burgess[18]

1. After electrophoresis, visualize the polypeptide bands with an ice-cold solution containing 0.25 *M* KCl and 1 m*M* DTT.

2. Excise each band and elute them separately in a buffer containing 50 m*M* Tris–Cl (pH 7.9), 0.1% (w/v) SDS, 0.1 m*M* EDTA, 1 m*M* DTT, and 0.2 *M* NaCl by incubating at room temperature overnight with gentle shaking.

3. Precipitate the protein by 80% (v/v) acetone, wash the pellet with acetone, and dry.

4. Resuspend the pellet in 2× buffer 4 containing bovine serum albumin (200 g/ml), and assay for DNA-binding activity.

Both polypeptides in the "B" component are renatured and bind to DNA after complementation with the "A" component.

[18] D. A. Hager and R. R. Burgess, *Anal. Biochem.* **109,** 76 (1980).

Southwestern Blot

1. After electrophoresis, transfer the protein from the gel to a nitrocellulose membrane.
2. Incubate the membrane with 5% (w/v) nonfat milk in binding buffer containing 25 m*M* HEPES (pH 7.9), 75 m*M* NaCl, 1 m*M* EDTA, 10% (v/v) glycerol, 0.05% (v/v) Nonidet P-40, 0.5 m*M* phenylmethylsulfonyl fluoride, and 0.5 m*M* DTT for 30 min at 4°.
3. Wash the membrane with 0.25% (w/v) nonfat milk in binding buffer, and then incubate with 100 fmol/ml ^{32}P-labeled (4000 cpm/fmol) double-stranded oligonucleotide used in the DNA-binding assay, poly[d(I-C)/d(I-C)] (5 μg/ml), 0.25% (v/v) nonfat milk, and purified "A" component (20 ng/ml) in binding buffer for 2 hr at 4°.
4. Wash the membrane for 15 min three times with 0.25% (v/v) nonfat milk in binding buffer at room temperature.
5. Detect the protein-bound DNA by autoradiography.

The two polypeptides in the "B" component bind to DNA with equal intensity. If the "A" component is omitted from the incubation mixture, none of the polypeptides forms a complex with DNA.

The polypeptides are digested with trypsin and the resulting peptides are separated by HPLC according to Stone *et al.*[19] Peptides are sequenced by automated Edman degradation, and the amino acid sequence information is utilized to isolate a cDNA clone corresponding to the polypeptides of the "B" component, designated CBF-B.[15] Recombinant CBF-B, which is expressed from the cDNA, binds to DNA after complementation with the chromatographic "A" component. Analysis of mouse CBF-B mRNA by Northern hybridization shows two mRNA species in different cell lines; one corresponds to full-length CBF-B and the other appears to be generated by alternate splicing.[20,21] Thus the two CBF-B polypeptides are most likely generated from the two species of mRNA.

"A" Component

Four major polypeptides of 45, 40, 32.5, and 32 kDa are present in the "A" component. Each polypeptide is eluted, renatured, and assayed for

[19] K. L. Stone, M. B. LoPresi, J. M. Crawford, R. De Angelis, and K. R. Williams, *in* "A Practical Guide to Protein and Peptide Purification for Microsequencing" (P. T. Matsudaira, ed.), pp. 31–48. Academic Press, New York, 1989.

[20] R. H. van Huijsduijnen, X. Y. Li, D. Black, H. Matthes, C. Benoist, and D. Mathis, *EMBO J.* **9,** 3119 (1990).

[21] X.-Y. Li, R. H. van Huijsduijnen, R. Mantovani, C. Benoist, and D. Mathis, *J. Biol. Chem.* **267,** 8984 (1992).

DNA binding as described above. Because our initial experiment showed that the 32.5- and 32-kDa doublet polypeptides bind DNA after complementation with the "B" component, tryptic peptides were isolated and sequenced. The amino acid sequences were utilized to isolate a cDNA corresponding to another CBF subunit, designated CBF-A. However, recombinant CBF-A expressed from the cDNA did not bind DNA after complementation with recombinant CBF-B. In contrast, when the two recombinant CBF subunits complemented the renatured 40-kDa species present in the "A" component, a specific complex with DNA was formed.[22] The 40-kDa polypeptide is the third CBF subunit, designated CBF-C, and a cDNA corresponding to this subunit was also isolated as described above. The CBF-A and CBF-C polypeptides, which are present in the chromatographic "A" component, are associated with each other in a tight complex, and the complex can be dissociated to individual subunits only by denaturing agents such as SDS or guanidine hydrochloride.[22]

Molecular Cloning of CCAAT-Binding Factor Subunits

Isolation of cDNAs. The cDNAs corresponding to each CBF subunit are isolated from a cDNA library according to the following method.[15,16,23] The sequences of two tryptic peptides (pep1 and pep2) are chosen for oligonucleotide synthesis. Sense and complementary antisense oligonucleotides corresponding to a portion of each peptide are used as primers in polymerase chain reaction (PCR) reactions. The sequences of these oligonucleotides are determined with the help of a codon usage table and also by the choice of either inosine residues or degenerate bases at certain places.[24] First-strand cDNA of poly(A)$^+$ RNA from rat liver is synthesized by reverse transcriptase using both oligo(dT) and random priming. Major DNA bands obtained after 40 cycles of amplification using two oligonucleotides primers are cloned by blunt-end ligation in Bluescript plasmid and sequenced. One of these PCR clones contained sequences adjacent to the primer sequences that encoded the expected amino acid sequences in pep1 and pep2. The PCR cDNA clones were then used to obtain cDNAs corresponding to the full sizes of the CBF polypeptides by screening a rat cDNA library constructed in λZAP (Stratagene, La Jolla, CA). After purification of the λ clone, the vector/insert was rescued from λ arms to the phagemid state according to the protocol developed by Stratagene.

[22] S. N. Maity, S. Sinha, E. C. Ruteshouser, and B. de Crombrugghe, *J. Biol. Chem.* **267,** 16574 (1992).

[23] S. Sinha, S. N. Maity, J. Lu, and B. de Crombrugghe, *Proc. Natl. Acad. Sci. U.S.A.* **92,** 1624 (1995).

[24] R. Lathe, *J. Mol. Biol.* **183,** 1 (1985).

Recombinant CCAAT-Binding Factor Subunits

Expression of CCAAT-Binding Factor Subunits. Each CBF subunit is expressed in *Escherichia coli* as a fusion polypeptide with glutathione *S*-transferase (GST) by cloning the cDNA of each subunit in frame with the coding sequence of GST in the bacterial expression vector, pGEX-2T or pGEX-4T3 (Pharmacia LKB).[22,23] The synthesis of fusion proteins is induced by adding 1 m*M* isopropyl-β-D-thiogalactopyranoside (IPTG) for 20 min at the midlog phase of bacterial growth, and the GST fusion proteins are purified from bacterial extracts using glutathione–agarose resin according to the method of Smith and Johnson.[25]

Each subunit is also expressed in *E. coli* without fusion with GST after the cDNA is introduced in the pET-23a vector (Novagen, Madison, WI), and the synthesis of the recombinant polypeptide is induced by IPTG in BL21(DE3)pLysS strain.[23] The recombinant subunits are purified over DEAE-cellulose followed by Mono S and Mono Q ion-exchange chromatography. Recombinant CBF-B binds to Mono S and is eluted at 350 m*M* NaCl. Recombinant CBF-A binds to Mono Q and elutes at 250 m*M* NaCl, whereas recombinant CBF-C is found in the flowthrough fractions of both Mono S and Mono Q columns.

The subunits are also expressed *in vitro* in a rabbit reticulocyte lysate after transcription/translation of the cDNA clones. The cDNAs of CBF-A and CBF-B are transcribed by either T3 or T7 RNA polymerase, depending on the orientation of the cDNA in Bluescript plasmid in the presence of the capping analog 7-methylguanosine(5)triphospho(5)-2-*O*-methylguanosine.[15,16] The resulting RNA is translated in the rabbit reticulocyte lysate (Promega). The CBF subunits are labeled by addition of [^{35}S]methionine in the translation reaction. The translation efficiency of the RNA produced from the isolated CBF-C cDNA clone is weak. To increase the efficiency of translation, the open reading frame of CBF-C is subcloned in the pCITE-2a (Novagen) vector, and then transcribed by T7 RNA polymerase followed by translation as described above.[23]

DNA Binding of Recombinant Subunits. A specific protein–DNA complex is formed when the DNA-binding reaction contains all three recombinant subunits (Fig. 3). No protein–DNA complex formation occurs with any of the three combinations of two recombinant subunits. All three subunits are present in the DNA–protein complex. Indeed, the mobility of the DNA–CBF complex is dependent on the size of each CBF subunit. When one CBF subunit is substituted with a homologous GST fused subunit, a DNA–protein complex is formed, which has a slower electrophoretic

[25] D. B. Smith and K. S. Johnson, *Gene* **67,** 31 (1988).

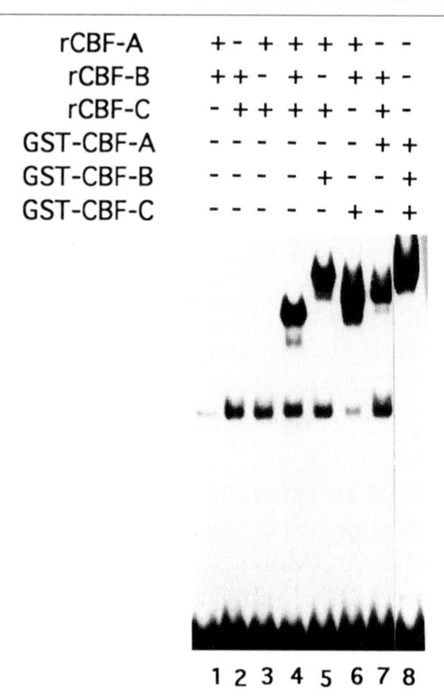

FIG. 3. DNA-binding activity of the three recombinant CBF subunits: gel-retardation assays using *E. coli*-synthesized CBF subunits. Bacteria transformed with PET-23a vectors containing the coding sequences of CBF-A, CBF-B, or CBF-C were induced with isopropyl-β-D-thiogalactopyranoside (IPTG). Two microliters of appropriate *E. coli* lysates was used as source of rCBF-A, rCBF-B, or rCBF-C as indicated. GST–CBF subunits were also synthesized in *E. coli* as fusion peptides with glutathione *S*-transferase (GST) and were purified from *E. coli* extracts using glutathione–agarose affinity resin. Ten nanograms of GST–CBF-A (lanes 7 and 8), 20 ng of GST–CBF-B (lanes 5 and 8), and 20 ng of GST–CBF-C (lanes 6 and 8) were used in the assays. No DNA binding occurred with any of the three combinations of two recombinant subunits (lanes 1–3). However, when three recombinant subunits were present, a CBF–DNA complex formed (lane 4). To test whether all three CBF subunits participated in this complex, each of these subunits was replaced in three separate reactions by its GST homolog, which adds 26 kDa to the size of each subunit. In each case (lanes 5–7), a slower mobility complex was produced, indicating that all three CBF subunits were present in the CBF–DNA complex. (Adapted from Ref. 23, with permission.)

mobility than that containing the nonfused subunit. Similar to native CBF, recombinant subunits do not form a complex with DNA containing a mutation in the CCAAT motif to CCAAA.[23]

Formation of CCAAT-Binding Factor Molecule. The CBF subunits interact with each other to form a ternary CBF complex in the absence of DNA binding.[23,26] The CBF subunit assembly follows a specific pathway

[26] S. N. Maity and B. de Crombrugghe, *J. Biol. Chem.* **267,** 8286 (1992).

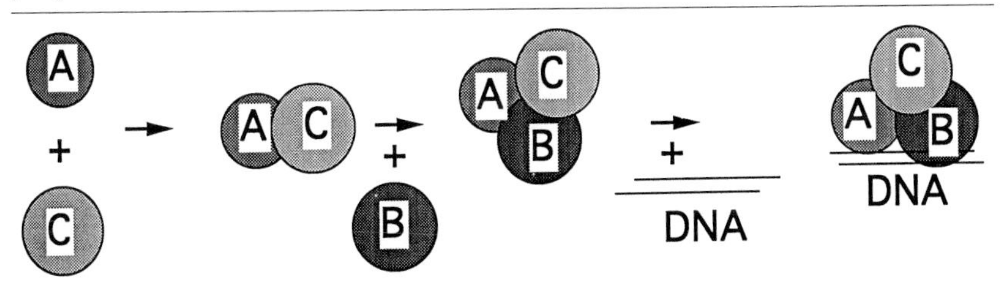

FIG. 4. Pathway for assembly of CBF subunits. The assembly of the CBF subunits is based on studies of subunit interaction between the CBF subunits and their mutants presented in Refs. 23 and 26a.

(Fig. 4). CBF-A and CBF-C first interact with each other to form a heterodimeric CBF-A/CBF-C complex. CBF-B does not interact with either CBF-A or CBF-C separately but interacts with the CBF-A/CBF-C complex to form a heterotrimeric CBF protein complex. Formation of the ternary complex is required for binding of CBF to DNA.

Segment of Amino Acid Sequence of Each CCAAT-Binding Factor Subunit Conserved during Evolution. Portions of the amino acid sequences of CBF-A, CBF-B, and CBF-C are evolutionarily conserved, showing a high degree of sequence identity with segments of yeast HAP3, HAP2, and HAP5, respectively (Fig. 5). In yeast, the products of these HAP genes are necessary for the expression of genes encoding components of the electron transport chains, and thus regulate respiratory function of this organism.[27–29] In each subunit the conserved portion is necessary for DNA binding.[26,26a]

CCAAT-Binding Factor to Activate Transcription of Several Eukaryotic Promoters

Native CBF, purified from rat liver extracts, activates transcription of several eukaryotic promoters, using an *in vitro* reconstitution assay with NIH 3T3 fibroblast nuclear extracts.[3] Activation of transcription requires specific binding of CBF to the promoter. Activation of transcription also occurs when the nuclear extracts is reconstituted with recombinant CBF-B and purified "A" component instead of the native CBF (Fig. 6).[30]

[26a] S. Sinha, I.-S. Kim, K.-Y. Sohn, B. de Crombrugghe, and S. N. Maity, *Mol. Cell Biol.* **16,** 328 (1996).

[27] S. Hahn, J. Pinkham, R. Weil, R. Miller, and L. Guarente, *Mol. Cell. Biol.* **8,** 655 (1988).

[28] J. L. Pinkham, J. T. Olesen, and L. Guarente, *Mol. Cell. Biol.* **7,** 578 (1987).

[29] D. S. McNabb, Y. Xing, and L. Guarente, *Genes Dev.* **9,** 47 (1995).

[30] F. Coustry, S. N. Maity, and B. de Crombrugghe, *J. Biol. Chem.* **270,** 468 (1995).

HOMOLOGY BETWEEN RAT CBF-A AND YEAST HAP3

```
CBF-A     53 REQDIYLPIANVARIMKNAIPQTGKIAKDAKECVQECVSEFISFITSEASERCHQEKR 110
             REQD +LPI NVAR+MKN +P + K++KDAKEC+QECVSE ISF+TSEAS+RC  +KR
HAP3      36 REQDRWLPINNVARLMKNTLPPSAKVSKDAKECMQECVSELISFVTSEASDRCAADKR 93

CBF-A    111 KTINGEDILFAMSTLGFDSYVEPLKLYLQKFR 142
             KTINGEDIL ++  LGF++Y E LK+YL K+R
HAP3      94 KTINGEDILISLHALGFENYAEVLKIYLAKYR 125
```

HOMOLOGY BETWEEN RAT CBF-B AND YEAST HAP2

```
CBF-B    260 YVNAKQYHRILKRRQARAKLEAEGKIPKERRKYLHESRHRHAMARKRGEGGRF 312
             YVNAKQY+RILKRR ARAKLE + +I +ER+ YLHESRH+HAM R RGEGGRF
HAP2     162 YVNAKQYYRILKRRYARAKLEEKLRISRERKPYLHESRHKHAMRRPRGEGGRF 214
```

HOMOLOGY BETWEEN RAT CBF-C AND YEAST HAP5

```
CBF-C     43 LPLARIKKIMKLDEDVKMISAEAPVLFAKAAQIFITELTLRAWIHTEDNKRRTLQRNDI 101
             LP ARIRK+MK DEDVKMISAEAP++FAKA  IFITELT RAW   E NKRRTLQ+ DI
HAP5     134 LPFARIRKVMKTDEDVKMISAEAPIIFAKACEIFITELTMRAWCVAERNKRRTLQKADI 192

CBF-C    102 AMAITKFDQFDFLI 115
             A A+ K D FDFLI
HAP5     193 AEALQKSDMFDFLI 206
```

FIG. 5. Amino acid sequence homologies between the CBF subunits and yeast HAP polypeptides. Each CBF subunit has been compared with the corresponding HAP polypeptide from *Saccharomyces cerevisiae,* using the BlastP program in GCG. Identical amino acid sequences are shown between the two sequences; +, conservative amino acid changes. CBF-A shows 83% sequence similarity with HAP3 and CBF-B shows 86% sequence similarity with HAP2 in the sequences that are shown.[15,16] These portions of the CBF-A and CBF-B sequences are identical between rat, mouse, and human. The segment of CBF-C between residues 42 and 115 shows 80% amino acid sequence similarity with HAP5.[23,29]

In Vitro Transcription Assay

1. Grow NIH 3T3 cells in Dulbecco's modified Eagle's medium supplemented with 10% (v/v) calf serum.

2. Harvest the cells at 80% confluence and prepare nuclear extracts as described by Dignam *et al.*,[31] except that leupeptin (10 g/ml) and pepstatin A (10 g/ml) are added to all buffers.

3. Incubate the nuclear extracts with the DNA affinity resin containing the CCAAT motif (1 vol of resin per 5 vol of extracts) for 30 min at 4° with gentle shaking. Centrifuge for 2 min in a microcentrifuge at top speed

[31] J. D. Dignam, R. M. Lebovitz, and R. G. Roeder, *Nucleic Acids Res.* **11,** 1475 (1983).

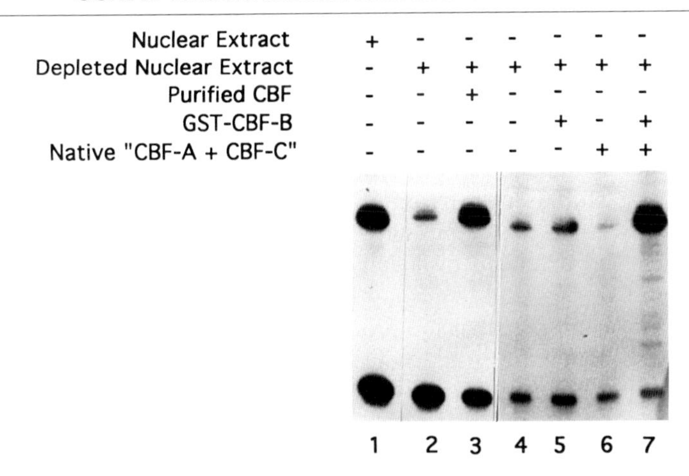

FIG. 6. Activation of transcription by CBF. Transcriptional activity of CBF was assayed using an *in vitro*-reconstituted transcription system with NIH 3T3 nuclear extracts.[3,30] The nuclear extracts were depleted of CBF by incubating with the DNA affinity resin containing a CBF-binding site. Two different promoter constructs, one containing $\alpha 2(1)$ collagen and the other containing $\alpha 1$(III) collagen sequences, were used for transcription. The CBF was purified from rat liver nuclear extracts by two cycles of DNA affinity chromatography and 180 ng of the purified protein was added to the depleted nuclear extracts (lane 3). The recombinant GST–CBF-B was synthesized and purified as described in Fig. 3, and 450 ng of the purified protein was added in the reactions (lanes 5 and 7). Native CBF-A + CBF-C ("A" component) was purified from rat liver nuclear extracts as described in Fig. 2, and 200 ng of the purified protein was added to the reactions (lanes 6 and 7). The CBF binds specifically with the $\alpha 2$(I) collagen promoter, but not with the $\alpha 1$(III) collagen promoter. Depletion of CBF from nuclear extracts resulted in substantial decrease in transcription of the $\alpha 2$(I) promoter but not of the $\alpha 1$(III) promoter (lanes 1 and 2). Addition of CBF to the depleted extracts increased specifically transcription of the $\alpha 2$(I) promoter (lane 3). Recombinant GST–CBF-B alone or nature CBF-A + CBF-C alone did not activate transcription (lanes 4–6). When the three subunits were reconstituted together, transcription of the $\alpha 2$(I) promoter increased sevenfold (lane 7).

at 4° and collect the supernatant, which contains nuclear extracts depleted of CBF.

4. Add 10 microliters of nuclear extracts to a final volume of 25 microliters containing 10 m*M* HEPES (pH 7.9), 10% (v/v) glycerol, 60 m*M* KCl, 5 m*M* $MgCl_2$, 2 m*M* spermidine, all four dNTPs (1 m*M* each), 300 ng of DNA template containing the mouse $\alpha 2$(I) collagen 350-bp proximal promoter with the CCAAT motif, and 150 ng of DNA template containing the mouse $\alpha 1$(III) collagen 100-bp proximal promoter as an internal control; this promoter does not contain a CCAAT motif in its sequence. Incubate the mixture at 30° for 1 hr.

5. Add 75 microliters of a solution containing 0.4 *M* sodium acetate (pH 5.2), 1% (w/v) SDS to the incubated mixture, extract the mixture with phenol–chloroform, and precipitate RNAs by ethanol.

6. Detect the specific RNAs by primer extension using two primers. Each primer contains a 21-nucleotide sequence corresponding to a part of the antisense strand downstream of the α2(1) or α1(III) promoter. Label the primers with [α-^{32}P]ATP using T4 polynucleotide kinase, hybridize with the RNAs, and extended by avian myeloblastosis virus (AMV) reverse transcriptase.[32]

7. Fractionate the extension product by 7 *M* urea, 8% (w/v) polyacrylamide gel electrophoresis in 1× Tris–borate–EDTA buffer, and analyze the gel by autoradiography.

Transcription of the α2(I) collagen promoter is reduced severalfold in CBF-depleted nuclear extracts compared to undepleted nuclear extracts, whereas transcription of α1(III) collagen promoter is similar in the two extracts. Addition of CBF to the depleted extracts resulted in stimulation of transcription of the α2(I) promoter, but not of the α1(III) promoter. Transcription of a mutant α2(I) promoter in which a mutation causes loss of CBF binding is not stimulated by CBF.

Amino acid sequences of both the amino-terminal part of CBF-B and the carboxyl-terminal part of CBF-C are rich in glutamine residues and hydrophobic amino acid residues.[15,23] These parts of both subunits are not necessary for DNA binding.[26,26a] In a heterologous *in vivo* transcription assay, the amino-terminal part of CBF-B, which was fused with the LexA DNA-binding domain, activated transcription of a promoter containing LexA-binding sites.[30] In an *in vitro* transcription assay, a recombinant CBF-B mutant deleted of this amino-terminal part was still able to activate after reconstitution with the "A" component. This suggest that CBF contains more than one activation domain, most likely one in CBF-B and another in CBF-C. The mechanism of transcription activation by CBF will be better understood in a reconstituted system using the three recombinant CBF subunits and their mutants, and in a more defined *in vitro* transcription system using purified or recombinant basal transcription components.

[32] M. P. Wickens, G. N. Buell, and R. T. Schmike, *J. Biol. Chem.* **253,** 2483 (1978).

[21] Purification, Assay, and Properties of RNA Polymerase I and Class I-Specific Transcription Factors in Mouse

By ANDREAS SCHNAPP and INGRID GRUMMT

In eukaryotes, ribosomal gene (rDNA) transcription is governed by a dedicated set of transcription factors and a specialized RNA polymerase, the DNA-dependent RNA polymerase I (Pol I). To characterize the biochemical properties of these factors it is necessary to purify them to homogeneity. The purified factors are required to analyze their function, to raise antibodies, as well as to isolate the genes encoding the proteins. It has been difficult, however, to isolate Pol I and the respective initiation and termination factors because they typically constitute less than 0.01% of the total cellular protein. In addition, most of the proteins are notoriously unstable and lose activity at low protein concentration or on freezing and thawing. This chapter deals first with a short introduction to the factors involved in initiation and termination of mouse rDNA transcription, and then describes the protocols used for the purification of the individual activities.

Background

The nomenclature for the protein factors required for rDNA transcription is confusing. Each laboratory has developed its own nomenclature and applied it to incompletely purified fractions rather than to pure components. Slight differences in the assay conditions used by different groups also lead to variations in results and thus to apparently different properties of the factors. Here we use the nomenclature TIF-I (for Pol I-specific transcription initiation factors), with letters appended to indicate different factors, to describe the basal factors required for transcription initiation of mouse rDNA. We use this nomenclature except for the upstream binding factor (UBF), a ubiquitous factor whose name is used by all laboratories. As with class II and III promoters, transcriptional initiation of class I genes involves the stepwise assembly of basal factors and Pol I on the ribosomal gene promoter to form a preinitiation complex. The first step in preinitiation complex formation involves binding of factor TIF-IB to the promoter. TIF-IB (the analogous human factor is called SL1) is a multiprotein complex that consists of the TATA-binding protein (TBP) and three TAFs, i.e., TBP-associated factors.[1,2] Binding of TIF-IB to the rDNA promoter is stabilized by cooperative interaction with UBF, another DNA-binding factor that belongs to the family of high mobility group (HMG) proteins. TIF-IB then recruits Pol I in association with two essential factors, termed TIF-IA and TIF-IC, to the template to form a preinitiation complex.[3] TIF-IA is a positive-acting factor whose amount or activity fluctuates in response to the nutritional state of the cell, and therefore plays a central role in the chain of events by which extracellular signals are transmitted from the cell surface to the nucleolus.[4] The second Pol I-associated factor, TIF-IC, serves a role both in initiation and elongation of transcription[5] and thus appears to be functionally homologous to the Pol II factor TFIIF. Termination of Pol I transcription is mediated by TTF-I, a 130-kDa DNA-binding protein that interacts with an 18-bp terminator element downstream of the murine rDNA transcription unit. The cDNA encoding TTF-I has been cloned and the recombinant protein has been shown to mediate transcription termination, i.e., stop of the elongation reaction and transcript release.[6]

Starting Material

The source of cells and the preparation of cell extracts represent critical steps in the establishment of a reconstituted transcription system, especially because these parameters influence the yield, activity, and purity of the individual proteins. Most important, extracts should be prepared from logarithmically growing cells because cellular rRNA synthetic activity fluctuates according to cell proliferation. Therefore, despite the expense and time consumption for the propagation of tissue culture cells, it is the method of choice for preparation of transcriptionally active extracts from which to purify the factors required to reconstitute a Pol I-dependent transcription system. Another important point concerns the origin of the cells that are used as starting material. The ribosomal transcription apparatus is specific to taxonomic orders, the promoter of one group being unrecognized by the transcription factors of another.[7] The disparity of Pol I promoter sequences is apparently in agreement with this species specificity. Because of this incompatibility of the transcription factor(s) of one species with the

[1] L. Comai, N. Tanese, and R. Tjian, *Cell* **68,** 965 (1992).
[2] D. Eberhard, L. Tora, J.-M. Egly, and I. Grummt, *Nucleic Acids Res.* **21,** 4180 (1993).
[3] A. Schnapp and I. Grummt, *J. Biol. Chem.* **266,** 24588 (1991).
[4] D. Buttgereit, G. Pflugfelder, and I. Grummt, *Nucleic Acids Res.* **13,** 8165 (1985).
[5] G. Schnapp, A. Schnapp, H. Rosenbauer, and I. Grummt, *EMBO J.* **13,** 4028 (1994).
[6] R. Evers, A. Smid, U. Rudloff, F. Lottspeich, and I. Grummt, *EMBO J.* **14,** 1248 (1995).
[7] I. Grummt, E. Roth, and M. Paule, *Nature (London)* **296,** 173 (1982).

rDNA promoter of a heterologous species, it is absolutely necessary that the cell extracts and the rDNA template originate from the same species. For humans, HeLa cells are commonly used. For mouse, cultured Ehrlich ascites cells, FM3A cells, or lymphoblastoma cells are convenient sources for the preparation of active extracts and purification of transcription factors. Because, at least in our hands, active extracts are obtained only from unfrozen cells, we scale up the cell culture to about 20–30 liters, and then harvest the cells and prepare the extract within the same day. The methods for preparation of cell extracts have been described in this series.[8] Usually nuclear or whole-cell extracts exhibit the highest transcriptional activity, but cytoplasmic extracts also contain significant amounts of Pol I and individual transcription factors and can be used for purification.

Experimental Procedures

Preliminary Considerations

Often a particular factor need not or cannot be obtained in a pure state, but it is essential that interfering activities (e.g., nucleases, proteases, and phosphatases) or contamination by other factors not be present. In our experience, attempting to purify most or all factors from a given amount of extract is an extremely frustrating endeavor. Instead, purifying so as to optimize yield and specific activity of a particular protein is likely to be more satisfying. Partially purified factors may be for many purposes sufficient to study their function. Moreover, the success of the purification is most dependent on appropriate assays. The assay should be specific, reliable, and sensitive. Quantitative accuracy is generally compromised, as each protein requires a different assay. The assays used to test the activity of the individual factors are described here, together with the fractionation procedure of the respective proteins.

Materials and Reagents

All chemicals should be of analytical-reagent grade. Most of the column matrices (e.g., DEAE-Sepharose CL-6B, CM-Sepharose, Q-Sepharose, protein A-Sepharose, ω-aminooctyl-agarose) and the FPLC columns (Mono Q HR 5/5, Mono S HR 5/5, Mono Q HR 10/10, S-Sepharose High-performance HP 16/10, and HiLoad 26/60 Superdex 200) are obtained from Pharmacia (Piscataway, NJ) Heparin-Ultrogel is from Serva (Heidelberg, Germany), and Bio-Rex 70 (200–400 mesh) is from Bio-Rad (Richmond,

[8] J. D. Dignam, *Methods Enzymol.* **182,** 194 (1990).

CA). Both HPLC columns (PEI-HPLC and GF-250) are purchased from Baker (Gross-Gerau, Germany).

Buffers

Buffers used for fast protein liquid chromatography (FPLC) and for high-performance liquid chromatography (HPLC) are degassed and filtered through a 0.22-μm pore size membrane (Sartorius, Göttingen, Germany).

Buffer A: 20 m*M* Tris–HCl (pH 7.9), 100 m*M* KCl, 0.2 m*M* EDTA, 0.5 m*M* dithioerythritol (DTE), 0.5 m*M* phenylmethylsulfonyl fluoride (PMSF), 20% (v/v) glycerol, plus KCl at the concentrations indicated by the numbers

Buffer AM: Same as buffer A, but with 5 m*M* $MgCl_2$

Buffer BM: Same as buffer AM, but with 20 m*M* *N*-2-hydroxyethylpiperazine-*N*′-2-ethanesulfonic acid (HEPES)–KOH instead of Tris–HCl

In Vitro Transcription Assay

The standard transcription assay contains, in a total volume of 25 μl, 10 to 100 ng of template DNA, 15 μl of either extract or purified transcription factors in 12 m*M* HEPES–KOH (pH 7.9), 0.1 m*M* EDTA, 0.5 m*M* DTE, 5 m*M* $MgCl_2$, 60–80 m*M* KCl, 10 m*M* creatine phosphate, 12% (v/v) glycerol, ATP, CTP, and UTP (0.66 m*M* each), 0.01 m*M* GTP, and 1–2 μCi of [α^{32}P]GTP. After incubation for 60 min at 30°, the RNA is purified by phenol–chloroform extraction, analyzed by gel electrophoresis, and visualized by autoradiography.

Fractionation of Extract Proteins

Step 1: Fractionation on DEAE–Sepharose

The first step in the fractionation procedure is chromatography on DEAE-Sepharose. This procedure is used to remove nucleic acids, to separate Pol I from the majority of Pol II and Pol III, and both to increase the specific activity and to concentrate Pol I and the initiation factors. Usually 5 mg of extract proteins is loaded per milliliter of resin in buffer AM-100 to a DEAE-Sepharose column. After binding, we wash the column and elute bound proteins with buffer A that does not contain $MgCl_2$. The use of a Mg^{2+}-free buffer at this early step of fractionation is essential if the growth-regulated factor TIF-IA is to be purified. Apparently the complex between Pol I and TIF-IA is dissociated under these conditions and both activities can be separated on the subsequent heparin column. On the

DEAE column, all proteins required for transcription initiation are eluted with a step of 280 m*M* KCl in buffer A. Compared to extracts, the DEAE-280 fraction exhibits a fivefold higher specific activity. As significant amounts of UBF remain bound to the column, residual UBF is eluted by a 1 *M* KCl step (DEAE-1000).

Step 2: Fractionation of DEAE-280 on Heparin–Ultrogel

To partially separate Pol I and different factor activities, the pooled DEAE-280 fractions (3 mg of protein/ml) are adjusted with buffer A to 200 m*M* KCl and fractionated on heparin–Ultrogel. The flowthrough fraction (H-200) is collected, the resin washed with buffer AM-250, and bound proteins are fractionated by step elution with buffer AM-400 and AM-600 to yield fractions H-400 and H-600, respectively. On this column, Pol I and individual factor activities are separated. None of the fractions alone or in combination with one another is sufficient to direct efficient transcription. Only the combination of H-200, H-400, and H-600 restores transcriptional activity. H-200 is the source of TIF-IA and TIF-IC. H-400 contains the majority of Pol I as well as significant amounts of UBF; H-600 fractions are the source for purification of TIF-IB.

Purification of RNA Polymerase I

Assay

The reactions contain 7.5 μl of the protein fractions, 15 μg of sheared calf thymus DNA, 12 m*M* HEPES–KOH, 0.12 m*M* EDTA, 85 m*M* KCl, 5 m*M* $MgCl_2$, 12% (v/v) glycerol, α-amanitin (100 μg/ml), 500 μM ATP, CTP and GTP, 2 μM UTP, and 1 μCi of [^{3}H]UTP in a final volume of 50 μl. After incubation for 20 min at 30°, reactions are terminated by the addition of 500 μl of cold 5% (w/v) trichloroacetic acid containing 40 m*M* sodium pyrophosphate. The precipitate is collected on glass fiber filters, washed three times with the same solution and once with ethanol, dried, and radioactivity is measured in a scintillation counter.

Fractionation Procedure

1. CHROMATOGRAPHY ON S-SEPHAROSE (FPLC)

a. Dialyze H-400 fractions against buffer BM-100.
b. Load on an S-Sepharose column (HP 16/10, 20 ml) at 10 mg of protein per milliliter of resin.

c. Elute with a 250-ml linear salt gradient from 150 to 450 m*M* KCl in buffer BM. RNA polymerase elutes together with UBF between 250 and 300 m*M* KCl.

2. Chromatography on Mono Q (FPLC)

a. Apply pooled fractions from S-Sepharose columns (60 mg of protein) in buffer AM-100 to a Mono Q column (HR 10/10, 8 ml).
b. Wash with three column volumes of buffer AM-200.
c. Elute with a 40-ml linear salt gradient from 200 to 500 m*M* KCl in buffer AM. RNA polymerase I activity elutes between 230 and 260 m*M* KCl, UBF at 400–450 m*M* KCl.

3. Size-Exclusion Chromatography on Superdex-200 (FPLC)

Fractionate 10 ml of Pol I-containing fractions (approximately 5–10 mg of protein) in buffer AM-150 on a Superdex-200 HiLoad column (26/60) at a flow rate of 2 ml/min.

On this 330-ml column, Pol I elutes between 120 and 150 ml. Because the protein is dilute and cannot be stored without unacceptable losses of activity, the active fractions should immediately be concentrated and further purified on a Mono Q column.

4. Ion-Exchange Chromatography on Mono Q

a. Apply active fractions from the Superdex-200 column to a Mono-Q column (HR 5/5) equilibrated with buffer AM-150.
b. Wash with buffer AM-200, and elute with a 10-ml linear salt gradient from 200 to 450 m*M* KCl. RNA polymerase I activity elutes between 230 and 260 m*M* KCl.

At this step of the purification procedure, Pol I is about 50% homogenous and can be stored in buffer AM-100 for several months without loss of activity.

Purification of the Upstream Binding Factor UBF

Assays

In Vitro Transcription. UBF stimulates transcription of rDNA in the reconstituted transcription system containing partially purified protein fractions. Note that transcription activation is observed only if the additional protein fractions are substantially free of UBF as estimated on immunoblots. Transcription is performed in the standard assay in the presence of 5–10 ng of template DNA, 5 μl of Pol I, 3 μl of TIF-IB (CM-400), 3 μl of TIF-IA/TIF-IC (MQ-230), and 0.5–5 ng of UBF1.

DNase I Footprint. Fifty-microliter assays contain 5 m*M* HEPES–KOH (pH 7.9), 25 m*M* KCl, 5 m*M* $MgCl_2$, 2.5 m*M* KF, 2% (v/v) polyvinyl alcohol, 5% (v/v) glycerol, and 1–2 ng of a 5′-labeled DNA fragment from the mouse rDNA enhancer. After incubation for 15 min at 30°, the DNA is digested with DNase I and the products are analyzed on a sequencing gel.[9]

Fractionation Procedure

UBF is much more abundant than Pol I, TIF-IA, TIF-IB, and TIF-IC, and therefore cross-contaminates most of the other fractions. In our hands, UBF is not absolutely required for mouse rDNA transcription but rather exerts a stimulatory function. In the presence of highly purified factor preparations, UBF stimulates transcription at most three- to fivefold. In crude fractions, however, 50- to 100-fold activation is observed. Thus UBF exerts most, if not all, of its stimulatory function by receiving repression exerted by a negative-acting factor that competes for binding to the rDNA promoter.[10] UBF is purified either from H-400 fractions after step 2 of the Pol I fractionation scheme or from DEAE-1000. The protocol below describes the purification from DEAE-1000 that contains none of the other activities, and therefore is a relatively inexpensive starting material that normally is discarded.

1. CHROMATOGRAPHY ON Q-SEPHAROSE

a. Apply dialyzed DEAE-1000 fractions to a Q-Sepharose column (7–8 mg of protein per milliliter of resin).
b. Wash with 3 vol of buffer AM-100 and step-elute bound proteins successively with buffer AM-300, AM-500, and AM-1000. Both UBF1 and UBF2, the two 97- and 94-kDa splice variants, elute in the 500 m*M* KCl step.

2. CHROMATOGRAPHY ON MONO Q (FPLC)

a. Apply dialyzed fractions from the Q-Sepharose column to a Mono Q column (HR 10/10).
b. Wash with buffer AM-100 and AM-300, and elute bound proteins with a linear salt gradient from 300 to 600 m*M* KCl. UBF1 and UBF2 elute between 430 and 480 m*M* KCl (MQ-450).
c. Dialyze UBF-containing fractions against AM-100 and quantify UBF on immunoblots.

[9] S. P. Bell, R. M. Learned, H.-M. Jantzen, and R. Tjian, *Science* **241,** 1192 (1988).
[10] A. Kuhn, R. Voit, V. Stefanovsky, R. Evers, M. Bianchi, and I. Grummt, *EMBO J.* **13,** 416 (1994).

3. Chromatography on Bio-Rex 70

a. Load MQ-450 fractions on a Bio-Rex 70 column (5 mg of protein per milliliter of resin). This step is also used if UBF is to be purified from H-400 fractions. In this case, use the MQ-450 fractions that elute after the peak of Pol I.
b. Wash with buffer AM-100, then with AM-300, and elute UBF with a linear gradient from 300 to 600 m*M* KCl. Fractions eluting at 360 m*M* KCl contain predominantly UBF2, whereas fractions eluting at around 550 m*M* KCl contain UBF1. At this step of purification, UBF is about 80–90% homogenous and can be stored at −80° for several months without significant loss of activity.

4. DNA Affinity Chromatography of UBF

If required, UBF can be purified to molecular homogeneity by purification on a sequence-specific DNA affinity column. For this, a double-stranded, concatenated oligonucleotide covering the upstream control element of the human rDNA promoter (5′ CAGGTGTCCGTGTCGC-GCGTCGCCTGGCCGGCGGCG 3′) is covalently bound to cyanogen bromide-activated Sepharose. Fractions from the Bio-Rex 70 column are preincubated for 30 min with nonspecific competitor DNA, and subjected to affinity chromatography as described.[11]

Purification of TIF-IB

Assay

The activity of TIF-IB is monitored under standard transcription conditions in a reconstituted system containing crude Pol I (H-400) and partially purified TIF-IA/TIF-IC (Q-Sepharose fractions). After complementation with TIF-IB, these fractions support high levels of specific transcription. This assay is quantitative and relatively easy to perform, and therefore this method is routinely used to follow TIF-IB activity during purification. Alternatively, the availability of antibodies against recombinant TBP and Pol I-specific TAFs may be used to detect TIF-IB on immunoblots. However, because TBP is also an integral component of TFIID and TFIIIB, TBP antibodies should be used only after the different TBP–TAF complexes have been chromatographically separated.

[11] J. T. Kadonaga, *DNA Protein Eng. Tech.* **2,** 82 (1990).

Fractionation Procedure

The purification of TIF-IB to molecular homogeneity is achieved by a combination of ion-exchange and affinity chromatography using monoclonal antibodies directed against TBP.[12] In protocol 1, the purification of TIF-IB is integrated in the overall fractionation scheme used for purification of Pol I and the initiation factors. Because, however, more than 50% of TIF-IB is lost during chromatography both on DEAE-Sepharose and CM-Sepharose, we use the purification scheme depicted in protocol 2 if high levels of homogeneous TIF-IB are needed.

Protocol 1

1. Chromatography of H-600 Fractions on CM-Sepharose

a. Apply H-600 fractions to a CM-Sepharose column (2.5 mg of protein per milliliter of resin) in buffer AM-100.
b. Wash with four column volumes of buffer AM-100 and step-elute bound proteins with AM-200 and AM-400. TIF-IB is present both in the CM-200 and CM-400 fraction. However, the CM-200 fraction contains significant amounts of UBF, whereas the CM-400 fractions contain TIF-IB, which is free of UBF.

2. Chromatography of CM-400 Fractions on Mono S (FPLC)

a. Load dialyzed CM-400 fractions on a Mono S column (HR 5/5, 1 ml).
b. Elute residual TFIID and TFIIIB by extensive washing with buffer BM-320.
c. Elute TIF-IB with a linear gradient from 320 to 700 m*M* KCl in buffer BM. TIF-IB elutes at 450 m*M* KCl. At this step of purification, TIFI-IB is highly active, practically free of TFIID and TFIIIB, and stable for several months at −80°.

3. Immunoaffinity Purification

a. All solutions contain the following set of protease inhibitors: aprotinin (5 μg/ml), pepstatin (5 μg/ml), leupeptin (5 μg/ml), benzamidine hydrochloride (3 m*M*), antipain (2 μg/ml), and chymostatin (2 μg/ml).
b. Cross-link monoclonal antibodies directed against TBP to protein A–Sepharose using standard procedures, and incubate pooled TIF-IB fractions with the antibody matrix for 5 hr at 4° in buffer AM-100 supplemented with 0.1% (v/v) Nonidet P-40 (NP-40) with gentle rotation.

[12] U. Rudloff, D. Eberhard, L. Tora, H. Stunnenberg, and I. Grummt, *EMBO J.* **13,** 2611 (1994).

c. Wash antibody–antigen complexes by successive treatment with buffer AM-700–0.1% (v/v) NP-40, then with buffer AM-100 supplemented with 150 m*M* guanidinium hydrochloride plus 0.1% (v/v) NP-40, and finally with buffer AM-300 containing 0.05% (v/v) NP-40.

d. Elute bound TBP complexes in buffer AM-300 containing epitope-specific peptide (2 mg/ml) and 0.05% (v/v) NP-40 for 3 hr at 4°. This eluate contains homogeneous TIF-IB and can be stored in this buffer for several months without loss of activity.

Protocol 2

1. Fractionation of Nuclear Extract Proteins on S-Sepharose Fast Flow

a. Undialyzed nuclear extracts are mixed with an equal volume of a saturated ammonium sulfate solution and proteins are precipitated at 4° for 30 min with gentle rotation.

b. The precipitate is collected by centrifugation and resuspended in buffer AM.

c. Proteins are applied to an S-Sepharose Fast Flow column equilibrated with buffer AM-100.

d. After washing, two step-elutions with buffer AM-320 and AM-700 are performed. TIF-IB elutes in the 700 m*M* salt step (S-700).

2. Chromatography of S-700 Fractions on Heparin–Ultrogel

a. Load S-700 fractions dialyzed against buffer AM-200 on a heparin–Ultrogel column.

b. Wash with buffer AM-200 and elute bound proteins at 400 and 700 m*M* KCl, respectively. The majority of TFIID elutes at 400 m*M* KCl, TIF-IB at 700 m*M* KCl (H-700).

3. Chromatography of H-700 Fractions on Mono S (FPLC)

a. Load pooled H-700 fractions on a Mono S column (HR 5/5, 1 ml) equilibrated with buffer BM-100.

b. Wash the resin with buffer BM-100, then with buffer BM-320, and elute TIF-IB with buffer BM-700. This procedure provides highly concentrated TIF-IB that is not contaminated by TFIID and TFIIIB.

4. Immunoaffinity Purification

The undialyzed Mono-S fractions are supplemented with 0.1% (v/v) NP-40 and affinity purified on a TBP antibody column as described above.

Purification of TIF-IA

Purification of TIF-IA has proven to be difficult. First, the cellular concentration of this growth-dependent factor is extremely low even in exponentially growing cells. Second, TIF-IA is associated with Pol I and dissociation of TIF-IA from Pol I is difficult to achieve and is normally not quantitative. Third, TIF-IA does not bind to negatively charged chromatographic residues, and therefore only anion-exchange matrices can be used. In addition, TIF-IA is labile and significant losses of activity occur in more purified fractions.

Assay

The assay for TIF-IA is based on the fact that this activity is absent in extracts from growth-arrested cells. Thus, fractions derived from exponentially growing cells are assayed for their ability to reconstitute transcriptional activity of extracts prepared from stationary, starved, or cycloheximide-treated cells. Thus, 7 μl of inactive extract is complemented with 5–8 μl of column fractions and assayed in the standard transcription system.[4] Alternatively, TIF-IA can be assayed in the reconstituted transcription system in the presence of highly purified Pol I, TIF-IB, UBF, and TIF-IC.[13]

1. Chromatography of H-200 Fractions on Q-Sepharose

a. Apply H-200 fractions, i.e., the flowthrough from the heparin column (step 2 of the fractionation procedure; see above), in buffer AM-100 to a Q-Sepharose column at a ratio of approximately 10 mg of protein per milliliter of resin.
b. Wash with buffer AM-100.
c. Step-elute bound proteins with AM-200 and AM-300. Both TIF-IA and TIF-IC are present in the 300 m*M* KCl step (fraction QS-300).

2. Chromatography of QS-300 Fractions on Mono Q (FPLC)

a. Apply QS-300 fractions in buffer AM-100 to a Mono Q column (HR 10/10) at 10 mg of protein per milliliter of resin.
b. Wash with buffer AM-200.
c. Recover bound proteins with a 60-ml linear gradient from 200 to 300 m*M* KCl. Both TIF-IA and TIF-IC elute at about 230 m*M* KCl.

3. Chromatography of Mono Q Fractions on ω-Aminooctyl–Agarose

a. Dialyze Mono Q fractions against buffer AM-75 and load on an ω-aminooctyl–agarose column.

[13] A. Schnapp, G. Schnapp, B. Erny, and I. Grummt, *Mol. Cell. Biol.* **13,** 6723 (1993).

b. Wash extensively with buffer AM-75.
c. Elute both TIF-IA and TIF-IC by a 150 mM KCl step (ωao-150 fractions).

4. Ammonium Sulfate Precipitation

a. Precipitate TIF-IA with 35% (w/v) ammonium sulfate from pooled ωao-150 fractions. Add $(NH_4)_2SO_4$ (0.194 g/ml) and 1 μl of 1 *N* NaOH per gram of $(NH_4)_2SO_4$.
b. Stir for 1 hr at 4°.
c. Collect precipitate by centrifugation at 13,000 rpm for 20 min in a Sorvall centrifuge (SS 34 rotor). The supernatant can be used for purification of TIF-IC.
d. Suspend the pellet (AS35%) in buffer AM-100 and dialyze against buffer AM-100.

5. Chromatography of the AS35% Fractions on a Polyethyleneimine Column

a. Load 10 ml of the AS35% fraction (10 to 20 mg of protein) onto a PEI-HPLC column (4.8 ml).
b. Wash with 10 ml of buffer AM-250 and elute bound proteins with a 50-ml linear gradient from 250 to 1000 m*M* KCl. The peak of TIF-IA activity is recovered at 700 m*M* KCl.

6. Size-Exclusion Chromatography on Superdex-200 (FPLC)

Apply active PEI fractions in buffer AM-120 to a Superdex-200 gel-filtration column (HiLoad 26/60, 330 ml) equilibrated with buffer AM-120, at a flow rate of 2 ml/min. The peak of TIF-IA activity elutes at 192 ml, corresponding to a native molecular mass of about 80 kDa.

7. Chromatography of Superdex-200 Fractions on Mono Q (Mono Q2)

a. Apply fractions eluting between 180 to 210 ml from the Superdex-200 column to a Mono-Q column (HR 5/5) equilibrated with buffer AM-120.
b. Wash with 10 ml of buffer AM-200.
c. Elute bound proteins with a 6-ml linear gradient from 200 to 300 m*M* KCl. TIF-IA elutes at 230 m*M* KCl.

Purification of TIF-IC

Both TIF-IC and TIF-IA are Pol I-associated factors and exert similar chromatographic properties. As with TIF-IA, TIF-IC is negatively charged, and therefore binds only to anion-exchange columns. With the fractionation procedure described below, partially purified TIF-IC preparations that are not cross-contaminated with other factor activities are obtained.

Assay

TIF-IC needs to be assayed in a reconstituted transcription system in the presence of purified factors. Importantly, highly purified Pol I and TIF-IA are required that are free of TIF-IC. We usually use 60 ng of template DNA, 1–2 μg of Pol I (Mono Q2), 2 μl of TIF-IA (PEI fraction), 1 μl of TIF-IB (Mono S), 1 μl of UBF (Bio-Rex 70), and 3–4 μl of TIF-IC.[5]

Fractionation Procedure

Steps 1 to 3 are identical to those described for the purification of TIF-IA.

4. Differential Ammonium Sulfate Precipitation

a. Precipitate TIF-IC from the supernatant of the 35% (w/v) ammonium sulfate precipitation (step 4 of TIF-IA preparation) by adding 65% (0.184 g/ml) ammonium sulfate (AS-65% fraction).
b. Spin again, resuspend the pellet in buffer AM-100, and dialyze.

5. Chromatography of AS-65% Fraction on a Polyethyleneiminie Column

a. Load the dialyzed AS-65% onto a PEI-HPLC column.
b. Wash with 10 ml of buffer AM-250 and elute bound proteins with a 50-ml linear gradient from 250 to 1000 m*M* KCl. The peak of TIF-IC activity is recovered at 550 m*M* KCl. Dialyze against buffer AM-120.

6. Size-Exclusion Chromatography on Superdex-200 (FPLC)

Apply PEI fractions onto a Superdex-200 gel-filtration column (HiLoad 26/60, 330 ml) equilibrated with buffer AM-120, at a flow rate of 2 ml/min. The peak of TIF-IC activity elutes after 220 ml, corresponding to a native molecular mass of about 60 kDa.

7. Chromatography of Superdex-200 Fractions on Mono Q (Mono Q2)

a. Apply fractions eluting between 210 and 235 ml from the Superdex-200 column to a Mono Q column (HR 5/5) equilibrated with buffer AM-120.
b. Wash with 10 ml of buffer AM-200.

c. Recover bound proteins with a 6-ml linear gradient from 200 to 300 m*M* KCl. TIF-IC elutes at 230 m*M* KCl.

Purification of TTF-I

Background

Although our understanding of the mechanisms of transcription initiation has progressed rapidly, the mechanisms involved in eukaryotic transcription termination are more poorly understood. By focusing on the events that produce the 3′ end of the ribosomal transcript made by murine Pol I, we have previously identified a protein, known as TTF-I (transcription termination factor). Murine TTF-I binds tightly to an 18-bp DNA element, eight repeats of which are located downstream of sequence encoding the mature 3′ end of the rRNA.[14] When mTTF-I is bound at this site, transcription elongation by Pol I is halted.

The isolation of TTF-I in sufficient amounts to obtain sequence information required for cloning of the cDNA encoding this termination factor turned out to be a difficult task because of the low cellular abundance and because of the heterogeneity of Sal box-binding proteins.[15] Using the purification procedure described below, we have purified a 50-kDa proteolytic version of TTF-I to apparent homogeneity, which enabled us to clone the cDNAs encoding murine and human TTF-I.[6] The availability of recombinant TTF-I, produced in *E. coli,* now provides an invaluable tool for mechanistic and structure/function studies on the process of transcription termination.

Assay

TTF-I activity is most easily detected by monitoring its specific DNA binding by electrophoretic mobility shift assays. Reactions of 25 μl contain 7.5 fmol of a double-stranded ^{32}P-labeled oligonucleotide (5′ CCCGGGATCCTTCGGAGGTCGACCAGTACTCCGGGCGAC 3′) containing 3′-terminal mouse rDNA sequences from +581 to +610 (with respect to the 3′ end of 28S rRNA), 12 m*M* Tris–HCl (pH 8.0), 100 m*M* KCl, 5 m*M* $MgCl_2$, 0.1 m*M* EDTA, 0.5 m*M* DTE, 8% (v/v) glycerol, 2 μg of poly(dI-dC), and 0.5 to 5 μl of protein fractions in buffer AM-100. After incubation for 30 min at room temperature, protein–DNA complexes are separated by electrophoresis on native 8% (w/v) polyacrylamide gels.

[14] I. Bartsch, C. Schöneberg, and I. Grummt, *Mol. Cell. Biol.* **8,** 3891 (1988).

[15] A. Smid, M. Finsterer, and I. Grummt, *J. Mol. Biol.* **227,** 635 (1992).

Alternatively, the termination activity is determined in a reconstituted transcription system. For this, we use artificial minigene templates, i.e., fusions between a 5′-terminal gene fragment including the rDNA promoter and a fragment from the 3′ end which contains one terminator. Ten to 50 ng of template DNA is incubated under standard transcription conditions in the presence of 8 μl of S-100 extract and different amounts (up to 5 μl) of fractions containing TTF-I. S-100 extracts contain low levels of TTF-I, and therefore the majority of RNA products are readthrough transcripts. In the presence of TTF-I terminated transcripts are produced whose length corresponds to the distance between the initiation site and the terminator element.[16]

Fractionation Procedure

Conventional Chromatography

1. Use only nuclear (not cytoplasmic) extracts as starting material.
2. Omit step 1 (involving the DEAE-Sepharose column) because about half of TTF-I is in the flowthrough.
3. Load nuclear extracts onto a heparin–Ultrogel A4-R column equilibrated with buffer AM-100.
4. Wash the column with buffer AM-280.
5. Step-elute TTF-I with buffer AM-450.
6. Dialyze against buffer BM-100 and apply to an S-Sepharose Fast Flow column.
7. Wash with BM-150 and step-elute with BM-400.
8. Dialyze pooled fractions against buffer AM-100 and apply to a Mono Q FPLC column.
9. Elute bound proteins with a linear gradient from 100 to 230 m*M* KCl. The peak of TTF-I activity elutes at 180 m*M* KCl.

DNA Affinity Chromatography

1. Prepare a DNA affinity matrix containing a 39-bp double-stranded Sal box oligonucleotide (5′ CCCGGGATCCTTCGGAGGTCGAC-CAGTACTCCGGGCGAC 3′) according to standard procedures.[11]
2. Dialyze active fractions from two Mono Q FPLC columns against buffer AM-80.
3. Incubate with nonspecific competitor DNA for 15 min at 4°. Next, add the affinity resin and incubate for 1 hr at 4°.

[16] I. Grummt, H. Rosenbauer, I. Niedermeyer, U. Maier, and A. Öhrlein, *Cell* **45,** 837 (1986).

4. Add 5–10 μg of V8 protease (Boehringer, Mannheim, Indianapolis, IN) per 1 mg of matrix-bound protein and incubate for 15 min at 30°. This proteolytic step converts the heterogeneous cellular Sal box-binding proteins into a single 50-kDa polypeptide.
5. Add 0.1% (v/v) Nonidet P-40 (NP-40), transfer the resin to a small column, and wash with buffer AM-80 plus 0.1% (v/v) NP-40.
6. Elute DNA-bound proteins in buffer AM plus 0.1% (v/v) NP-40 at 0.25 and 1 *M* KCl. The bulk of TTF-I elutes at 1 *M* KCl.
7. For purification to homogeneity, it is suggested that one perform a second round of affinity chromatography using concatenated and biotinylated Sal box oligonucleotides bound to streptavidin Dynabeads (Dyual, Oslo, Norway).

Conclusions

As stated at the beginning of this chapter, it is not possible to present a single step-by-step protocol for the purification of all factors required for initiation and termination of Pol I transcription in high yield and purity. However, using the fractionation procedure outlined here, it is possible to use the same starting material to purify, at least partially, all components involved in Pol I initiation. The chromatographic procedures outlined here have allowed us to reconstitute a Pol I transcription system with highly purified factors that mimicks all steps in basal transcription, i.e., initiation, elongation, and termination. In the near future we anticipate that it will be possible to reconstitute Pol I transcription from recombinant factors alone. Such a system would be invaluable both for the elucidation of basic mechanisms of transcription and for structure–function studies.

Acknowledgments

We acknowledge the numerous students and postdoctoral fellows who have assisted in the development of these assays and fractionation procedures over the years. This work has been supported for many years by the Deutsche Forschungsgemeinschaft.

[22] RNA Polymerase III and Class III Transcription Factors from *Saccharomyces cerevisiae*

By JANINE HUET, NATHALIE MANAUD, GIORGIO DIECI, GÉRALD PEYROCHE, CHRISTINE CONESA, OLIVIER LEFEBVRE, ANNY RUET, MICHEL RIVA, and ANDRÉ SENTENAC

There are three distinct RNA polymerases in eukaryotic cell nuclei.[1] RNA polymerase III transcribes class III genes, which encode small RNA molecules involved in translation and RNA processing, such as tRNAs, 5S rRNA, and U6 snRNA. These genes display a great diversity of promoter organization, with intragenic and/or extragenic elements.[2,3] RNA polymerase III is specifically recruited at the transcription start sites via a cascade of protein–protein–DNA interactions involving various transcription factors. Remarkably, complex assemblies are formed that involve more than 25 polypeptides and cover the entire transcription units. The transcriptional components and the process of transcription complex formation on various class III genes have been best analyzed in *Saccharomyces cerevisiae*.[3–5] Transcription factor IIIB (TFIIIB) is the critical factor that is ultimately responsible for RNA polymerase III recruitment. The proper assembly of TFIIIB upstream of the transcription start site is performed by TFIIIC. TFIIIC is a multisubunit protein that recognizes two promoter elements, the A and B blocks, mainly found in tRNA genes, as well as preformed TFIIIA–5S rDNA complexes. TFIIIA and TFIIIC, therefore, play the role of assembly factors, while TFIIIB acts as the proper initiation factor.[6] The general TATA box-binding protein (TBP) is part of TFIIIB and is essential for transcription complex formation. Another essential component of

[1] A. Sentenac, *CRC Crit. Rev. Biochem.* **18,** 31 (1985).

[2] I. M. Willis, *Eur. J. Biochem.* **212,** 1 (1993).

[3] R. J. White, *in* "RNA Polymerase III Transcription." Molecular Biology Intelligence Unit, R. G. Landes Company, Austin, TX, 1994.

[4] P. Thuriaux and A. Sentenac, *in* "The Molecular and Cellular Biology of the Yeast *Saccharomyces*: Gene Expression" (J. Broach, J. R. Pringle, and E. W. Jones, eds.), Vol. 2, p. 1. Cold Spring Harbor Laboratory Press, Cold Spring Harbor, NY, 1992.

[5] G. A. Kassavetis, C. Bardeleben, B. Bartholomew, B. R. Braun, C. A. P. Joazeiro, M. Pisano, and E. P. Geiduschek, *in* "Transcription: Mechanisms and Regulation" (R. C. Conaway and J. W. Conaway, eds.), p. 107. Raven, New York, 1994.

[6] G. A. Kassavetis, B. R. Braun, L. H. Nguyen, and E. P. Geiduschek, *Cell* **60,** 235 (1990).

TFIIIB is $TFIIIB_{70}$, or BRF, which is structurally and functionally homologous to TFIIB.[3]

In this chapter we describe the purification of yeast RNA polymerase III and of yeast factors TFIIIA, TFIIIB, and TFIIIC. RNA polymerase III (≈700 kDa, 17 subunits) and TFIIIC (≈600 kDa, 6 subunits) can be purified to near homogeneity as stable multisubunit assemblies. In contrast, TFIIIB can be easily dissociated into three components (TBP, $TFIIIB_{70}$, and B″). TFIIIA, $TFIIIB_{70}$ and TBP can be obtained in active form as recombinant proteins made in *Escherichia coli*. Twenty components of the yeast class III transcription machinery have already been cloned and mutagenized.[3,4]

Assay Methods

Reconstituted Specific Transcription Assay

The reconstituted system contains, in 40 μl, 20 m*M* *N*-2-hydroxyethylpiperazine-*N*′-2-ethanesulfonic acid (HEPES, pH 7.9), 90 m*M* KCl, 5 m*M* $MgCl_2$, 1 m*M* dithiothreitol, 0.1 m*M* EDTA, 10% (v/v) glycerol, 7 units of RNasin (40,000 units/ml; Promega, Madison, WI), 0.6 m*M* ATP, CTP, and GTP, 0.03 m*M* UTP, and 10 μCi of [α-^{32}P]UTP (400 Ci/mmol), 0.1–0.15 μg of pUC-Glu plasmid DNA,[7] TFIIIC (0.2 μg of DEAE-Sephadex fraction or 0.05 μg of DNA affinity fraction), TFIIIB (1.5 μg of heparin–Ultrogel fraction or 0.25 μg of Cibacron-Blue fraction), and 50 ng of purified RNA polymerase III. The TFIIIB fraction can be replaced by a combination of recombinant TBP (10 ng), recombinant $TFIIIB_{70}$ (50 ng), and partially purified B″ component [0.5 μg of Bio-Rex fraction (Bio-Rad, Richmond, CA); see below]. The best order of addition is DNA, TFIIIC, and TFIIIB, followed by RNA polymerase III. For transcription of the yeast 5S RNA gene, recombinant yeast TFIIIA is added (5 ng). After a 45-min incubation at 25°, the reaction is stopped by addition of 14 μl of a stop solution [*E. coli* tRNA (1.8 mg/ml), 2% (w/v) sodium dodecyl sulfate (SDS), and proteinase K (0.5 mg/ml)]. After 10 min at 65°, the RNA is recovered by precipitation with 3 vol of ethanol in the presence of 0.3 *M* ammonium acetate. The dried pellet is dissolved in 15 μl of sample buffer [90% (v/v) formamide, Tris–borate–EDTA (TBE) buffer, 0.5% (w/v) bromphenol blue, 0.5% (w/v) xylene cyanole] and subjected to electrophoresis in a

[7] O. Feldman, J. Olah, and H. Friedenreich, *Nucleic Acids Res.* **9,** 2949 (1981).

6% (w/v) polyacrylamide gel containing 8 *M* urea at 200 V for 90 min in TBE buffer. The gel is soaked in fixing buffer [acetic acid–ethanol–water, 1:2:7 (v/v)], dried on 3MM Whatman (Clifton, NJ) paper, and exposed to X-ray film at −70° with an intensifying screen for autoradiography.

Single-Round Transcription

The following procedure is adapted from Kassavetis *et al.*,[8] with minor modifications. The yeast SUP4 $tRNA^{Tyr}$ gene in plasmid pRB1[9] is preincubated for 30 min at 25° with transcription components, in a final volume of 35 μl, to form stable preinitiation complexes. Transcription is initiated by adding 5 μl of a solution providing 0.4 m*M* ATP, 0.4 m*M* CTP, 0.03 m*M* UTP (molecular biology grade; Boehringer-Mannheim, Indianapolis, IN) and 15 μCi of [α-^{32}P]UTP. In the absence of GTP, a stable ternary complex is formed that is stalled 17 nucleotides downstream of the transcription initiation site. After 1 min, elongation is allowed to resume by the addition of 0.4 m*M* GTP, together with heparin (Sigma, St. Louis, MO) to 0.3 mg/ml, to inhibit reinitiation. Incubation is continued for 3 min, then the reaction is blocked and RNA products purified as described above, with the exception that ethanol precipitation is performed twice.

Nonspecific Transcription Assay

A standard incubation mixture (50 μl) contains 65 m*M* Tris-HCl (pH 8.0), 1 m*M* dithiothreitol, 5 m*M* $MgCl_2$, 2.5 m*M* $MnCl_2$, 0.1 *M* ammonium sulfate, 1 m*M* ATP, 0.5 m*M* [α-^{32}P]UTP (20 to 60 cpm/pmol), and 1 μg of poly[d(A-T)]. The reaction is started by addition of 1μg of purified RNA polymerase III or 10 μl of heparin–Ultrogel fraction. After 30 min of incubation at 30°, the reaction is stopped by addition of 2 ml of ice-cold 5% trichloroacetic acid (w/v). Acid-insoluble radioactivity is collected on a Millipore (Bedford, MA) filter HAWP 025 (0.45-μm pore size) and washed with about 30 ml of cold 5% (w/v) trichloroacetic acid. The filter is dried and radioactivity is counted in a scintillation counter. Under these conditions, the specific activity of the purified enzyme is 3500 units/mg.

[8] G. A. Kassavetis, D. L. Riggs, R. Negri, L. H. Nguyen, and E. P. Geiduschek, *Mol. Cell. Biol.* **9,** 2551 (1989).

[9] R. E. Baker, O. S. Gabrielsen, and B. D. Hall, *J. Biol. Chem.* **261,** 5275 (1986).

One unit of activity corresponds to the incorporation of 1 nmol of UMP in 60 min at 30°.

Gel-Retardation Assay

To monitor the TFIIIC–DNA interaction, a 206-bp *Bam*HI–*Eco*RI DNA fragment from pUC-Glu plasmid, carrying the yeast $tRNA_3^{Glu}$ gene (3–10 fmol; ≈3000–15,000 cpm) is incubated in a 20-μl reaction mixture containing carrier pBR322 DNA [we use a 1:1 (w/w) ratio of protein and carrier DNA] to prevent nonspecific retention of the probe on top of the gel, 20 m*M* Tris–HCl (pH 8), 1 m*M* EDTA, 150 m*M* KCl, bovine serum albumin (2 mg/ml), and 10% (v/v) glycerol. The binding reaction is initiated by addition of TFIIIC (1–2 μl of column fractions). After a 10-min incubation at 25°, DNA–TFIIIC complexes are analyzed by gel electrophoresis.[9] The gel employed to resolve TFIIIC–DNA complexes from free DNA is 11 × 13 × 0.1 cm in size and consists of 5% (w/v) polyacrylamide (acrylamide:bisacrylamide, 20:0.24, v/v) in a buffer of 20 m*M* Tris–HCl (pH 8), 1 m*M* EDTA, and 5% (v/v) glycerol. The electrode buffer is the same except that it lacks glycerol and mercaptoethanol is added to a final concentration of 5 m*M*. Gels are run at 200 V at 4°. Before loading the samples, gels are prerun for 1 hr; then the buffer in the electrode reservoir is changed. As bromphenol blue dissociates TFIIIC–DNA complexes, the dye is loaded in empty wells to monitor the electrophoresis. When the bromphenol blue reaches the bottom of the gel (≈90 min), electrophoresis is stopped, the gel is incubated for 30 min in 10% (v/v) acetic acid–20% (v/v) ethanol to precipitate nucleic acids, dried, and autoradiographed overnight on a Kodak (Rochester, NY) X-Omat AR film.

Buffers

TE* buffer: 50 m*M* Tris–HCl (pH 8), 0.1 m*M* EDTA
TBE buffer: 90 m*M* Tris–borate, 2 m*M* EDTA (pH 8)
Extraction buffer: 0.2 *M* Tris–HCl (pH 8), 1 m*M* EDTA, 10 m*M* mercaptoethanol, 10 m*M* $MgCl_2$, 0.4 *M* ammonium sulfate, 10% (v/v) glycerol, freshly supplemented with 2 m*M* phenylmethylsulfonyl fluoride (PMSF) in ethanol solution
Buffer I: 20 m*M* Tris–HCl (pH 8), 0.5 m*M* EDTA, 10 m*M* mercaptoethanol, 10% (v/v) glycerol
Buffer II: Buffer I containing 20% (v/v) glycerol
Buffer III: 20 m*M* Tris–HCl (pH 8), 0.5 m*M* EDTA, 1 m*M* dithiothreitol, 25% (v/v) glycerol, 0.05% (v/v) Nonidet P-40
HEG buffer: 20 m*M* HEPES (pH 7.9), 0.2 m*M* EDTA, 0.5 m*M* PMSF, 0.5 m*M* dithiothreitol (DTT), 20% (v/v) glycerol

The ammonium sulfate concentration of buffers I, II, III, and HEG is indicated in parentheses in the following sections.

RNA Polymerase III

Large-Scale Purification

RNA polymerase III has been purified from various species, including human,[10] mouse,[11,12] frog,[13,14] silkworm,[15] *Drosophila*,[16] wheat,[17,18] and yeast.[19–22] The purification procedure described here was designed to purify simultaneously from a yeast crude extract the RNA polymerase III transcription factors TFIIIB and TFIIIC and the three forms of RNA polymerase (Fig. 1). This method is an adaptation of that developed by Huet *et al.*[22] and Ruet *et al.*[23]

Preparation of Cell Extract. Yeast cells are grown in a fermentor, harvested in exponential phase (4 OD_{600}), and stored at −80°. Frozen cells are thawed in a water bath at 30°, washed with water once, and then washed with extraction buffer. After centrifugation at 8000 rpm for 10 min at 4° in a JA10 rotor (Beckman, Fullerton, CA), cells (150 g) are resuspended in 150 ml of extraction buffer, frozen at −70° in dry ice and ethanol, and disrupted in an Eaton press. The efficiency of this cell-breaking method is approximately 90% as estimated by cell counting with an hemocytometer. The cell extract is centrifuged at 40,000 rpm for 1 hr at 4° in a 45 Ti Beckman rotor after dilution with 300 ml of extraction buffer. The supernatant is filtered through gauze to remove the white lipid layer and either stored (up to 6 months) at −80° after freezing in liquid nitrogen or directly submitted to ammonium sulfate precipitation. Protein concentration of the crude

[10] J. A. Jaehning, P. S. Woods, and R. G. Roeder, *J. Biol. Chem.* **252,** 8762 (1977).
[11] V. E. F. Sklar, L. B. Schwartz, and R. G. Roeder, *Proc. Natl. Acad. Sci. U.S.A.* **72,** 348 (1975).
[12] V. Sklar and R. G. Roeder, *J. Biol. Chem.* **251,** 1064 (1976).
[13] D. R. Engelke, B. S. Shastry, and R. G. Roeder, *J. Biol. Chem.* **258,** 1921 (1983).
[14] R. G. Roeder, *J. Biol. Chem.* **258,** 1932 (1983).
[15] V. Sklar, J. Jaehning, P. Gage, and R. G. Roeder, *J. Biol. Chem.* **251,** 3794 (1976).
[16] E. Gundelfinger, H. Saumweber, A. Dallendorfer, and H. Stein, *Eur. J. Biochem.* **111,** 395 (1980).
[17] M. Teissere, P. Penon, Y. Azou, and J. Ricard, *FEBS Lett.* **82,** 77 (1977).
[18] J. Jendrisak, *Plant Physiol.* **67,** 438 (1981).
[19] P. Valenzuela, G. L. Hager, F. Weinberg, and W. J. Rutter, *Proc. Natl. Acad. Sci. U.S.A.* **73,** 1024 (1976).
[20] G. L. Hager, M. J. Holland, and W. J. Rutter, *Biochemistry* **16,** 1 (1977).
[21] T. M. Wandzilak and R. W. Benson, *Biochemistry* **17,** 426 (1978).
[22] J. Huet, M. Riva, A. Sentenac, and P. Fromageot, *J. Biol. Chem.* **260,** 15304 (1985).
[23] A. Ruet, S. Camier, W. Smagowicz, A. Sentenac, and P. Fromageot, *EMBO J.* **3,** 343 (1984).

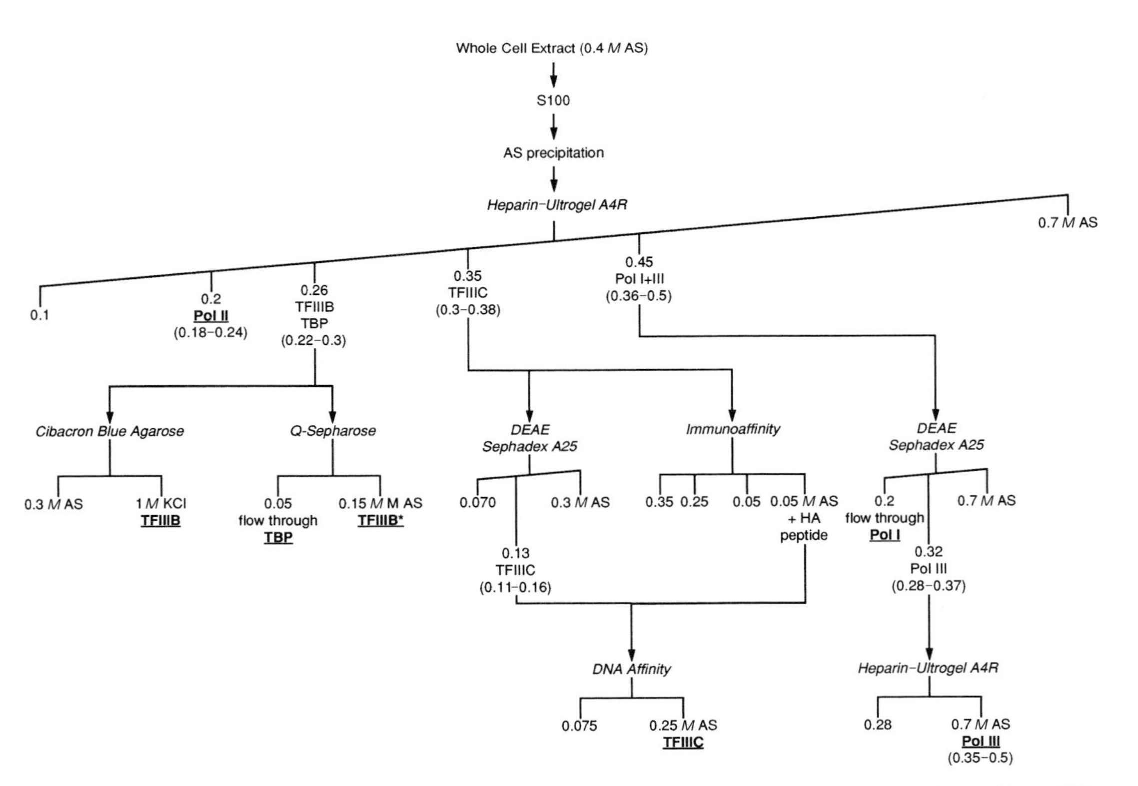

FIG. 1. Purification scheme for RNA polymerase III and its transcription factors. AS, Ammonium sulfate; S100, 100,000 *g* supernatant; Pol, RNA polymerase; TF, transcription factor. All active fractions are stored at −80°.

extract, estimated by the Bradford method,[24] is around 30 to 35 mg/ml. All the subsequent steps are performed at 4°, and 1 m*M* PMSF is added to all buffers immediately before use.

Ammonium Sulfate Precipitation. Ammonium sulfate is ground in a mortar or a Waring blender to facilitate its rapid dissolution and is progressively added to the crude extract (35 g/100 ml). After complete solubilization of the salt, the solution is slowly stirred for 30 min and allowed to stand on ice for 1 hr or overnight. The precipitate is collected by centrifugation at 40,000 rpm for 30 min in a 45 Ti rotor (Beckman). The supernatant is discarded and a second centrifugation is performed at 20,000 rpm for 10 min to eliminate thoroughly, by aspiration, the residual ammonium sulfate solution. The pellets are dissolved in the minimal volume of buffer I (200 ml) and the solution is diluted with buffer I until the ammonium sulfate concentration reaches 0.1 *M* (950 ml final).

Heparin–Ultrogel Chromatography. Proteins (11.5 g) are chromatographed on a heparin column (Fig. 2). Heparin–Ultrogel A4R (BioSepra; 350 ml) equilibrated in buffer I (0.1 *M*) is allowed to settle in a Büchner funnel (12.5-cm diameter) with sintered glass (No. 2) and covered with a sheet of Whatman 3MM paper to avoid perturbance of the resin bed during adsorption and washing steps. The protein solution, supplemented with 1 m*M* PMSF, is allowed to flow through the resin without suction. Only 15% of the proteins are adsorbed. The resin is washed extensively with 6 liters of buffer I (0.1 *M*) until the absorbance at 280 nm of the flowthrough reaches 0.05. The heparin–Ultrogel cake is then resuspended and poured into a column ($12.5 \text{ cm}^2 \times 28$ cm) that is developed at a flow rate of 100 ml/hr with 1.7 liters of a linear gradient of ammonium sulfate from 0.1 to 0.7 *M* in buffer I. Fractions (10 ml) are collected and assayed for the presence of transcription factors and RNA polymerases. RNA polymerase II, assayed in a nonspecific assay using poly(rC) as template,[25] is eluted first at 0.2 *M* ammonium sulfate. TFIIIB- and TFIIIC-containing fractions are eluted at 0.26 and 0.34 *M* ammonium sulfate, respectively (see Fig. 1). Note that the slope of the elution gradient is critical to avoid contamination of TFIIIC by trailing fractions of TFIIIB. TFIIIB fractions contain TBP as determined by Western blotting with anti-TBP antibodies (gift of J.-M. Egly, IGBMC, Illkirch, France). RNA polymerases I and III are coeluted at a higher salt concentration (optimum of 0.45 *M* ammonium sulfate) but RNA polymerase III activity can be specifically assayed in the presence of 100 m*M* ammonium sulfate.

[24] M. M. Bradford, *Anal. Biochem.* **72,** 248 (1976).

[25] A. Ruet, A. Sentenac, P. Fromageot, B. Winsor, and F. Lacroute, *J. Biol. Chem.* **255,** 6450 (1980).

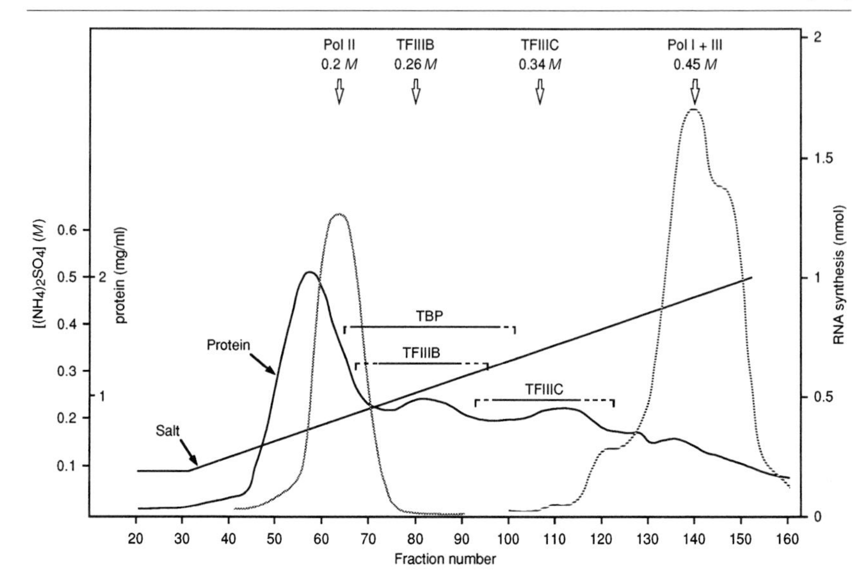

FIG. 2. Heparin–Ultrogel chromatography of yeast RNA polymerase III and transcription factors. Proteins are adsorbed on heparin–Ultrogel A4R and eluted by an ammonium sulfate gradient as described in text. Protein is monitored by the Bradford assay.[24] Fractions are assayed in 10-μl aliquots for RNA polymerase activity (dotted line) with poly(rC) as template in the case of RNA polymerase II or with poly[d(A-T)] for RNA polymerases I and III. TFIIIB is assayed in a reconstituted, specific transcription assay, and TFIIIC in a gel-retardation assay. TBP, visualized by Western blotting, overlaps the peak of TFIIIB activity.

DEAE-Sephadex Chromatography. Heparin–Ultrogel fractions containing RNA polymerases I and III are pooled (117 mg) and dialyzed for 2 hr against 1 liter of buffer I containing 1 m*M* PMSF until the ionic strength corresponds to 0.2 *M* ammonium sulfate and loaded at a flow rate of 100 ml/hr on a DEAE-Sephadex A25 column (5 cm^2 $\times$ 16 cm) equilibrated with buffer II (0.2 *M*). Owing to the instability of the RNA polymerase III in purified fractions, chromatography is performed in the presence of 20% (v/v) glycerol. RNA polymerase I flows through the column, which is washed with 600 ml of buffer II (0.2 *M*). RNA polymerase III is eluted with a linear gradient of 260 ml from 0.2 to 0.7 *M* ammonium sulfate in buffer II. Active enzyme fractions, eluted at 0.32 *M* ammonium sulfate, are pooled (1.9 mg), diluted to 0.28 *M* amonium sulfate with buffer II, and concentrated on a 1.8-ml heparin–Ultrogel A4R column (Poly-Prep columns; Bio-Rad) equilibrated with buffer II (0.28 *M*). After adsorption (12 ml/hr), the column is washed successively with 4 ml of buffer II (0.28 *M*) and 0.6 ml of buffer II (0.3 *M*) and the RNA polymerase III is stepwise eluted with buffer II (0.7 *M*). One to 2 μl of fractions (300 μl) are used to assay the enzyme activity in the nonspecific assay. About 1.5–2 mg of RNA polymerase III at a concentration reaching 1.8 mg/ml is usually obtained, starting from 150 g of cells. This procedure can be scaled down to purify RNA polymerase III from mutant strains, starting from a lower amount of cells (30 g). In that case, the cell crude extract is directly loaded on the heparin column.

The subunit composition of the purified enzyme has been described.[22] The enzyme contains 17 subunits [C160, C128, C82, C53, C40, C37, C34, C31, C27, C25, C23, C19, C17, C14.5, C11, and C10 (α and β)], named according to their apparent size derived from SDS-gel electrophoresis. The enzyme preparation shown in Fig. 3 contained one additional polypeptide

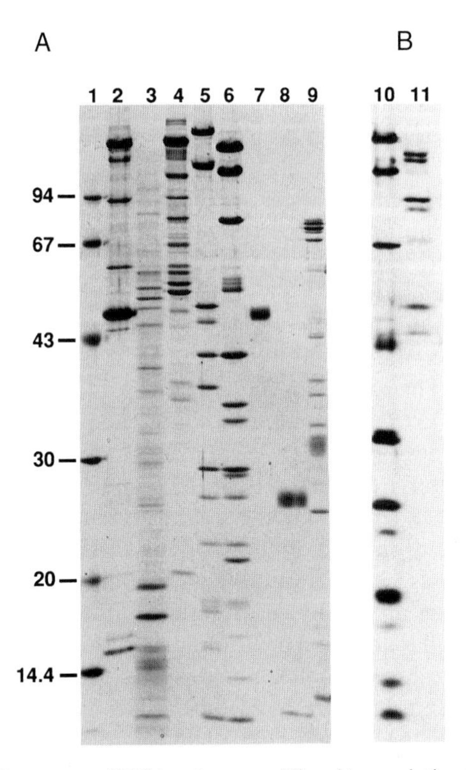

FIG. 3. Polypeptide content of RNA polymerase III and transcription factors. Proteins are analyzed by SDS–PAGE using a 13% (w/v) gel.[68] Proteins are stained by Coomassie blue (A) or with silver (B).[30] Lane 1, molecular weight markers; lane 2, Cibacron Blue fraction (5 μg); lane 3, B″ fraction (7.2 μg); lane 4, TFIIIB* (5 μg); lane 5, RNA polymerase I (7 μg); lane 6, RNA polymerase III (9 μg); lane 7, rTFIIIA (0.6 μg); lane 8, rTBP (0.1 μg); lane 9, rTFIIIB$_{70}$ (Ni^{2+}-NTA agarose) (2 μg); lane 10, RNA polymerase III (1.5 μg); lane 11, immunoaffinity-purified TFIIIC (0.05 μg).

of 13 kDa. Note that the migration of C37 polypeptide is variable.[26] Most RNA polymerase III subunits have been cloned and mutagenized.[4]

Microscale Purification

A micropurification procedure based on the large-scale method was developed starting from 10 g of cells. This procedure allows purification of RNA polymerase III in 1 day, using high-performance liquid chromatography (HPLC)-grade resins.

Heparin Hyper D Chromatography. Crude extract (20 ml) is diluted to 0.25 *M* ammonium sulfate with buffer I and directly loaded at 1 ml/min on a heparin Hyper D (BioSepra) column ($3\ cm^2 \times 7$ cm). At that ammonium sulfate concentration, RNA polymerase II passes through the column and only a small amount of RNA polymerase I is retained. The resin is washed with 200 ml of buffer I (0.365 *M*) at 3 ml/min; a linear gradient of ammonium sulfate from 0.365 to 0.8 *M* in 180 ml of buffer I is then applied at 3 ml/min. Fractions (2 ml) are assayed on 5-μl aliquots. The peak of RNA polymerase III activity eluted at 0.48 *M* ammonium sulfate. Note that when the crude extract is loaded to 0.1 *M* ammonium sulfate, RNA polymerases I, II, and III can be separated and are eluted at 0.35, 0.2, and 0.48 *M*, respectively.

Mono Q Chromatography. Fractions containing RNA polymerase III are pooled (20 ml) in a disymmetrical way because trailing fractions of the activity peak are contaminated by proteins that are not eliminated in the subsequent steps. The solution is diluted slowly, to avoid enzyme precipitation, with buffer I to 0.2 *M* ammonium sulfate and applied on a 1-ml Mono Q column (Pharmacia, Piscataway, NJ) at 1 ml/min. The column is washed with 60 ml of buffer I (0.28 *M*), then developed in 30 min with a 15-ml gradient from 0.28 to 0.6 *M* ammonium sulfate in buffer I. RNA polymerase III activity is tested on 5-μl aliquots of the different fractions (200 μl). Maximum activity is eluted around 0.36 *M* ammonium sulfate. Only the most active fractions are pooled (six fractions). The enzyme is then concentrated on a 100-μl heparin Hyper D column (Poly Prep columns; Bio-Rad) as described above except that the flow rate is 50 μl/min and the fraction size is 50 μl. Fifteen micrograms of pure RNA polymerase is commonly obtained. Owing to the FPLC-grade resins and FPLC or HPLC chromatographs used, the different chromatography steps are highly reproducible and RNA polymerase III activity tests can be omitted once the system has been calibrated.

[26] B. Bartholomew, D. Durkovich, G. A. Kassavetis, and E. P. Geiduschek, *Mol. Cell. Biol.* **13,** 942 (1993).

TFIIIC

Two additional chromatographic steps are necessary to purify TFIIIC to near homogeneity after the heparin–Ultrogel A4R column (see Large-Scale Purification, above). In the original procedure,[27] TFIIIC is further purified by DEAE-Sephadex, then by DNA–Sepharose chromatography. This procedure allows the purification of TFIIIC to near homogeneity but a significant amount of TFIIIC–DNA binding activity is lost. We also describe another procedure based on immunoaffinity purification that improves the purification yield about 10-fold.

DEAE-Sephadex Chromatography

The most active TFIIIC fractions from the heparin column (see Fig. 2) are pooled (≈140–180 ml; 100 mg of protein) and dialyzed for 2 hr against 1 liter of buffer I containing 1 m*M* PMSF to lower the salt concentration. An additional dilution to 70 m*M* salt is performed with buffer I (with 1 m*M* PMSF) before loading the proteins on a DEAE-Sephadex A-25 (Pharmacia) column ($2\ cm^2 \times 10\ cm$) at a flow rate of 100 ml/hr. The column is washed with 100 ml of buffer II (0.07 *M*). Proteins are eluted by a 200-ml gradient from 0.07 to 0.3 *M* ammonium sulfate in buffer II. Proteins are collected in 2-ml fractions. The active fractions are eluted between 0.11 and 0.16 *M* salt (≈25–30 ml; 0.1 mg of protein per milliliter). This step is efficient as it eliminates nearly 90 to 95% of the proteins.

DNA–Sepharose Chromatography

The last purification step of TFIIIC is by DNA affinity chromatography.[28] The DNA consists of one to seven repeats of a synthetic, 81-bp DNA fragment corresponding to the entire mature sequence of the yeast $tRNA_3^{Glu}$ gene (the two TFIIIC-binding sites are underlined):

```
GATCCTCCGATATAGTGTAACGGCTATCACATCACGCTTTCACCGTGGAGACCGGGGTTCGACTCCCCGTATCGGAGTACG
    GAGGCTATATCACATTGCCGATAGTGTAGTGCGAAAGTGGCACCTCTGGCCCCAAGCTGAGGGGCATAGCCTCATGCCTAG
            -----------                                 -------------
              Block A                                      Block B
```

To prepare the DNA affinity column, the two gel-purified complementary 81-mers (50 μg each) are combined in TE buffer [10 m*M* Tris–HCl (pH 8), 1 m*M* EDTA], incubated for 10 min at 100°, and gradually cooled for about 1 hr to room temperature to allow annealing. After precipitation,

[27] O. S. Gabrielsen, N. Marzouki, A. Ruet, A. Sentenac, and P. Fromageot, *J. Biol. Chem.* **264,** 7505 (1989).

[28] R. N. Swanson, C. Conesa, O. Lefebvre, C. Carles, A. Ruet, E. Quemeneur, J. Gagnon, and A. Sentenac, *Proc. Natl. Acad. Sci. U.S.A.* **88,** 4887 (1991).

the annealed DNA is kinased. The pellet is resuspended in a 300-μl reaction volume containing 50 mM Tris–HCl (pH 7.6), 1 mM EDTA, 10 mM $MgCl_2$, 5 mM dithiothreitol, 1 mM spermidine, 3 mM ATP, 1 μCi of [γ-32P]ATP, and T4 polynucleotide kinase (300 units). After 2 hr of incubation at 37°, the DNA is purified by phenol–chloroform extraction and ethanol precipitation. The DNA pellet is redissolved in 66 mM Tris–HCl (pH 7.5), 10 mM $MgCl_2$, 15 mM dithiothreitol, 1 mM spermidine, 4 mM ATP, and 15 units of T4 DNA ligase (300 μl). After 3 hr of incubation at 15°, 3 μl of ligated DNA is analyzed by agarose gel electrophoresis. The majority of the ligation products should be four to seven repeats of double-stranded 81-mer. After phenol–chloroform extraction and ethanol precipitation, the DNA is redissolved in water (100 μl) to be covalently coupled to CNBr-Sepharose 4B (Pharmacia). The DNA is coupled to CNBr-activated Sepharose 4B (Pharmacia) precisely as described by Kadonaga and Tjian.[29] One gram of CNBr-activated Sepharose 4B ($\approx$4-ml final gel volume) is washed successively on a sintered glass funnel (porosity 3) with 100 ml of 1 mM HCl, 200 ml of H_2O, and 100 ml of 10 mM potassium phosphate, pH 8 (never dry the resin). The coupling reaction is performed at 4° overnight in a screw-cap plastic tube on a rotating wheel with activated resin resuspended in 4 ml of 10 mM potassium phosphate (pH 8) and 100 μg of DNA. The resin is collected again on a sintered glass funnel and washed with 100 ml of H_2O, then with 50 ml of 1 M ethanolamine hydrochloride (pH 8), under slight suction. The resin is transferred to a screw-cap plastic tube and incubated with 2 ml of 1 M ethanolamine hydrochloride (pH 8) on a rotating wheel for 4 hr at room temperature to block the remaining active groups. The resin is then successively washed on a sintered glass funnel under slight suction with 50 ml of 10 mM potassium phosphate (pH 8), 50 ml of 1 M potassium phosphate (pH 8), 10 ml of 1 M KCl, 10 ml of 1 M NaCl, 50 ml of H_2O, and 50 ml of buffer III (0.075 M). Sodium azide is added at 0.02% (w/v) to prevent bacterial contamination. Usually the resin can be used up to 20 times.

A 1-ml DNA–Sepharose column is used per milligram of protein. Active TFIIIC fractions from DEAE-Sephadex ($\approx$25–30 ml; 2.5–3 mg of protein) are diluted to 75 mM salt with buffer III containing 0.5 mM PMSF and adsorbed at room temperature at a flow rate of 30 ml/hr. The flowthrough is reloaded on the column at a flow rate of 15 ml/hr. The column is washed with three column volumes of buffer III (0.09 M), and TFIIIC is eluted with buffer III (0.25 M) in fractions of 250 μl. Addition of Nonidet P-40 (NP-40) to buffer III decreases the nonspecific adsorption of proteins but interferes with the Bradford protein concentration measurement.[24] There-

[29] J. T. Kadonaga and R. Tjian, *Proc. Natl. Acad. Sci. U.S.A.* **83,** 5889 (1986).

fore, protein purification is estimated by gel-retardation assay and by silver staining after separation by SDS-polyacrylamide gel electrophoresis of the proteins.[30] Typically, the final fractions contain seven major polypeptides of 50, 60, 91, 95, 131, 138, and 170 kDa.[28] The 170-kDa polypeptide is unrelated to TFIIIC and corresponds to topoisomerase II, as evidenced by amino acid sequence determination.

Immunopurification of TFIIIC

The sequence encoding the epitope YPYDVPVYA[31] was introduced after the initiation codon of *TFC1*, the gene for the 95-kDa subunit of TFIIIC, by site-directed mutagenesis as described previously.[32] Centromeric plasmid harboring the epitope-tagged allele of *TFC1* was used to transform a yeast diploid strain in which one of the chromosomal copies of *TFC1* was disrupted. After sporulation, tetrad analysis was performed and a haploid strain (named YCS7) containing a chromosomal disrupted copy of *TFC1* as well as the modified allele on a plasmid was selected.

Purification of tagged TFIIIC is performed from 150 g of YCS7 cells as described for wild-type TFIIIC (this procedure can also be applied to 30 g of cells). All heparin–Ultrogel A4R fractions containing TFIIIC activity are pooled (200–260 ml, 0.8–1.2 mg of protein per milliliter) and chromatographed on an immunoaffinity column (instead of DEAE-Sephadex).

Preparation of the Immunoaffinity Column. The monoclonal antibody 12CA5[31] (2.8 mg) is covalently coupled with 20 m*M* dimethylpimelimidate (Sigma) to 7 ml of protein A–Sepharose CL-4B beads (Sigma) as described by Harlow and Lane.[33] After extensive washing with 0.2 *M* glycin hydrochloride (pH 3), the column is equilibrated in HEG buffer (350 m*M*). After use, the column is regenerated with 100 ml of 5 *M* NaCl, then with 100 ml of glycin hydrochloride (pH 3, 0.2 M), and conserved at 4° in phosphate-buffered saline (PBS) containing 0.01% (w/v) sodium azide. The resin can be used about five times without a significant loss of epitope-tagged TFIIIC binding.

Immunopurification of Epitope-Tagged TFIIIC. Active heparin–Ultrogel A4R fractions (100 ml) containing 0.1% (v/v) NP-40 are incubated in an IBF 80 column (5 cm^2) (BioSepra) overnight at 4° under gentle end-

[30] H. Blum, H. Beier, and H. J. Gross, *Electrophoresis* **8,** 93 (1987).

[31] I. A. Wilson, H. L. Niman, R. A. Houghten, A. R. Cherenson, M. L. Connolly, and R. A. Lerner, *Cell* **37,** 767 (1984).

[32] C. Conesa, R. N. Swanson, P. Schultz, P. Oudet, and A. Sentenac, *J. Biol. Chem.* **268,** 18047 (1993).

[33] E. Harlow and D. Lane, eds., "Antibodies: a laboratory manual" Cold Spring Harbor Laboratory, Cold Spring Harbor, New York, (1988).

over-end rocking with 7 ml of immunoaffinity resin equilibrated in HEG buffer (350 mM). After incubation, the flowthrough is removed by gravity flow and the column is washed by gravity flow with 100 ml of HEG buffer (250 mM) and 0.1% (v/v) NP-40, then with 120 ml of HEG buffer (50 mM). Elution is performed stepwise at 4° by adding 2 ml of HEG buffer (50 mM) and 0.1 mg/ml of a synthetic peptide corresponding to the HA epitope. Every 15 min, 2 ml is collected by gravity flow and replaced by 2 ml of the same elution buffer. Thirty fractions are collected and assayed for TFIIIC–DNA binding activity. This stepwise elution is more efficient than a slow, continuous elution process.

DNA Affinity Chromatography. Immunoaffinity fractions containing TFIIIC activity (≈40 ml; 5–10 μg of protein per milliliter) are pooled, brought to 75 mM ammonium sulfate with HEG buffer (500 mM), and loaded at 10–15 ml/hr on a 1-ml DNA–Sepharose column. The flowthrough is immediately loaded on a second 1-ml DNA–Sepharose column. The two columns are then washed with 10 ml of HEG buffer containing 75 mM ammonium sulfate and 0.05% (v/v) NP-40. Elution is performed stepwise with 500 μl of HEG buffer (250 mM) and 0.05% (v/v) NP-40 (10 fractions are collected), then with 500 μl of HEG buffer (500 mM) and 0.05% (v/v) NP-40 (5 fractions are collected). Elution fractions are collected in siliconized microtubes saturated with insulin (0.01 mg/ml). Fractions containing TFIIIC–DNA binding activity are pooled (5–6 ml; ≈2.5 μg of protein per milliliter). Purified TFIIIC is composed of six polypeptides of 138, 131, 95, 91, 60, and 50 kDa.[28] The sample shown has an additional polypeptide of 80 kDa (Fig. 3). The three largest subunits of yeast TFIIIC have been cloned.[4]

Xenopus TFIIIC purified to near homogeneity is also a multisubunit factor.[34] Human TFIIIC can be resolved into two components, TFIIIC1 and TFIIIC2.[35] Highly purified TFIIIC2 is a multisubunit protein.[36–38]

TFIIIB

The partial purification of TFIIIB from yeast has been previously described by Kassavetis *et al.*[8] TFIIIB activity was resolved in two components; B′, containing TBP and $TFIIIB_{70}$ (or BRF); and B″, containing an active

[34] H. J. Keller, P. J. Romaniuk, and J. M. Gottesfeld, *J. Biol. Chem.* **267,** 18190 (1992).

[35] S. K. Yoshinaga, P. A. Boulanger, and A. J. Berk, *Proc. Natl. Acad. Sci. U.S.A.* **84,** 3585 (1987).

[36] S. Yoshinaga, N. D. L'Etoile, and A. Berk, *J. Biol. Chem.* **264,** 10726 (1989).

[37] N. D. L'Etoile, M. L. Fahnestock, Y. Shen, R. Aebersold, and A. J. Berk, *Proc. Natl. Acad. Sci. U.S.A.* **91,** 1652 (1994).

[38] G. Lagna, R. Kovelman, J. Sukegawa, and R. G. Roeder, *Mol. Cell. Biol.* **14,** 3053 (1994).

component of 90 kDa. Both are required to reconstitute transcription.[39] Recombinant TBP (rTBP), recombinant $TFIIIB_{70}$, and fraction B″ have been shown to be necessary and sufficient to reconstitute yeast TFIIIB activity.[40] TBP and $TFIIIB_{70}$ genes have been cloned[41–47] and the corresponding polypeptides can now be easily produced in a recombinant, highly active form. In addition, an easy procedure has been described for the preparation of a highly active B″ fraction devoid of TBP and $TFIIIB_{70}$ activities.[40] In addition, Margottin *et al.*[48] have resolved TBP from TFIIIB by gel filtration. This procedure could not be scaled up, therefore a new method to prepare TFIIIB devoid of TBP (TFIIIB*) was developed by Huet and Sentenac.[49]

TFIIIB

The procedure of Moenne *et al.*[50] is described. Ten milliliters (7 mg) of the most active TFIIIB fractions from the heparin–Ultrogel A4R column (see Fig. 1) is adjusted to 0.3 *M* ammonium sulfate with buffer I and loaded at a flow rate of 10 ml/hr on a Cibacron Blue 3GA–agarose (Sigma) column (1.5 ml) equilibrated with buffer I (0.3 *M*). Ninety percent of the proteins pass through the column and, after washing, TFIIIB is eluted stepwise with 1 *M* KCl in buffer I. Fractions (300 μl) are collected and 0.25 to 0.5 μg is used to assay TFIIIB activity in the reconstituted transcription assay. TFIIIB fraction is contaminated by a major 49-kDa protein (EF1α) that can be removed by Q-Sepharose (Pharmacia) chromatography (see below). Note that TFIIIC is retained at 0.3 *M* ammonium sulfate on Cibacron Blue and therefore care must be taken when pooling the TFIIIB fractions from the

[39] G. A. Kassavetis, B. Bartholomew, J. A. Blanco, T. E. Johnson, and E. P. Geiduschek, *Proc. Natl. Acad. Sci. U.S.A.* **88,** 7308 (1991).

[40] G. A. Kassavetis, C. A. P. Joazeiro, M. Pisano, E. P. Geiduschek, T. Colbert, S. Hahn, and J. A. Blanco, *Cell* **71,** 1055 (1992).

[41] B. Cavallini, I. Faus, H. Matthes, J. M. Chipoulet, B. Winsor, J. M. Egly, and P. Chambon, *Proc. Natl. Acad. Sci. U.S.A.* **86,** 9803 (1989).

[42] S. Hahn, S. Buratowski, P. A. Sharp, and G. Guarente, *Cell* **58,** 1173 (1989).

[43] M. Horikoshi, C. K. Wang, H. Fujii, J. A. Cromlish, P. A. Weil, and R. G. Roeder, *Nature (London)* **341,** 299 (1989).

[44] M. C. Schmidt, C. C. Kao, R. Pei, and A. J. Berk, *Proc. Natl. Acad. Sci. U.S.A.* **86,** 7785 (1989).

[45] S. Buratowski and H. Zhou, *Cell* **71,** 221 (1992).

[46] T. Colbert and S. Hahn, *Genes Dev.* **6,** 1940 (1992).

[47] A. López-De-León, M. Librizzi, K. Puglia, and I. Willis, *Cell* **71,** 211 (1992).

[48] F. Margottin, G. Dujardin, M. Gérard, J. M. Egly, J. Huet, and A. Sentenac, *Science* **251,** 424 (1991).

[49] J. Huet and A. Sentenac, *Nucleic Acids Res.* **20,** 6451 (1992).

[50] A. Moenne, S. Camier, G. Anderson, F. Margottin, J. Beggs, and A. Sentenac, *EMBO J.* **9,** 271 (1990).

heparin–Ultrogel A4R column to avoid TFIIIC contamination. At that ionic strength, RNA polymerase II, which can also contaminate the TFIIIB pool, is not retained.

*TFIIIB**

TFIIIB fractions from the heparin–Ultrogel A4R column (45 mg) are chromatographed on a Q-Sepharose column (Pharmacia) (2 cm^2 × 7 cm; 14 ml) equilibrated in buffer II (50 m*M*) containing 0.1 m*M* PMSF. The column is washed with the same buffer (200 ml), and the proteins are eluted stepwise with buffer II (0.15 *M*). Two microliters of each fraction (1 ml) is assayed for transcription activity on U6 snRNA gene (SNR6)[51] or $tDNA_3^{Glu}$ and for the presence of TBP by Western blotting. TBP is found in the flowthrough with 90% of the proteins, while TFIIIB* activity is recovered in the step-eluted fractions. Aliquots (0.8 μg) of eluted fractions must be supplemented with 1.2 μg of the flowthrough or 30 ng of purified rTBP to transcribe *SNR6*. Active fractions are pooled after analysis of their protein content by SDS–PAGE and stored at −80°.

B″

The following procedure for the preparation of crude B″ starts from the pellet fraction of whole-cell lysates after ammonium sulfate extraction. It is adapted from Kassavetis *et al.*,[40] with minor modifications. Pellets deriving from 60 g of cells are resuspended in 320 ml of Bx buffer [40 m*M* Tris–HCl (pH 8), 1 m*M* EDTA, 10% (v/v) glycerol, 10 m*M* mercaptoethanol] plus protease inhibitors containing 1 *M* KCl and 6 *M* urea, by blending for 5 sec in a Waring blender at low speed at 4°. The slurry is centrifuged at 15,000 *g* for 45 min. After a five-fold dilution with Bx buffer (final volume, 1.3 liters) the supernatant is applied to a 100-ml Bio-Rex 70 (Bio-Rad) column (flow rate, 60 ml/hr) equilibrated with Bx buffer plus 0.2 *M* KCl. After washing with 250 ml of the same buffer plus 0.25 *M* KCl, bound proteins are eluted with 0.5 *M* KCl. A transcriptionally active fraction (6 ml) is generated containing 3.6 mg of protein.

Recombinant TBP

Several procedures for the purification to homogeneity of recombinant yeast TBP have been described.[52,53] We usually follow the one described in detail by Burton *et al.*[54] High-level expression in *E. coli* is obtained by

[51] D. A. Brow and C. Guthrie, *Nature* (*London*) **334,** 213 (1988).
[52] P. M. Lieberman, M. C. Schmidt, C. C. Kao, and A. J. Berk, *Mol. Cell. Biol.* **11,** 63 (1991).
[53] P. Reddy and S. Hahn, *Cell* **65,** 349 (1991).
[54] N. Burton, B. Cavallini, M. Kanno, V. Moncollin, and J.-M. Egly, *Protein Expr. Purif.* **2,** 432 (1991).

the use of the pET expression system,[55] and the recombinant protein is purified to homogeneity through sequential DEAE-Sephacel, heparin–Ultrogel A4R, and heparin TSK-5PW chromatographies. Human[56,57] and *Drosophila*[58] TBPs can also be easily produced in a recombinant form.

Recombinant TFIIIB$_{70}$

The following procedure for the purification of recombinant TFIIIB$_{70}$ polypeptide is based on that described by Colbert and Hahn,[46] with several modifications. A standard preparation starts from a 2-liter culture of *E. coli* BL21(DE3) transformed with plasmids pLysS[55] and pSH360, which contain the BRF1-6His fusion in the Novagen expression vector pET11d (a gift of S. Hahn, Fred Hutchinson Cancer Research Center, Seattle, WA). Cells are grown in LB medium supplemented with ampicillin (100 μg/ml) and chloramphenicol (34 μg/ml) at 30° to an A_{600} value of 0.7. Then isopropyl-β-D-thiogalactopyranoside (IPTG) is added to 0.55 m*M*, and the incubation continued for 3 hr. Cells are harvested by centrifugation at 6000 *g* for 15 min at 4°, washed with 500 ml of cold TE* buffer, then resuspended in 150 ml of lysis buffer [50 m*M* sodium phosphate (pH 8), 0.3 *M* NaCl, 5 m*M* mercaptoethanol] plus protease inhibitors (0.5 m*M* PMSF, 0.5 m*M* benzamidine, 1 μ*M* leupeptin, 1 μ*M* pepstatin). After one freeze–thawing cycle, lysozyme (Sigma) is added to a final concentration of 0.2 mg/ml, and the cell suspension is incubated on ice for 30 min. Aliquots of 75 ml are then sonicated three or four times (10 sec each time) at power 5, 60% cycle (Vibracell; Bioblock Scientific, Illkirch, France) and centrifuged for 30 min at 15,000 *g*. The supernatant is recovered and filtered over gauze. Ni^{2+}-NTA agarose (4.5 ml) (Qiagen, Chatsworth, CA), equilibrated in lysis buffer plus 10% (v/v) glycerol and protease inhibitors, is then added. Binding is performed in rotating 30-ml centrifuge tubes for 30 min at 16°. The resin with bound BRF1-6His is sedimented by centrifugation (5 min at 3000 rpm in a Beckman JA-20 rotor). Most of the supernatant is discarded and the slurry is transferred into a 1-cm diameter column. The column (4.5 ml) is washed at 4° at gravity flow with 70 ml of lysis buffer plus 10% (v/v) glycerol and protease inhibitors, then with 40 ml of the same buffer containing 20 m*M* imidazole. Bound proteins are eluted with the same buffer plus 250 m*M* imidazole. Active fractions are pooled (7.5 ml; 5.5 mg of protein) and dialyzed for 3 hr against 1 liter of dialysis buffer [25 m*M* Tris–HCl (pH 8), 0.2 m*M* EDTA, 15% (v/v) glycerol, 0.12 *M* KCl, 5 m*M* mercaptoethanol, 0.5 m*M* benzamidine]. At this stage, the TFIIIB$_{70}$ polypeptide is about

[55] F. W. Studier, A. H. Rosenberg, and J. J. Dunn, *Methods Enzymol.* **185,** 60 (1990).
[56] M. G. Peterson, N. Tanese, B. F. Pugh, and R. Tjian, *Science* **248,** 1625 (1990).
[57] A. Hoffmann and R. G. Roeder, *Nucleic Acids Res.* **19,** 6337 (1991).
[58] T. Hoey, B. D. Dynlacht, M. G. Peterson, B. F. Pugh, and R. Tjian, *Cell* **61,** 1179 (1990).

20–30% pure. Further purification can be obtained by chromatography on heparin–Ultrogel (BioSepra) at an input of 0.5 mg of protein per milliliter of resin. Proteins are bound at 0.12 *M* KCl in dialysis buffer containing protease inhibitors, and eluted with a KCl linear gradient (0.12–0.5 *M*). $TFIIIB_{70}$ elutes around 0.25–0.3 *M* KCl. Heparin can be efficiently replaced by a Mono S column (Pharmacia), as described in the original paper.[46] Following SDS–PAGE, the 70-kDa polypeptide represents 60–70% of the stainable material, the remaining 30–40% being constituted by smaller polypeptides recognized by anti-BRF antiserum and by [^{35}S]-labeled TBP.[59]

TFIIIA

TFIIIA from *Xenopus laevis* was the first eukaryotic transcription factor to be purified to homogeneity.[60] Human TFIIIA has also been purified to homogeneity from HeLa cells.[61] Wang and Weil[62] have described a procedure for the purification to apparent homogeneity of endogenous yeast TFIIIA, with a low yield probably due to the low *in vivo* abundance of the protein. Since the cloning of the yeast TFIIIA gene,[63,64] an easier, high-yield procedure for the preparation of recombinant yTFIIIA has been published.[65] High-level expression of the protein in *E. coli* depends on the combined use of a T7 RNA polymerase expression system[55] and of a plasmid carrying the *E. coli* gene *argU* encoding a minor Arg(AGA/AGG) tRNA species.[66] A 10-ml culture of LB broth plus ampicillin (200 μg/ml) and kanamycin (50 μg/ml) is inoculated with the *E. coli* BL21 (DE3) strain transformed with pETIIIA[65] and pUBS520[66] plasmids, and grown to an OD_{600} of 0.6. This culture is diluted to 2 liters, growth is continued at 37° to an OD_{600} of 0.6, then 50 μM $ZnSO_4$ and 0.6 m*M* IPTG are added, and the culture incubated at 30° for an additional 4 hr. Harvested cells are rinsed with TE* buffer, resuspended in TE* containing 0.1 *M* KCl, 0.1 m*M* $ZnSO_4$, 5 m*M* mercaptoethanol plus protease inhibitors [leupeptin and pepstatin (1 μ*M* each), 0.5 m*M* benzamidine, and 0.1 m*M* PMSF], and

[59] J. Huet, C. Conesa, N. Manaud, N. Chaussivert, and A. Sentenac, *Nucleic Acids Res.* **22,** 3433 (1994).

[60] D. R. Engelke, S. Y. Ng, B. S. Shastry, and R. G. Roeder, *Cell* **19,** 717 (1980).

[61] K. H. Seifart, L. Wang, R. Waldschmidt, D. Jahn, and E. Wingender, *J. Biol. Chem.* **264,** 1702 (1989).

[62] C. K. Wang and P. A. Weil, *J. Biol. Chem.* **264,** 1092 (1989).

[63] J. Archambault, C. A. Milne, K. T. Schappert, B. Baum, J. D. Friesen, and J. Segall, *J. Biol. Chem.* **267,** 3282 (1992).

[64] N. A. Woychik and R. A. Young, *Proc. Natl. Acad. Sci. U.S.A.* **89,** 3999 (1992).

[65] S. Ottonello, A. Ballabeni, C. Soncini, and G. Dieci, *Biochem. Biophys. Res. Commun.* **203,** 1217 (1994).

[66] U. Brinkmann, R. E. Mattes, and P. Buckel, *Gene* **85,** 109 (1989).

frozen. After thawing and addition of KCl, glycerol, and lysozyme to 0.3 *M*, 10% (v/v), and 0.2 mg/ml, respectively, the sample is kept for 30 min at 4°, then for 5 min at 30°, and sonicated as described for $TFIIIB_{70}$. After centrifugation at 15,000 *g* for 30 min, the clarified supernatant is adjusted to 80% ammonium sulfate saturation, centrifuged at 30,000 *g* for 70 min, and the pellet dissolved in 25 ml of TEG buffer [TE* plus 10% (v/v) glycerol, 0.1 m*M* $ZnSO_4$, 5 m*M* mercaptoethanol, and protease inhibitors]. The sample is dialyzed against TEG buffer to a conductivity value equivalent to that of 0.15 *M* KCl, then applied to a 45-ml Bio-Rex 70 (Bio-Rad) column equilibrated in TEG buffer containing 0.15 *M* KCl. After washing with 2 vol of TEG plus 0.2 *M* KCl, bound proteins are eluted with 0.75 *M* NaCl. TFIIIA-containing fractions are pooled, concentrated by ammonium sulfate precipitation as described above, resuspended in 8 ml of TEG buffer, and loaded onto a 430 ml Sephacryl S-300 HR (Pharmacia) gel-filtration column ($5\ cm^2 \times 90$ cm) equilibrated with TEG buffer plus 0.25 *M* NaCl. The 51-kDa TFIIIA polypeptide elutes at an apparent molecular weight of 150,000, probably due to structural asymmetry.[62] TFIIIA-containing fractions are pooled and loaded on a 1-ml heparin–Ultrogel A4R (BioSepra) column equilibrated in TEG buffer and eluted with a 0.3–1.0 *M* NaCl linear gradient in TEG (15 column volumes). Yeast TFIIIA, essentially free of contaminants, elutes around 0.6 *M* NaCl. A new, one-step purification of rTFIIIA on magnetic beads of phosphocellulose has been developed.[67]

Note Added in Proof

The gene encoding the 90 kDa component of B″ has been cloned.[69,70] Recombinant B''_{90} (or $TFIIIB_{90}$), produced in *E. coli* cells, directed accurate transcription of tRNA genes in combination with recombinant $TFIIIB_{70}$ and recombinant TBP.

Acknowledgments

This work was supported in part by a grant from the CEE biotechnology program. G.D. was supported by an EMBO long-term fellowship. We thank Steve Hahn for plasmid pSH360 and for anti-BRF antibodies, Simone Ottonello for a sample of pure ryTFIIIA, and Jean-Marc Egly for anti-TBP antibodies.

[67] P. A. Risøen, K. Struksnes, A. H. Myrset, and O. S. Gabrielsen, *Protein Expr. Purif.* **6,** 272 (1995).

[68] U. K. Laemmli, *Nature (London)* **227,** 680 (1970).

[69] G. A. Kassavetis, S. T. Nguyen, R. Kobayashi, A. Kumar, E. P. Geiduschek, and M. Pisano, *Proc. Natl. Acad. Sci. U.S.A.* **92,** 9786 (1995).

[70] J. Rüth, C. Conesa, G. Dieci, O. Lefebvre, A. Dusterhoft, S. Ottonello, and A. Sentenac, *EMBO J.* **15,** 1941 (1996).

[23] Basal and Activated *in Vitro* Transcription in Plants by RNA Polymerase II and III

By HAO FAN and MASAHIRO SUGIURA

Developed plant organs and tissues are highly differentiated in shape and function and contain distinct constituents, especially different forms of plastids. In different organs and tissues, genes are differentially transcribed under the control of *cis*-elements and *trans*-factors. Thus, development of basal and activated plant *in vitro* transcription systems is critical for elucidating the biochemical mechanism of gene transcription and its regulation in plants. Tobacco (*Nicotiana tabacum*) nongreen cultured BY-2 cells are a good source for this purpose because this cell line has the fastest growth rate among the plant cell lines reported[1] and the tobacco plant provides a powerful tool with which to analyze plant gene transcription *in vivo* by the well-established transgenic technique. As BY-2 cells do not differentiate in response to light or any other stimuli, they give a basal *in vitro* transcription system that lacks tissue-specific and light-responsive factors. By combination of the basal transcription system and active fractions from organs and tissues in which genes in question are specifically transcribed, *cis*-elements and *trans*-factors involved in their transcription are expected to be detected *in vitro*. We describe the procedures for preparation of basal and activated *in vitro* transcription systems for RNA polymerase II (Pol II)-dependent genes. The basal system also supports RNA polymerase I- and III-dependent transcription.[2,3]

Preparation of Basal *in Vitro* Transcription System

Cell Culture

Tobacco BY-2 cells are grown in modified MS medium as described,[4] except that KH_2PO_4 and inositol are added to 540 and 100 mg/liter, respectively. Fifteen milliliters of a 7-day-old culture is transferred to 500 ml of fresh medium in a 2-liter flask and shaken at 130 rpm in the dark at 27°. BY-2 cells from the 2-liter culture (ca. 200 g wet weight) are harvested on

[1] T. Nagata, Y. Nemoto, and S. Hasezawa, *Int. Rev. Cytol.* **132,** 1 (1992).
[2] H. Fan and M. Sugiura, *EMBO J.* **14,** 1024 (1995).
[3] H. Fan, K. Yakura, M. Miyanishi, M. Sugita, and M. Sugiura, *Plant J.* **8,** 295 (1995).
[4] T. Nagata, K. Okada, I. Takebe, and C. Matsui, *Mol. Gen. Genet.* **184,** 161 (1981).

one layer of Miracloth at mid-log phase (86 hr after inoculation), transferred to two 500-ml centrifuge tubes, and washed once with 700 ml of 0.4 *M* mannitol solution (pH 5.4) followed by centrifugation at 400 *g* for 2 min at 4°.

Protoplast Preparation

Protoplasts are prepared by the method described.[5] Pelleted cells in two tubes are digested in a final volume of 500 ml each of enzyme solution [1% (w/v) Onozuka-RS cellulase (Yakult Honsha Company, Tokyo, Japan) and 0.1% (w/v) pectolyase Y-23 (Kikkoman Company, Tokyo, Japan) in MS medium containing 0.4 *M* mannitol, pH 5.4] at 30° for about 45 min (checked occasionally under a microscope). Protoplasts are collected by centrifugation at 200 *g* for 5 min at 4° and gently washed three times with 700 ml each of ice-cold 0.4 *M* mannitol solution.

Nuclear Isolation

Loosely packed protoplasts (200–220 ml) are suspended in 1 vol of ice-cold nuclear isolation buffer [30 m*M* *N*-2-hydroxyethylpiperazine-*N*′-2-ethanesulfonic acid HEPES–KOH (pH 7.9), 36% (w/v) Ficoll 400 (Pharmacia, Uppsala, Sweden), 2 m*M* NaF, 6 m*M* dithiothreitol (DTT), 1 m*M* EGTA, 0.4 m*M* EDTA, 0.3 m*M* spermine, 1 m*M* spermidine (spermine and spermidine can be replaced by 4 m*M* $MgSO_4$), 1 m*M* phenylmethylsulfonyl fluoride (PMSF), 1 m*M* benzamide, pepstatin A (3 μg/ml), and leupeptin (2 μg/ml)]. The suspension is immediately vacuum filtered through one layer of nylon mesh of 20-μm pore size (type HD-20; Schweiz, Seidengazefabrik AG, Switzerland). The filtrate is again vacuum filtered as described above, transferred to one 500-ml tube, and centrifuged at 2500 *g* for 15 min at 2°. The supernatant and yellow floating layer are carefully removed by a water aspirator, and the remaining yellow layer inside the tube is wiped off with tissue paper. The precipitated nuclei are resuspended in 0.5× nuclear isolation buffer (gradual addition of 400 ml) with a paintbrush and centrifuged at 2500 *g* for 10 min at 2°. The pellet is resuspended as described above in 40 ml of 0.5× nuclear isolation buffer and transferred into a 50-ml centrifuge tube; nuclei are collected by centrifugation at 2500 *g* for 10 min at 2°.

Preparation of Nuclear Extracts

Purified nuclei (ca. 1.2 g = 1.2 ml) are immediately suspended in 2 vol of ice-cold nuclear extraction buffer [25 m*M* HEPES–KOH (pH 7.9), 25%

[5] Y. Nemoto, S. Kawano, S. Nakamura, T. Mita, T. Nagata, and T. Kuroiwa, *Plant Cell Physiol.* **29,** 167 (1988).

(v/v) glycerol, 4 m*M* $MgSO_4$, 0.2 m*M* EGTA, 5 m*M* DTT, pepstatin A (3 μg/ml), leupeptin (2 μg/ml), and 1 m*M* PMSF. Lysis is carrier out by dropping in 4 *M* ammonium sulfate to a final concentration of 0.42 *M*. Highly viscous lysate is slowly (ca. 2 rpm) rotated with a tube rotator at 4° for 30 min, transferred carefully into two 5-ml open ultracentifuge tubes, and centrifuged at 80,000 rpm [Beckman (Fullerton, CA) TL100.1 rotor] for 1 hr at 2°. The supernatant above the cloudy pellet is carefully collected immediately after centrifugation and dialyzed twice on ice for 3 hr against 500 ml each of dialysis buffer [20 m*M* HEPES–KOH (pH 7.9), 20% (v/v) glycerol, 100 m*M* potassium acetate, 4 m*M* $MgSO_4$, 0.2 m*M* EGTA, 0.1 m*M* EDTA, 2 m*M* DTT, and 0.5 m*M* PMSF]. The dialyzed extract is aliquoted (150 μl each), frozen in liquid nitrogen, and stored at −80° for up to 1 year. Typically, 3 ml of the nuclear extract with 6 to 8 mg of protein per milliliter, as determined by the Bradford dye-binding assay, is obtained from 200 g of BY-2 cells.

Comments

To obtain active extracts, the time of preparation should be kept as short as possible, generally within 8 hr. It is important to use a well-maintained BY-2 culture, as RNA synthesis activity is the highest and RNase activity is relatively low at the mid-log phase (86 hr after inoculation). It is absolutely essential that isolated nuclei are pure and intact. As the nuclear envelope is permeable to small solutes and proteins, the leakage of necessary factors is inevitable during the isolation of nuclei. We used 18% (w/v) Ficoll (final concentration) to prevent this rapid loss of nuclear constituents during nuclear isolation (there were no active nuclear extracts with Ficoll lower than 15%, w/v). Nuclear extracts inactivated by dephosphorylation cannot be rescued by the addition of protein kinase C, and we therefore add 1 m*M* NaF in nuclear isolation buffers to inhibit phosphatase activities. Conditions of BY-2 cells and others vary from preparation to preparation; we routinely obtain an active extract in three of five trials.

RNA Polymerase II-Dependent *in Vitro* Transcription

Preparation of DNA Templates

The promoter region (−1470 to +23) of the tobacco β-1,3-glucanase gene[6] is subcloned into Bluescript KS$^+$ (Stratagene, La Jolla, CA) at *Hin*dIII

[6] M. Ohme-Takagi and H. Shinshi, *Plant Mol. Biol.* **15,** 941 (1990).

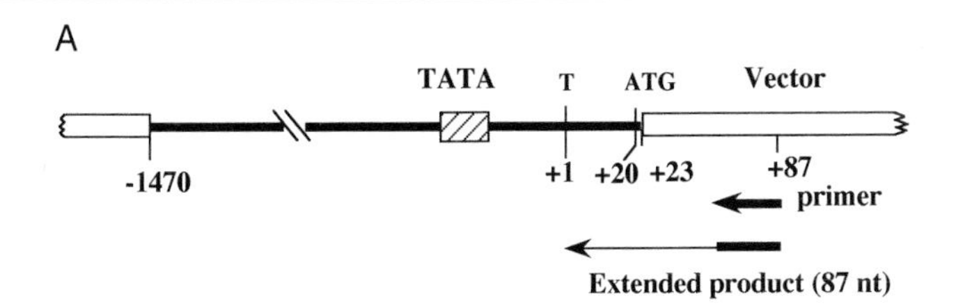

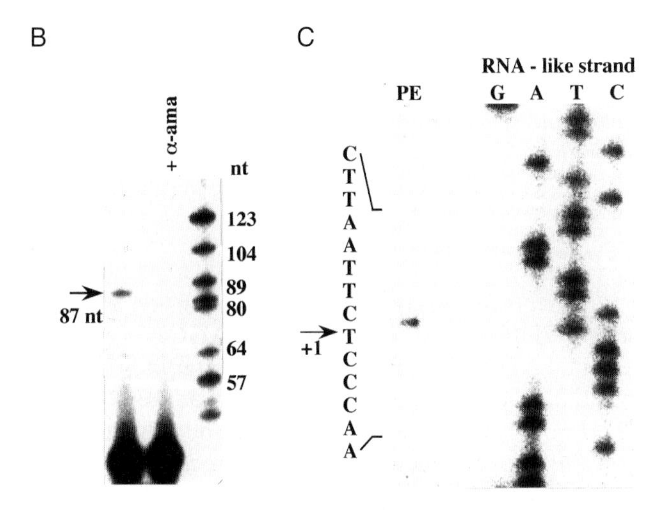

FIG. 1. *In vitro* transcription of the tobacco β-1,3-glucanase gene with the BY-2 nuclear extract. (A) Schematic view of DNA template and primer extension assay.[2] A short, boldface arrow indicates primer T7DKS. A long arrow shows the expected extended product (87 nt) calculated from the primer (+87) to the *in vivo* transcription start site (+1). (B) Detection of *in vitro* transcripts by 8% (w/v) PAGE. The expected extended product is indicated by an arrow (87 nt). No product was observed in the presence of α-amanitin (α-ama; 20 μg/ml). Size markers are from a pBR322 *Hae*III digest. (C) Determination of the transcription initiation site *in vitro*.[2] The primer extension product was compared with DNA sequence ladders generated from the same primer. The RNA start site is shown as +1. [Reprinted from H. Fan and M. Sugiura, *EMBO J.* **14,** 1024 (1995), with permission.]

and *Bam*HI sites and verified by sequence analysis (Fig. 1A). Plasmid is amplified in *Escherichia coli* XL1-Blue, isolated by the alkaline lysis method, and purified further by CsCl gradient centrifugation. Closed circle plasmid DNA (1 μg/μl) is used as templates.

In Vitro Transcription Reaction

A typical reaction mixture is constructed in 50 μl by adding (in the following order) 5 μl of 10× transcription buffer [200 m*M* HEPES–KOH (pH 7.9), 300 m*M* potassium acetate, 30 m*M* $MgSO_4$, and 15 m*M* DTT], 1 pmol of DNA (ca. 4 μg), 20 U of pancreatic ribonuclease inhibitor (TaKaRa Company, Kyoto, Japan) and RNase-free distilled H_2O to 22 μl (mixed), and then 25 μ of the nuclear extract (thawed on ice), mixed, and incubated on ice for 10 min. Two microliters (12.5 m*M* each) of the four RNase-free NTPs (Pharmacia, Piscataway, NY) and 1 μl of 50 m*M* *S*-adenosyl-L-methionine (SAM; Sigma, St. Louis, MO) are added to initiate transcription. After incubation for 30 min at 25 or 28°, the reaction is stopped by adding 200 μl of stop solution [20 m*M* EDTA, 0.2 *M* NaCl, 1% sodium dodecyl sulfate (SDS), glycogen (50 mg/ml), and 1 μl of proteinase K (20 mg/ml)] and incubated at 37° for 10 min. Next, 200 μl of 0.3 sodium acetate is added and the solution is extracted twice with 500 μl of phenol–chloroform–isoamyl alcohol (25:24:1, v/v) and once with chloroform–isoamyl alcohol (24:1, v/v). The resulting aqueous phase is recovered.

Assay of in Vitro Transcripts

In vitro transcripts are assayed by primer extension essentially as described.[7] Primer T7DKS (5′GGGCGAATTGGAGCTCCACCGCGGTG-GCGGCCGCTCTAGA 3′), which is complementary to the Bluscript KS^+ sequence (position 644–683) and purified by high-performance liquid chromatography (HPLC), is labeled at the 5′ end and diluted to 10 fmol/μl. Transcripts in the above, aqueous phase are precipitated with 2.5 vol of ethanol after addition of 10 μl (100 fmol) of labeled T7DKS, and collected immediately by centrifugation at 14,000 rpm for 15 min at 20°. Centrifugation at 20° is preferred for reducing background. The pellet is washed with 1 ml of 75% (v/v) ethanol, dried at room temperature, and dissolved in 10 μl of reverse transcriptase buffer [50 m*M* Tris–HCl (pH 8.3), 75 m*M* KCl, 3 m*M* $MgCl_2$, 10 m*M* DTT]. To anneal the primer, the solution is heated at 75° for 5 min, then incubated at 60° for 30 min and cooled to room temperature. Next, 10 μ of solution containing a 2 m*M* concentration of each of the four dNTPs, 100 μl of actinomycin D, 200 U of Moloney murine leukemia virus (M-MLV) reverse transcriptase (U.S. Biochemical, Cleveland, OH), and 20 U of RNase inhibitor in reverse transcriptase buffer is added and incubated at 37° for 1 hr. The reaction is terminated by the addition of an equal volume of loading buffer [50 m*M* EDTA, 1% bromophenol blue (BPB), 1% (w/v) xylene cyanole, 95% (v/v) formamide] and denatured at 95° for 3 min. Half of the sample is resolved on a denatured 8% (w/v) polyacrylamide gel 19:1, v/v) and the remaining half is loaded on a 6% (w/v) polyacrylamide sequencing gel and run in parallel with DNA

[7] J. T. Kadonaga, *J. Biol. Chem.* **265,** 2624 (1990).

sequencing ladders generated by a double-stranded DNA (dsDNA) PCR (polymerase chain reaction) sequencing method (GIBCO-BRL, Gaithersburg, MD), using the same DNA template and primer. After electrophoresis, ^{32}P-labeled products are assayed either by using a bio-imaging analyzer BAS-2000 (Fuji Photo Film, Tokyo, Japan) or by exposing to X-ray film.

The *in vitro* reaction with the β-glucanase gene produced a single clear band of predicted size and the initiation site (T) has been verified to be identical to that determined *in vivo*[6] (Fig. 1B and C). The transcription is DNA template dependent and sensitive to a low concentration of α-amanitin (20 μg/ml).

Comments

The present reaction conditions are optimized using the β-glucanase gene. Mn^{2+} cannot be substituted for Mg^{2+}; instead the addition of Mn^{2+} results in nonspecific transcription. The optimal reaction temperature and time are 25 to 30° and 30 min, respectively. Incubation for longer than 45 min causes product degradation. The amount of DNA template is optimal at 1 pmol in a 50-μl reaction mixture containing about 150 μg of nuclear proteins, and saturated at about 1.5 pmol. The addition of SAM apparently reduces the degradation of the transcripts. Transcription efficiency is about 0.001 transcript per template.[2] Reaction conditions and transcription efficiency depend on promoters, and optimization is necessary for individual genes.

We have measured transcription activity by primer extension analysis of transcripts synthesized *in vitro* as this assay yields transcriptional start sites at a nucleotide level and prevents artifactual results sometimes observed when using the G-free cassette. In addition, the use of primers complementary to vector sequences in circle DNA templates greatly reduces background due to nonspecific transcription from endogenous DNA incompletely removed by centrifugation of the nuclear extracts.

Activated *in Vitro* Transcription System by Supplementing Leaf Nuclear Extract

Nuclear genes encoding components of the photosynthesis apparatus are actively transcribed in green leaves under illumination but poorly in dark-grown plants and other nonphotosynthetic organs. Northern blot analysis showed that the gene (*rbcS*) for the small subunit of ribulose-bisphosphate carboxylase is inactive in dark-grown tobacco BY-2 cells, but actively expressed in tobacco leaves.[2] We then carried out *in vitro* transcription with the *rbcS3C* promoter from tomato, considering that tomato is close to tobacco (both belong to the Solanaceae family).

In Vitro Transcription of rbcS

Two promoter constructs in Bluescript KS$^+$ from tomato *rbcS3C*,[8] a full-length promoter (pRBC-1657, −1657 to +44), and a truncated promoter (pRBC-174, −174 to +1), have been tested. *In vitro* transcription and transcript assays are as described above. As shown in Fig. 2 (lanes 7 and 10), transcription from the *rbcS3C* promoters cannot be detected with the BY-2 nuclear extract, as expected.

Leaf Nuclear Extracts

We prepare a nuclear extract from light-grown tomato leaves. Sterilized tomato seeds (*Lycopersicon esculentum*) are germinated on fiberglass soaked with 1:1000 diluted Hyponex in a growth chamber under a 16-hr light/8-hr dark cycle at 28°. Green cotyledons (along with the one-quarter of the hypocotyl) are collected from 7-day-old seedling and frozen in liquid nitrogen. The preparation of nuclei from tomato leaves is essentially as described,[9] except that the isolation buffer consists of 20 m*M* HEPES–KOH (pH 7.0), 18% (w/v) sucrose, 1% (w/v) polyvinylpyrrolidine, 10 m*M* NaF, 5 m*M* DTT, 0.1% (v/v) Triton X-100, 0.5 m*M* EDTA, 1 m*M* EGTA, 0.15 m*M* spermine, 0.5 m*M* spermidine, pepstatin A (3 μg/ml), leupeptin (2 μg/ml), 1 m*M* PMSF, and 1 m*M* benzamide. The preparation of nuclear extracts is as described above. Only 0.4 ml of a nuclear extract with 5 mg of protein per milliliter is obtained from 200 g of tomato leaves after concentration of the dialysate with Centricut-10 (Kurabo Company, Osaka, Japan).

Activated in Vitro Transcription Reaction

In vitro transcription and transcript assays are as described above except that a volume of 100 μl is used. The reaction mixture (constituents are doubled) is preincubated with the leaf nuclear extract (30 or 40 μl) on ice for 10 min, followed by the addition of 30 μl of BY-2 nuclear extract and incubation for a further 10 min on ice prior to the addition of NTPs (4 μl) and SAM (2 μl) to start the reaction.

As shown in Fig. 2, the BY-2 system supplemented with a leaf extract drove accurate transcription from the *rbcS* promoter, detected as a clear 100-nucleotide band by primer extension assay (lanes 3 and 4). The transcription was α-amanitin sensitive (lanes 5 and 6). The *in vitro* transcription initiation site was confirmed to be exactly the same as *in vivo*.[2,10] The leaf

[8] M. Sugita and W. Gruissem, *Proc. Natl. Acad. Sci. U.S.A.* **84,** 7104 (1987).

[9] T. Manzara, P. Carrasco, and W. Gruissem, *Plant Cell* **3,** 1305 (1991).

[10] P. Carrasco, T. Manzara, and W. Gruissem, *Plant Mol. Biol.* **21,** 1 (1993); W. C. Taylor, *Annu. Rev. Plant Physiol.* **40,** 211 (1989).

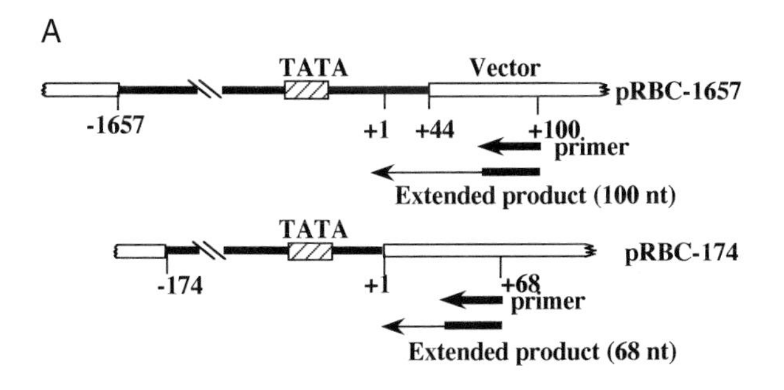

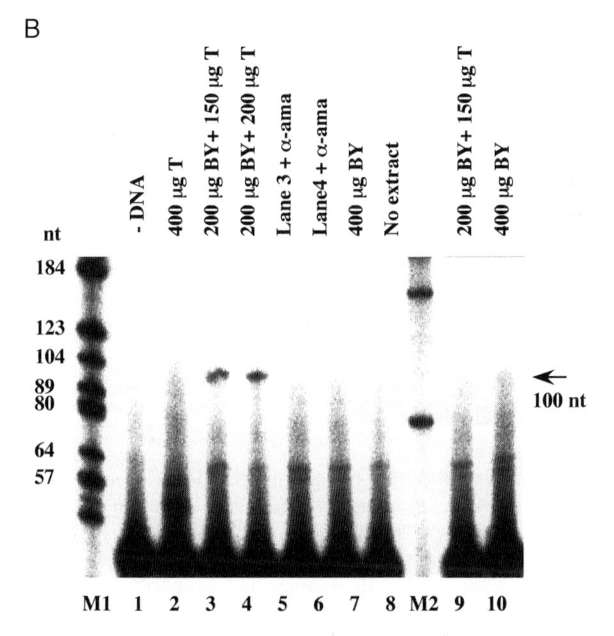

FIG. 2. *In vitro* transcription of the tomato *rbcS3C* gene.[2] (A) Schematic view of DNA templates and primer extension assay. A short, boldface arrow indicates primer T7DKS. The long arrows show the expected extended products [pRBC-1657 (100 nt) and pRBC-174 (68 nt)] calculated from the primer to the *in vivo* transcription start site (+1). (B) Detection of *in vitro* transcripts by 8% (w/v) PAGE. A full-length promoter (pRBC-1657) (lanes 1 to 8) and a truncated promoter (pRBC-174) (lanes 9 and 10) were transcribed by the tobacco BY-2 nuclear extract (BY) with or without the tomato leaf nuclear extract (T). Expected extended products are indicated by an arrow (100 nt). α-Amanitin (α-ama) was added at a final concentration of 20 μg/ml. Size markers are from a pBR322 *Hae*III digest (M1) and ϕX174-RF-DNA *Hinc*II digest (M2). [Reprinted from H. Fan and M. Sugiura, *EMBO J.* **14,** 1024 (1995), with permission.]

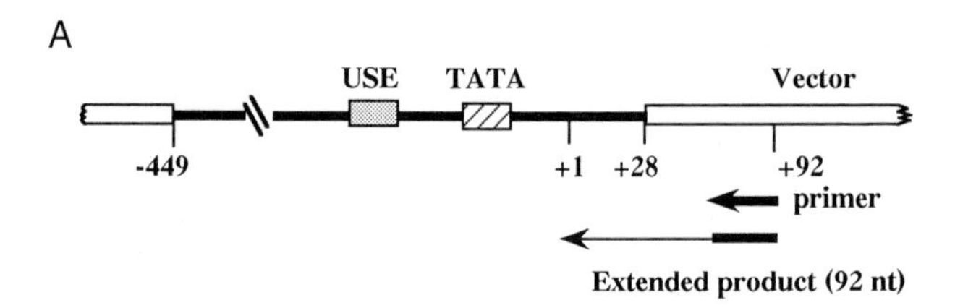

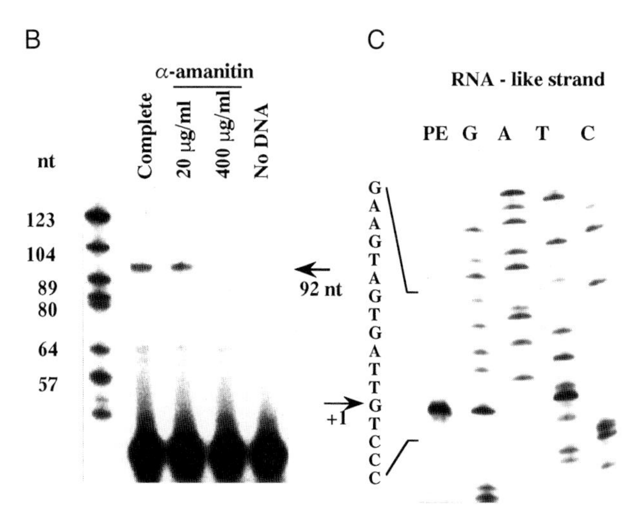

FIG. 3. *In vitro* transcription of the *Arabidopsis* U6 snRNA gene with the BY-2 nuclear extract.[2] (A) Schematic view of DNA template and primer extension assay. A short, boldface arrow indicates primer T7DKS. A long arrow shows the expected extended product (92 nt) calculated from the primer (+92) to the *in vivo* transcription start site (+1). (B) Detection of *in vitro* transcripts by 8% (w/v) PAGE. α-Amanitin was added as indicated. The expected product is indicated by an arrow (92 nt). (C) Determination of the transcription initiation site *in vitro*. The *in vitro* transcript was initiated at G residue (+1) as *in vivo*. [Reprinted from H. Fan and M. Sugiura, *EMBO J.* **14,** 1024 (1995), with permission.]

nuclear extract itself did not support *rbcS* transcription (lane 2), probably because of low Pol II activity compared with that of BY-2 cells. The truncated promoter pRBC-174 was not transcribed by the BY-2 extract with or without the leaf extracts (lanes 9 and 10), indicating that sequences upstream of −174 are necessary for green leaf factor-dependent transcription.

RNA Polymerase III-Dependent *in Vitro* Transcription

RNA polymerase III-dependent *in vitro* transcription is carried out with an *Arabidopsis* U6 small nuclear RNA (snRNA) gene as template. This gene contains an upstream sequence element (USE) and a −30 TATA box, which are essential for transcription by Pol III.[11] The promoter region (−449 to +28) is subcloned in Bluescript KS^+ and the circular plasmid DNA is used as template. *In vitro* transcription and transcript assays are as described above. As shown in Fig. 3, the *in vitro* transcription produces a dense band of expected extended product (92 nucleotides), found to be the same initiation site as that *in vivo*.[12] The U6 gene transcription is resistant to a low concentration of α-amanitin (20 μg/ml) but is inhibited by its high concentration (400 mg/ml), which is known to be a feature of Pol III activity. The optimal Mg^{2+} concentration for the U6 snRNA gene is at 3.6 to 4.6 m*M*. Transcription efficiency of the U6 gene is highest among plant genes tested to date.

[11] F. Waibel and W. Filipowicz, *Nature* (*London*) **346,** 199 (1990).
[12] F. Waibel and W. Filipowicz, *Nucleic Acids Res.* **18,** 3451 (1990).

Section V

Genetic Analysis of Transcription and Its Regulation

[24] Analysis of Two-Component Signal Transduction Systems Involved in Transcriptional Regulation

By REGINE HAKENBECK and JEFFRY B. STOCK

Regulation of gene expression in bacteria in response to environmental signals is in large part controlled by phosphorylation of a superfamily of cytoplasmic proteins termed response regulators (for reviews, see Refs. 1–4). These proteins share a conserved N-terminal sequence of approximately 125 amino acids that has been termed the receiver domain. Any two response regulators exhibit 20–30% amino acid sequence identity within this region, and contain several essentially invariant residues, including an aspartate residue near the N-terminal border of the domain, an aspartate near the center, and a lysine near the C-terminal border.[5]

Response regulators that function to control transcription act via C-terminal extensions that encode domains that interact with RNA polymerase and DNA. Phosphorylation of the receiver domain activates the C-terminal DNA-binding/transcriptional regulation domain causing the repression or activation of target genes. Some receiver sequences do not follow the common structural features and may employ different modes of activation; they have been referred to as unorthodox response regulators. The receiver domain can also occur as a single independent domain (e.g., CheY, Spo0F, PleC), two may be linked in tandem (e.g., *Myxococcus xanthus* FrzZ or *Caulobacter crescentus* PleD), or attached to domains with functions that are unrelated to transcription (e.g., CheB). In addition, receiver domains are sometimes linked to histidine kinase domains (see below).

The principal sources of phosphoryl groups for most response regulators are phosphohistidine residues in a second superfamily of proteins termed the histidine kinases. Because each member of the kinase family tends to function preferentially as a phosphodonor for a specific response regulator, the proteins have been considered as cognate pairs termed two-component signal transduction systems.[6] This designation should not be taken literally,

[1] J. B. Stock, A. M. Stock, and J. M. Mottonen, *Nature* (*London*) **344,** 395 (1990).
[2] J. B. Stock, A. J. Ninfa, and A. M. Stock, *Microbiol. Rev.* **53,** 450 (1989).
[3] J. S. Parkinson and E. C. Kofoid, *Annu. Rev. Genet.* **26,** 71 (1992).
[4] J. A. Hoch and T. J. Silhavy (eds.), "Two-Component Signal Transduction." American Society for Microbiology, Washington, DC, 1995.
[5] K. Volz, *Biochemistry* **32,** 11741 (1993).
[6] L. M. Albright, E. Huala, and F. M. Ausubel, *Annu. Rev. Genet.* **23,** 311 (1989).

however, because the phospho donor specificity for each response regulator is not absolute and there are generally auxiliary components besides the histidine kinases that function to specifically control the phosphorylation of a given receiver domain.

Histidine protein kinases are characterized by a conserved domain of approximately 200 amino acids. Any two histidine kinases exhibit 20–30% sequence identity within this region, and almost all members of the family share a characteristic set of signature sequences termed the H-, N-, D-, F-, and G-boxes.[1,3] Histidine kinases catalyze the transfer of γ-phosphoryl groups from MgATP to the N-3 position in substrate histidine residues that are located within their conserved H-box regions.[7–9] The highly conserved residues within the N-, D-, F-, and G-boxes are thought to form an active site for ATP binding and phospho-transfer catalysis. Histidine kinases generally function as homodimers that autophosphorylate *in trans* with the kinase domain of one subunit catalyzing the phosphorylation of the H-box histidine in the other subunit.[10–15] Variable flanking sequences N-terminal to the conserved kinase domain uniquely define a given member of the family. When, as is often the case, these variable regions contain hydrophobic membrane-spanning sequences, the proteins are thought to function as transmembrane receptors whose kinase activity is modulated in response to extracellular signals.

Response regulators and their associated histidine kinases are common proteins. They are found in all eubacterial strains that have been examined as well as in the archaea, and have been described in eukaryotes.[16] It has been estimated in *Escherichia coli* that there may be as many as 50 different response regulators and a comparable number of different histidine kinases, and there is no reason to expect that enteric bacteria are unusual in this respect. Because this represents approximately 2% of the *E. coli* genome, it is not surprising that two-component systems are so frequently encountered when bacterial regulatory loci are subjected to sequence analysis, and an increasing number of novel two-component systems of unknown function

[7] D. L. Roberts, D. W. Bennett, and S. A. Forst, *J. Biol. Chem.* **269,** 8728 (1994).

[8] J. F. Hess, R. B. Bourret, and M. I. Simon, *Nature (London)* **336,** 139 (1988).

[9] A. J. Ninfa and R. L. Bennett, *J. Biol. Chem.* **266,** 6888 (1991).

[10] Y. Yang and M. Inouye, *J. Mol. Biol.* **231,** 335 (1993).

[11] Y. Yang and M. Inouye, *Proc. Natl. Acad. Sci. U.S.A.* **88,** 11057 (1991).

[12] E. G. Ninfa, M. R. Atkinson, E. S. Kamberov, and A. J. Ninfa, *J. Bacteriol.* **175,** 7024 (1993).

[13] S. Q. Pan, T. Charles, S. Jin, Z.-L. Wu, and E. W. Nester, *Proc. Natl. Acad. Sci. U.S.A.* **90,** 9939 (1993).

[14] R. V. Swanson, R. B. Bourret, and M. I. Simon, *Mol. Microbiol.* **8,** 435 (1993).

[15] A. J. Wolfe and R. C. Stewart, *Proc. Natl. Acad. Sci. U.S.A.* **90,** 1518 (1993).

[16] R. V. Swanson, L. A. Alex, and M. I. Simon, *Trends Biochem. Sci.* **19,** 485 (1994).

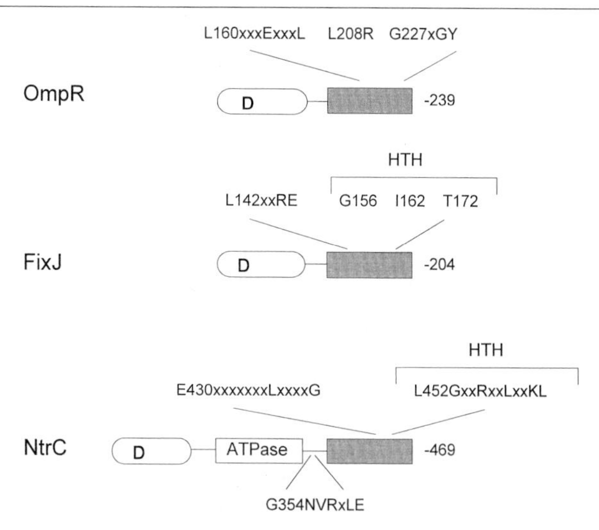

FIG. 1. Signature sequences in DNA-binding domains. For comparison of the DNA-binding domains, the sequences were aligned using the pileup program. D, Receiver domains. The numbering corresponds to the amino acid sequences of *E. coli* OmpR, *Rhizobium meliloti* FixJ, and *E. coli* NtrC, whose domain structures are shown schematically. Amino acid residues that are conserved in 90% or more of the proteins analyzed are shown. HTH, Putative helix–turn–helix motif.

are being revealed by this method. Here we discuss the experimental approaches that have been used to assess the function of novel two component transcriptional regulation systems following the discovery of a new response regulator or kinase gene.

Structure–Function Analysis

Depending on their associated C-terminal DNA-binding/transcriptional activation domain, the majority of response regulators fall into three distinct subfamilies designated OmpR, FixJ, and NtrC in accord with the name of their best characterized member.[5,17,18] Figure 1 schematically depicts the domain structure of the prototype of each of these families, and response regulators and other proteins with homologous DNA-binding domains are shown in Fig. 2.

Each member within a given subfamily controls a different set of promoters, and almost identical response regulators can function in different

[17] G. M. Pao and M. H. Saier, Jr., *J. Mol. Evol.* **40,** 136 (1995).

[18] S. Da Re, S. Bertagnoli, J. Fourment, J.-M. Reyrat, and D. Kahn, *Nucleic Acids Res.* **22,** 1555 (1994).

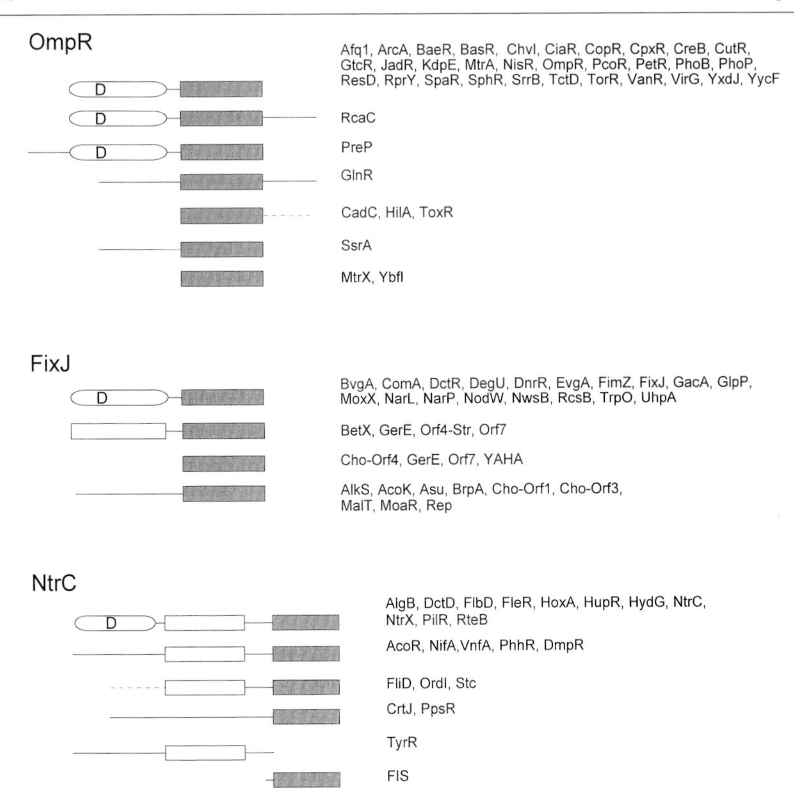

FIG. 2. Modular organization of subfamilies of response regulators and related proteins. Proteins are organized into families of related transcriptional regulation domains. Using the C-terminal DNA-binding domain of the prototype of each of the response regulator subfamilies (*E. coli* OmpR, *Rhizobium meliloti* FixJ, and *E. coli* NtrC), members of each subfamily were identified by using the BLAST network service of the National Center for Biotechnology Information (Bethesda, MD). D, Receiver domains; DNA-binding domains are shaded. Dashed lines indicate protein extensions of variable length.

species to produce different responses. For instance, in *E. coli*, OmpR controls the porin composition of the outer membrane in response to medium osmolarity whereas in *Shigella* it also functions to control virulence.[19] Similarly, the nitrogen response regulator NtrC of *Bradyrhizobium japonicum* does not control *nifA* where *Rhizobium melilota* and *Azorhizobium caulinodans* NtrC contribute to *nifA* regulation but in different ways.[20]

Members of the OmpR family of response regulators contain a conserved DNA-binding domain of approximately 150 amino acids that is linked to the C terminus of a generic receiver domain. OmpR homologs

[19] M. L. Bernardini, A. Fontaine, and P. J. Sansonetti, *J. Bacteriol.* **172,** 6274 (1990).
[20] H.-M. Fischer, *Microbiol. Rev.* **58,** 352 (1994).

can repress and/or activate target genes, depending on features of the promoters with which they interact. *E. coli* genes that are regulated by members of the OmpR family are generally transcribed by σ^{70} RNA polymerase holoenzyme.[21] OmpR proteins appear to function, at least in part, by making direct contact with the C-terminus of the α subunit of RNA polymerase.[22,23] Although the structure of the OmpR-type DNA-binding domain has not been determined, helix–turn–helix consensus sequences have been discerned in many members of this family. In their dephosphorylated states, the receiver domains of OmpR-type regulators appear to interfere with DNA binding and/or transcriptional control. Presumably, phosphorylation of the N-terminal phospho-accepting domain in OmpR and its homologs activates the associated C-terminal DNA-binding transcriptional regulation domain by relieving the inhibitory effect of the N-terminal domain. Proteins (e.g., *Vibrio* ToxR) that lack a receiver domain but nevertheless have an OmpR-type DNA-binding domain are known. DNA sequences encoding homologs to the OmpR-type DNA-binding domain as a single protein unit of unknown function also exist.

The FixJ family is characterized by a conserved C-terminal domain of approximately 100 residues that does not exhibit significantly sequence similarity to the DNA-binding domain of the OmpR family, but nevertheless appears to function similarly.[18] FixJ response regulators control transcription from *E. coli* σ^{70} promoters and deletion of their receiver domains can produce constitutively active trascription factors.[18,24] A putative helix–turn–helix DNA-binding motif near their C-termini within a region homologous to the transcriptional activator domain of σ factors has been described.[25] The occurrence of homologs of the DNA-binding domain independent of a receiver module is best exemplified in this family. This class of DNA-binding domains is also shared by another large protein family involved in regulation of diverse physiological processes, the LuxR proteins,[26] and occurs without or with additional domains of variable length in many proteins from a wide variety of bacteria (Fig. 2).

Members of the NtrC family of response regulators have two domains that are linked to the N-terminal receiver domain: a C-terminal DNA-binding domain and a middle domain that functions as an ATPase. The

[21] M. Drummond, P. Whitty, and J. Wootton, *EMBO J.* **5,** 441 (1986).

[22] L. A. Pratt and T. J. Silhavy, *J. Mol. Biol.* **243,** 579 (1994).

[23] L. A. Pratt and T. J. Silhavy, *in* "Two-Component Signal Transduction" (J. A. Hoch and T. J. Silhavy, eds.), p. 105. American Society for Microbiology, Washington, D.C., 1995.

[24] A. Galiner, A.-M. Garnerone, J.-M. Reyrat, D. Kahn, J. Batut, and P. Boistard, *J. Biol. Chem.* **269,** 23784 (1994).

[25] D. Kahn and G. Ditta, *Mol. Microbiol.* **5,** 987 (1991).

[26] G. P. C. Salmond, B. W. Bycroft, G. S. A. B. Stewart, and P. Williams, *Mol. Microbiol.* **16,** 615 (1995).

middle domain has a glycine-rich consensus sequence termed a Walker box, the ATP-binding motif characteristic of many ATPases, and the C-terminal DNA-binding domain contains a helix–turn–helix DNA-binding motif.[21] The DNA-binding domain is homologous to the FIS protein, which is related to eukaryotic enhancer-binding proteins.[27] Members of the NtrC family control transcription from promoters that use the σ^{54} form of RNA polymerase. Whereas activation by members of the OmpR and FixJ families seems to involve recruitment of polymerase, members of the NtrC family facilitate isomerization from closed to open transcription initiation complexes.[28] The transition depends on the central domain ATPase activity, which is activated by phosphorylation of the N-terminal phospho-accepting domain in the absence of DNA.[29,30] Unphosphorylated NtrC is a homodimer, and the ATPase activity of the middle domain depends on the formation of tetramers or higher order multimers.[30–32] In addition to ATPase-specific motifs, a short region between the DNA-binding domain and the ATPase domain is homologous to a sequence conserved in DnaA proteins (Fig. 1), which has been tentatively related to a common function of both protein families, local unwinding of DNA.[33–34]

The *Bacillus subtilis* Spo0A protein contains a DNA-binding domain that is unrelated to those of the three above-mentioned families. The only potential Spo0A DNA-binding domain homologs detected in the present search were VirB, a transcriptional activator of *Shigella* virulence genes, and a protein associated with yeast mitochondrial ribosomes. Spo0A is a central component of the phosphorelay system in sporulating *Bacillus*.[35] Spo0A binds to specific DNA sites termed 0A boxes, and this interaction is increased by phosphorylation.[36]

The sequences of the receiver domains that are found in these various classes of response regulators do not fall into subfamilies corresponding to their associated transcriptional regulation domains. Generally, conservation of this domain is slightly higher within families than across family lines, but exceptions have been noted.[5,17] Unusual receiver domains have been

[27] A. K. North, K. E. Klose, K. M. Stedmann, and S. Kustu, *J. Bacteriol.* **175,** 4267 (1993).

[28] D. L. Popham, D. Szeto, J. Keener, and S. Kustu, *Science* **243,** 629 (1989).

[29] S. Austin and R. Dixon, *EMBO J.* **11,** 2219 (1992).

[30] D. Weiss, J. Batut, K. E. Klose, J. Keener, and S. Kustu, *Cell* **67,** 155 (1991).

[31] J. Feng, M. R. Atkinson, W. R. McCleary, J. B. Stock, B. L. Wanner, and A. J. Ninfa, *J. Bacteriol.* **174,** 6061 (1992).

[32] V. Weiss, F. Claverie-Martin, and B. Magasanik, *Proc. Natl. Acad. Sci. U.S.A.* **89,** 5088 (1992).

[33] A. Roth and W. Messer, *EMBO J.* **14,** 2106 (1995).

[34] E. V. Koonin, *Nucleic Acids Res.* **20,** 1997 (1992).

[35] K. Trach, D. Burbulys, M. Strauch, J.-J. Wu,N. Dhillon, R. Jonas, C. Hanstein, P. Kallio, M. Perego, T. Bird, G. Spiegelman, C. Fogher, and J. A. Hoch, *Res. Microbiol.* **142,** 815 (1991).

[36] D. Burbulys, K. A. Trach and J. A. Hoch, *Cell* **6,** 545 (1991).

identified in which the highly conserved aspartate near the N-terminal border of the phospho-accepting domain and the conserved lysine near the C-terminal border are missing. In these instances (e.g., FlbD in *Caulobacter crescentus*[37] and FrzG in *Myxococcus xanthus*[38]), it is not clear that the central aspartate is phophorylated, and preliminary results suggest that the proteins may be active in the absence of phosphorylation.[39]

Many response regulators are phosphorylated by acetyl phosphate or other low molecular weight phosphodonors independently of their associated histidine kinases.[40,41] Well-characterized examples include CheY, PhoB, OmpR, NtrC, and FixJ.[31,42–45] However, there are also well-defined examples, such as CheB, where acetyl phosphate will not serve as a phosphodonor.[43] No feature of the sequence of a receiver domain has been discerned that allows one to deduce its phosphodonor specificity. Although phosphorylation by acetyl phosphate can be physiologically significant, there is no example of a response regulator that is clearly designed to function independently of a histidine kinase.

There are no subfamilies of *E. coli* histidine kinases in the same sense that there are subfamilies of response regulators. Kinases with homologous sequences outside the conserved ATP-binding ad histidine phosphorylation domains generally represent different isoforms of the same protein. For instance, motile bacteria generally have a histidine kinase termed CheA that functions in sensory motor regulation and chemotaxis; and all CheA kinases have conserved domains outside the histidine kinase region that mediate interactions with response regulators and auxiliary signal transduction components that are specialized for chemotaxis. Thus the relatively variable flanking regions can be used to design oligonucleotide primers for polymerase chain reaction (PCR) amplification of corresponding genes in different species. This approach was used to isolate a *cheA* kinase in archaea, and thereby provided considerable information about chemotaxis signal transduction in these cells.[46] In the absence of a well-characterized variant from another species, the sequences of novel histidine kinases provide

[37] G. Ramakrishnan and A. Newton, *Proc. Natl. Acad. Sci. U.S.A.* **87,** 2369 (1990).

[38] W. R. McCleary, M. J. McBride, and D. R. Zusman, *J. Bacteriol.* **172,** 4877 (1990).

[39] A. K. Benson, G. Ramakrishnan, N. Ohta, J. Feng, A. J. Ninfa, and A. Newton, *Proc. Natl. Acad. Sci. U.S.A.* **91,** 4989 (1994).

[40] W. R. McCleary, J. B. Stock, and A. J. Ninfa, *J. Bacteriol.* **175,** 2793 (1993).

[41] B. L. Wanner, *J. Bacteriol.* **174,** 2053 (1992).

[42] W. R. McCleary and J. B. Stock, *J. Biol. Chem.* **269,** 31567 (1994).

[43] G. S. Lukat, W. R. McCleary, A. M. Stock, and J. B. Stock, *Proc. Natl. Acad. Sci. U.S.A.* **89,** 718 (1992).

[44] J. M. Reyrat, M. David, C. Blonski, P. Boistard, and J. Batut, *J. Bacteriol.* **175,** 6867 (1993).

[45] I. Schroder, C. D. Wolin, R. Cavicchioli, and R. P. Gunsalus, *J. Bacteriol.* **176,** 4985 (1994).

[46] J. Rudolph and D. Oesterhelt, *EMBO J.* **14,** 667 (1995).

little information as to either mechanism of sensory regulation or their phosphodonor specificity.

Most histidine kinases are integral membrane proteins with two or more hydrophobic membrane-spanning sequences N terminal to the kinase domain. It has generally been assumed that stimulatory ligands bind to the extracytoplasmic aspect of these membrane-spanning regions to cause conformational changes that modulate kinase activity. In no case has a regulatory ligand been identified, however; and in several instances a more complicated sensory apparatus has been implicated. A good example is provided by PhoR, a histidine kinase that regulates the expression of the *E. coli* Pho regulon including *phoA*, the gene that encodes alkaline phosphatase (for reviews see Refs. 2 and 47). PhoR appears to have an internal hydrophobic transmembrane sequence that connects a periplasmic loop to a histidine kinase domain within the cytoplasm. PhoR functions to sense levels of phosphate causing *phoA* repression when phosphate is abundant and *phoA* activation when phosphate is scarce. One might suppose that phosphate in the periplasm simply binds to the extracytoplasmic domain to regulate kinase activity. The situation is far more complex, however; the activity of PhoR is not regulated by phosphate directly, but rather by the activity of an ABC transporter termed PST (phosphate-specific transporter), that functions both in phosphate uptake and sensing. Although PhoR has been extensively investigated both genetically and biochemically, the mechanism by which its activity is regulated by PST remains a mystery.

VanS provides another good illustration of the problem of identifying the molecular mechanism of kinase regulation. Like PhoR, VanS appears to be a transmembrane receptor. It functions to regulate the induction of expression of vancomycin resistance genes when enterococci are exposed to this antibiotic.[48] Induction is also observed with a completely different antibiotic, moenomycin, indicating that the antibiotics are not directly sensed but that inhibition of peptidoglycan biosynthesis, the common property of both antibiotics, may be the critical determinant of a regulatory signal.[49]

Histidine kinases function specifically to control the level of phosphorylation of cognate receiver domains. Frequently, kinase and regulator genes are organized in an operon. Invariably these proteins are members of a cognate pair. There are numerous instances, however, when a kinase–regulator pair is independently expressed. Moreover, aside from their loca-

[47] B. L. Wanner, *J. Cell. Biochem.* **51,** 47 (1993).

[48] M. Arthur, C. Molinas, and P. Courvalin, *J. Bacteriol.* **174,** 2582 (1992).

[49] S. Handwerger and A. Kolokathis, *FEMS Microbiol. Lett.* **70,** 167 (1990).

tion in a common operon, there is no feature of the sequence of a given kinase that allows one to discern anything concerning the regulator with which it interacts. A particularly good example of this is provided by the chemotaxis kinase, CheA, which typically functions with two response regulators, CheY and CheB. Despite the fact that phosphotransfer from phospho-CheA to CheY and CheB occurs at comparable rates, the sequences of CheY and CheB are not especially similar.[5,17,20] Another example is provided by the *E. coli* histidine kinases PhoR and CreC (formerly PhoM). These two kinases both donate phosphoryl groups to PhoB.[50] Whereas PhoR and PhoB are encoded together in a single operon, *creC* is linked to another response regulator gene, *creB*, at a completely different locus. Thus, whereas linked kinase–regulator genes produce proteins with linked functions, one cannot assume that this is the whole story. Finally, as illustrated by both the Che and Pho systems, although kinases and regulators function as pairs, more often than not there are more than two components involved.

Several histidine kinases have been shown to catalyze the dephosphorylation as well as the phosphorylation of their cognate regulators. For instance, the kinase EnvZ facilitates both phosphorylation and dephosphorylation of OmpR. Genetic and biochemical studies suggest that under a given set of osmotic conditions the transcriptional regulatory function of EnvZ is controlled by the relative balance between these two activities.[51] In other systems the phosphatase activity of a histidine kinase depends on its interaction with auxiliary signal transduction components. The best characterized example is NtrB, the histidine kinase that regulates transcription of the gene that encodes glutamine synthetase, *glnA*, as well as several other nitrogen-regulated genes. NtrB mediates both the phosphorylation and dephosphorylation of its cognate response regulator, NtrC, and the dephosphorylation reaction is regulated by uridylylation of a regulatory protein called P_{II} (GlnB).[52] In still other systems the kinase does not function as a phosphatase, but rather a completely distinct protein serves this role. The best characterized example is the Che system in which the dephosphorylation of phospho-CheY has been shown to be catalyzed by a protein termed CheZ independently of the kinase CheA.[43,53,54] Another example is the phosphorelay system that controls sporulation in *B. subtilis*, in which a family of proteins termed Rap ("response regulator aspartate phospha-

[50] M. Amemura, K. Makino, H. Shinagawa, and A. Nakata, *J. Bacteriol.* **172,** 6300 (1990).

[51] S. A. Forst and D. L. Roberts, *Res. Microbiol.* **145,** 363 (1994).

[52] E. S. Kamberov, M. R. Atkinson, P. Chandran, and A. J. Ninfa, *J. Biol. Chem.* **269,** 28294 (1994).

[53] J. F. Hess, K. Oosawa, N. Kaplan, and M. I. Simon, *Cell* **53,** 79 (1988).

[54] Y. Blat and M. Eisenbach, *Biochemistry* **33,** 902 (1994).

tases") has been identified.[55] No feature of the sequences of kinases has been discerned that allows one to distinguish whether or not they have phosphatase activity.

Genetics of Two-Component Regulation

Two-component systems have commonly been identified through the sequencing of loci where mutations lead to the aberrant regulation (reduced or constitutive expression) of a particular target gene such as *phoA* or *glnA* in *E. coli*[56–58] or to defects in complex regulatory processes such as genetic competence in *B. subtilis* (reviewed in Ref. 59), cell cycle progression in *C. crescentus*,[60,61] fruiting body formation in *M. xanthus*,[62] or susceptibility to β-lactam antibiotics in *Streptococcus pneumoniae*.[63] Once such a component has been recognized, further mutational analysis can be used to understand its role *in vivo*. Interacting proteins can be identified through the isolation and analysis of pseudorevertant and bypass suppressor mutations. These techniques are greatly facilitated by the fact that most two-component systems are not essential at least under laboratory conditions.

Clearly, the situation that pertains when the regulation of a known gene is being examined is much simpler than that for a complex process for which target genes are not defined. In such cases one has a means to readily assay the effect of mutations on the activity of a target promoter using reporter gene constructs, and methods to select and/or screen for defects in regulation can also readily be developed. Nevertheless, whereas the actual mutation in the response regulator or the histidine kinase can easily be identified, interpretation of the phenotype is often difficult because these interactive proteins are involved in several functions that may directly or indirectly result in aberrant regulation.

One common class of response regulator mutations contains defects in the C-terminal DNA-binding domain. Such mutants are often unable to express the target gene or exhibit constitutive expression. In every system

[55] M. Perego, C. Hanstein, K. M. Welsh, T. Djavakhishvili, P. Glaser, and J. A. Hoch, *Cell* **79,** 1047 (1994).

[56] A. Torriani and F. Rothman, *J. Bacteriol.* **81,** 835 (1961).

[57] K. Makino, H. Shinagawa, M. Amemura, and A. Nakata, *J. Mol. Biol.* **190,** 37 (1986).

[58] M. R. Atkinson and A. J. Ninfa, *J. Bacteriol.* **174,** 4538 (1992).

[59] D. Dubnau, *Microbiol. Rev.* **55,** 395 (1991).

[60] T. Lane, A. Benson, G. B. Hecht, G. J. Burton, and A. Newton, *in* "Two-Component Signal Transduction" (J. A. Hoch and T. J. Silhavy, eds.), p. 403. American Society for Microbiology, Washington, D.C., 1995.

[61] H. R. Horvitz and I. Herskowitz, *Cell* **68,** 237 (1992).

[62] S. K. Kim, D. Kaiser, and A. Kuspa, *Annu. Rev. Microbiol.* **46,** 117 (1992).

[63] E. Guenzi, A. M. Gase, M. A. Sicard, and R. Hakenbeck, *Mol. Microbiol.* **12,** 505 (1994).

that has been examined this phenotype corresponds to the null phenotype, in which the gene that encodes the regulator has been deleted, and it usually corresponds to the phenotype that would be expected from the response regulator if it could not be phosphorylated. Caution should be taken in interpreting mutations in the DNA-binding/transcriptional regulation domains of response regulators, however. These transcription factors often interact with several promoters and frequently have multiple DNA-binding sites that differentially affect a single promoter; examples include OmpR and *Bordetella* BvgA.[64,65] In NtrC response regulators, mutations in the DNA-binding domain as well as in the ATPase domain can affect DNA binding.[66]

The effects of mutations in the receiver domain are even more difficult to interpret. Because the dephosphorylated form of this domain generally functions to interfere with the activity of the C-terminal DNA-binding domain to which it is attached, deletion of the receiver domain or any mutation that interferes with its proper folding may lead to a constitutively activated regulator. This is a likely explanation for the transcriptional activation that is seen for some regulators when mutations are engineered within the phospho-accepting active site region.[67] A list of spontaneous mutations in receiver domains and associated phenotypes has been compiled by Volz.[5]

The interpretation of the phenotypic effects of mutations in histidine kinases is also not straightforward. The problems stem primarily from two sources: the phosphatase activities that are frequently associated with kinases, and the existence of alternative phosphodonors. Mutants with deletions in a given kinase frequently exhibit patterns of regulated gene expression that, to a first approximation, seem similar to those of wild type. In the best studied cases, regulation of gene expression by OmpR, NtrC, and PhoB in *E. coli*, it has been shown that (1) the corresponding kinases, EnvZ, NtrB, and PhoR, function as phosphatases; and (2) the regulators are all subject to phosphorylation by acetyl phosphate as well as by other noncognate histidine kinases.[31,38,43] Thus, in the absence of the relatively tight regulation of phosphorylation and dephosphorylation elicited by its cognate kinase, the level of phosphorylation of a given receiver domain can depend on alternative sources of phosphorylation such as noncognate histidine kinases or acetyl phosphate that fluctuate so to mimic superficially wild-type regulation. In fact, it is most common for the phenotype of a kinase deletion to fit what one would expect of a leaky mutation in the

[64] T. Mizuno, M. Kato, Y. Jo, and S. Mizushima, *J. Biol. Chem.* **263,** 1008 (1988).

[65] S. Stibitz, *J. Bacteriol.* **176,** 5615 (1994).

[66] S. C. Porter, A. K. North, A. B. Wedel, and S. Kustu, *Genes Dev.* **7,** 2258 (1993).

[67] J. Delgado, S. Forst, S. Harlocker, and M. Inouye, *Mol. Microbiol.* **10,** 1037 (1993).

response regulator DNA-binding domain rather than what would be expected for a mutation that was completely deficient in phosphorylation.

Point mutations in histidine kinase genes can also lead to a wide range of phenotypes. Missense mutations of the codon for the phospho-accepting histidine usually lead to the simultaneous loss of both kinase and phosphatase activity[68,69] but can also exclusively eliminate the kinase activity as in the $His_{139}Asn$ mutation in NtrB.[52] Phosphatase-specific mutations have been characterized in EnvZ,[51] and mutations found at homologous sites in other kinases have been described.[63,70] In general, phosphorylation-specific mutations tend to cluster within the highly conserved residues associated with the H-, N-, D-, F-, and G-boxes whereas phosphatase-specific mutations tend to result from alterations in the H-box as well as within more variable upstream regions.[51] Phosphatase mutations can be identified phenotypically as well because they generally lead to patterns of expression that are the opposite of those seen in null mutations of the corresponding response regulator. These phenotypes correspond roughly to the patterns of expression caused by overproduction of a response regulator. In contrast, the high-level expression of a histidine kinase frequently has no obvious phenotypic consequence.

The lessons learned from studies of the regulation of specific genes can be applied to the dissection of more complex regulatory networks in which gene targets are less well defined. One feature of such systems that needs to be kept in mind is the way quantitative changes in the level of phosphorylation of a response regulator can lead to qualitatively different phenotypic outputs. This feature of two-component regulation is best exemplified by the OmpR/EnvZ system of *E. coli*, in which a relatively simple signal transduction mechanism operates to produce a complex output.[51] A single histidine kinase, EnvZ, acts to phosphorylate and dephosphorylate a single response regulator, OmpR, in response to changes in medium osmolarity. OmpR is required for the expression of the two major *E. coli* outer membrane porins, OmpF and OmpC. A shift to high osmotic strength medium results in repression of *ompF* and preferential expression of *ompC*. Phosphorylation of OmpR alters its affinity for DNA, and it appears to be the intracellular concentration of phospho-OmpR that is responsible for both activation and repression of the *omp* promoters.[22,71] Mutations in OmpR can confer constitutive *ompC* expression (R2 class) or *ompF* expression

[68] S. Forst, J. Delgado, and M. Inouye, *Proc. Natl. Acad. Sci. U.S.A.* **86,** 6052 (1989).
[69] M. M. Igo and T. J. Silhavy, *J. Bacteriol.* **170,** 5971 (1988).
[70] G. B. Hecht, T. Lane, N. Ohta, J. M. Sommer, and A. Newton, *EMBO J.* **14,** 3915 (1995).
[71] H. Aiba, F. Nakasai, S. Mizushima, and T. Mizuno, *J. Biochem.* **106,** 5 (1989).

(R3 class).[64,72] Neither porin is expressed in *ompR* deletion mutations, or in EnvZ-deficient strains under conditions in which levels of acetylphosphate are low. Finally, further increases in phosphorylation that occur in mutants in which the phosphatase activity of EnvZ is differentially lost or in wild-type cells exposed to procaine, led to pleiotropic effects in several other systems, most notably the repression of genes that are required for maltose utilization.[73] Thus, the simplest of two-component regulatory systems has been shown to produce at least four distinct phenotypic outputs: low-level porin expression, *ompF* expression, *ompC* expression, and a pleiotropic phenotype. These effects are mediated by multiple OmpR-binding sites upstream of the relevant promoters, and perhaps more important, by direct protein–protein interactions between OmpR and RNA polymerase.

Biochemical Approaches

Despite the relative complexity of two-component regulatory systems, biochemical techniques have been developed that can readily be used to confirm and extend results obtained from molecular genetic approaches. The critical technology involves the isolation of individual components and the subsequent reconstitution of partial reactions such as kinase autophosphorylation, receiver domain phosphorylation and dephosphorylation, DNA binding, and transcription.

The overproduction and purification of response regulators has been relatively straightforward. These proteins are usually abundant cytoplasmic constituents with levels typically in the micromolar range, and they tend to be well behaved even when expressed at extremely high intracellular concentrations. Any of the currently available multicopy expression vectors that are in common use for protein overproduction in *E. coli* can be used to routinely obtain 10–100 mg of purified response regulator protein from a few liters of cells. In general it has not been necessary to resort to any molecular genetic tricks such as His-tagging or MBP (maltose binding protein) fusions to facilitate the purification process; all that is required to obtain pure native protein are a few relatively simple chromatographic steps.

In principle, the DNA-binding properties of both the phosphorylated and dephosphorylated forms of purified response regulators can be characterized using standard techniques for examining protein–nucleic acid interactions. *In vivo* as well as *in vitro* DNA protection assays have been per-

[72] J. M. Slauch, and T. J. Silhavy, *J. Mol. Biol.* **210,** 281 (1989).

[73] S. Tokishita, A. Kojima, H. Aiba, and T. Mizuno, *J. Biol. Chem.* **266,** 6780 (1991).

formed in a variety of two component regulatory systems. It is clear from these studies that promoter recognition is a complex process that involves interactions with both RNA polymerase holoenzyme and DNA. OmpR protects large regions of DNA, more than 60 bp, at both the *ompF* and *ompC* promoters *in vivo* and *in vitro*.[64,74,75] Although homologous repeated sequence motifs were found within these regions, a recognition sequence could not be deducted from these analyses. In contrast, footprinting experiments using PhoB clearly revealed binding to a conserved sequence termed a *pho* box.[76] However, although PhoP of *B. subtilis* is 40% identical to the *E. coli* PhoB protein, *B. subtilis* Pho promoter sequences do not exhibit the *pho* box consensus.[77] In the case of NtrC-dependent promoters, transcriptional activation depends on protein–protein interactions between σ^{54} RNA polymerase holoenzyme and the ATPase domain of the activator. The interaction of NtrC proteins with DNA is not absolutely required, and seems to function solely to increase local concentrations of the activator. Moreover, no σ^{54}-dependent promoter has been identified that does not require an NtrC-like transcription factor for its activation.

The phosphorylated forms of response regulators can be obtained using small molecule phosphodonors such as acetyl phosphate. The ability of acetyl phosphate to phosphorylate a given regulator can be evaluated using acetyl [^{32}P]phosphate prepared as described previously.[43,78] Alternatively, one can use unlabeled acetyl phosphate and measure its rate of hydrolysis by the $FeCl_3$ procedure.[79] In either case a kinetic analysis of the reaction should provide estimates of both the rate of phosphorylation and the intrinsic phospho-response regulator phosphatase activity. In cases in which acetyl phosphate can serve as a donor, millimolar concentrations are generally required to obtain substantial levels of phosphorylation. In several instances it has been shown that acetyl phosphate does not serve as a phosphodonor.[43] In these cases phosphoramidate can be used.[40,80] No procedure has been reported for the synthesis of radiolabeled phosphoramidate, however, so one must generally resort to an indirect measure of phosphorylation such as a probe for a phosphorylation-induced change in response regulator conformation, transcriptional activation, and, in the case of NtrC family members, ATPase activity.[31]

[74] K. Tsung, R. Brissette, and M. Inouye, *J. Biol. Chem.* **264,** 10104 (1989).

[75] S. Maeda and T. Mizuno, *J. Biol. Chem.* **263,** 14629 (1988).

[76] K. Makino, H. Shinagawa, M. Amemura, S. Kimura, and A. Nakata, *J. Mol. Biol.* **203,** 85 (1988).

[77] R. S. Chestnut, C. Bookstein, and F. M. Hulett, *Mol. Microbiol.* **5,** 2181 (1991).

[78] W. Langert, M. Meuthen, and K. Mueller, *J. Biol. Chem.* **226,** 21608 (1991).

[79] E. R. Stadtman, *Methods Enzymol.* **3,** 228 (1957).

[80] J.-M. Reyrat, M. David, J. Batut, and P. Boistard, *J. Bacteriol.* **176,** 1969 (1994).

Biochemical analysis of histidine kinases tends to be more problematic. These frequently are integral membrane proteins that cannot be purified without using complex detergent solubilization and phospholipid reconstitution procedures. Moreover, each kinase has a distinct sensory domain, and the functioning of such regions remains an unsolved problem, even in the most studied examples. Because of these difficulties, the usual approach has been to construct genetically engineered fragments with the extracytoplasmic and membrane-spanning regions deleted. When these fragments are overproduced they frequently partition, at least to some extent, into particulate aggregates or inclusion bodies that can then be washed with high salt and detergent to obtain relatively pure denatured protein.[81] It has generally been possible to solubilize these denatured forms of the kinases using the urea or guanidinium hydrochloride, which can then be removed to obtain soluble domains with kinase activity. Purification and, more important, refolding can be facilitated by appending a His-tag to the engineered fragments. Using this approach the unfolded protein in urea or guanidinium hydrochloride can be bound to a Ni^{2+} column and washed free of denaturant, and subsequently eluted with imidazole. This procedure favors formation of native structures, presumably because during the renaturation process protein monomers bound to the column are not free to interact with one another to form aggregates. In some cases, an intermediate step has been added to enhance the purity of the final product. The unfolded protein bound to the Ni^{2+} columns is first eluted using a pH gradient in the presence of a denaturant. Whereas most nonspecifically bound proteins elute around pH 6, His-tagged protein will elute at pH 4.

Kinase autophosphorylation activity has typically been assayed by incubating isolated protein with [γ-^{32}P]ATP, quenching the reaction in sodium dodecyl sulfate (SDS), subjecting the product to SDS-polyacrylamide gel electrophoresis (PAGE) according to the method of Laemmli, and measuring ^{32}P incorporation by standard phosphoimaging or fluorometric techniques.[43] Because phosphohistidine groups are stable in alkali but relatively unstable in acid, procedures such as trichloroacetic acid (TCA) precipitation or washes of polyacrylamide gels in acetic acid should be avoided or at least minimized.

To measure phosphotransfer between proteins, one generally first incubates the kinase with [^{32}P]ATP for sufficient time to allow the autophosphorylation reaction to go to equilibrium (generally only a few minutes are required). Addition of cognate response regulators to such mixtures

[81] J. K. Krueger, A. M. Stock, C. E. Schutt, and J. B. Stock, *in* "Protein Folding: Deciphering the Second Half of the Genetic Code" (L. M. Gierasch and J. King, eds.), p. 136. American Association for the Advancement of Science, Washington, D.C., 1990.

results in a rapid transfer of phosphoryl groups. The reaction can be followed by SDS–PAGE, but it should be kept in mind that the response regulator phosphoaspartate groups are relatively labile, especially under alkali conditions such as those encountered during SDS–PAGE by the Laemmli procedure. It is therefore desirable to minimize time and temperature during electrophoresis protocols.

In reactions in which response regulator phosphorylation is analyzed in the presence of MgATP and a histidine kinase, there are at least three reactions that must be considered: kinase autophosphorylation, phosphotransfer from kinase to response regulator, and phospho-response regulator dephosphorylation. The combined effect is ATP hydrolysis to ADP plus P_i. This reaction can conveniently be measured spectrophotometrically at 340 nm by following the oxidation of NADH in a coupled system with phosphoenolpyruvate, pyruvate kinase, and lactate dehydrogenase.[31] By adjusting the concentrations of the histidine kinase and response regulator in such reactions one can generally obtain conditions in which the rate is independent of response regulator concentration and the kinase autophosphorylation rate is limiting, or alternatively, the rate is independent of kinase concentration so that the phospho-response regulator dephosphorylation rate is limiting. Such kinetic analyses can be quite complex, however, especially when one takes into account the formation of complexes between the kinase and response regulator and the possibility that the kinase functions as a phospho-response regulator phosphatase. In studies of NtrC and its homologs the approach cannot be used because of the high ATPase activity of the middle domain of these proteins. Many of these problems can be obviated by using acetyl phosphate as a phosphodonor instead of the protein kinase, or by examining rates of phosphotransfer using a phosphorylated kinase preparation in which the ATP has been removed prior to addition of the phospho-accepting response regulator proteins.

Before the discovery that small molecules could be used as response regulator phosphodonors it was common to employ well-characterized heterologous kinases such as CheA for this purpos. In every case that has been examined noncognate phosphorylated kinases such as phospho-CheA, phospho-NtrB, and phospho-EnvZ were capable of supplying phosphoryl groups to any response regulator, although the rates of phosphotransfer are low compared to those obtained with cognate kinase–regulator pairs.[82] This approach could provide a useful alternative for the phosphorylation of response regulators that cannot use acetyl phosphate as a donor.

There are several potential artifacts that should be kept in mind when

[82] M. M. Igo, A. J. Ninfa, J. B. Stock, and T. J. Silhavy, *Genes Dev.* **3,** 1725 (1989).

using the types of procedures outlined above. Phosphoimidazoles such as the phosphoryl groups found in histidine kinases are potent phosphodonors, and under some conditions protein phosphorylations may be observed that are not physiologically relevant. For instance, mutant variants of CheY that lack the phospho-accepting aspartate, Asp-57, are nevertheless phosphorylated by phospho-CheA.[83] The phosphorylated residue is a serine or threonine, presumably Thr-87, whose side chain is located near Asp-57 in the wild-type protein. It seems likely that this phosphotransfer results from high local concentration of phosphohistidine that occurs when CheY forms a complex with phospho-CheA. Similarly, secondary phosphorylation sites observed in OmpR with a mutated Asp-55, the primary phosphate acceptor site, may not be physiologically relevant.[67] It can also be expected that under some conditions in which proteins are denatured by heating in SDS, side reactions may result in a transfer of phosphoryl groups from phosphohistidines to other side chains. Another possible source of misleading results can come from contaminating ATPase activities. These are of major concern when one is directly analyzing ATP hydrolysis using the coupled assay. In addition, because of the high energy of the phospho-histidine group, any significant conversion of ATP and ADP tends to shift the equilibrium toward low levels of kinase autophosphorylation.

It is clear that reversible protein–protein interactions play a central role in two-component function. Chemical cross-linking has been used to analyze such contacts, but with some notable exceptions the results obtained by this method have generally not been convincing. Most of these experiments have been performed using relatively high concentrations of proteins under conditions in which nonspecific aggregation is likely. Another approach has been to measure the adsorption of one component to a second component that is linked to a surface. Because of multivalent interactions, however, quantitative estimates of binding constants that have been determined by this approach tend to be orders of magnitude higher than the solution values.[84] Nevertheless, this method seems to give good qualitative indications of physiologically relevant interactions. This is especially true when mutants are available that can be used as controls.

Homotypic interactions between histidine kinase domains have been shown in several studies to be a central feature of the kinase autophosphorylation mechanism. It is apparent that the kinase domain in one monomer serves to phosphorylate a substrate histidine in a second monomer, and soluble histidine kinases such as CheA and NtrB have been shown to

[83] R. B. Bourret, J. F. Hess, and M. I. Simon, *Proc. Natl. Acad. Sci. U.S.A.* **87,** 41 (1990).

[84] J. E. Ladbury, M. A. Lemmon, M. Zhou, J. Green, M. C. Botfield, and J. Schlessinger, *Proc. Natl. Acad. Sci. U.S.A.* **92,** 3199 (1995).

be dimers under physiological conditions.[53,85,86] It seems likely that dimer interactions play a role in the regulation of kinase activity in histidine kinases that function as membrane receptors, just as they do in the regulation of eukaryotic tyrosinase kinases such as the epidermal growth factor (EGF) receptor.[87]

An attractive hypothesis for the function of the response regulator domains in the regulation of transcription has been that they serve as phosphorylation-dependent dimerization domains. There is little evidence to support this supposition, however. NtrC is a stable dimer that, on phosphorylation, is converted to a tetramer, but it is not clear that the N-terminal regulatory domain directly participates in this transition.[68,88] Preliminary reports of phosphorylation-induced oligomerization of members of the OmpR family have not been confirmed.

A recurrent problem in the analysis of the quarternary structures of two-component proteins has stemmed from attempts to use molecular sieve chromatography to estimate their size. The proteins are frequently composed of multiple domains tethered by flexible linker regions and they tend to run at aberrantly high molecular weights compared to the more compact globular proteins that are routinely used as standards. Moreover, one cannot assume that changes in quarternary structure will lead to changes in chromatographic mobility, and conversely, changes in interactions between domains could easily lead to changes in mobility without affecting quarternary structure.

The three-dimensional structure of the receiver domain has been determined by X-ray crystallographic and nuclear magnetic resonance (NMR) techniques.[89–95] The domain has a doubly wound α/β fold with a central

[85] J. A. Gegner and F. W. Dahlquist, *Proc. Natl. Acad. Sci. U.S.A.* **88,** 750 (1991).

[86] A. J. Ninfa, S. Ueno-Nishio, T. P. Hunt, B. Robustell, and B. Magasanik, *J. Bacteriol.* **168,** 1002 (1986).

[87] C.-H. Heldin, *Cell* **80,** 213 (1995).

[88] D. S. Weiss, K. E. Klose, T. R. Hoover, A. K. North, S. C. Porter, A. B. Wedel, and S. Kustu, *in* "Transcriptional Regulation" (S. L. McKnight and K. R. Yamamoto, eds.), p. 667. Cold Spring Harbor Laboratory Press, Cold Spring Harbor, NY, 1992.

[89] A. M. Stock, J. M. Mottonen, J. B. Stock, and C. E. Schutt, *Nature (London)* **337,** 745 (1989).

[90] K. Volz and P. Matsumura, *J. Biol. Chem.* **266,** 15511 (1991).

[91] A. M. Stock, E. Martinez-Hackert, B. F. Rasmussen, A. H. West, J. B. Stock, D. Ringe, and G. A. Petsko, *Biochemistry* **32,** 13375 (1993).

[92] L. Bellsolell, J. Prieto, L. Serrano, and M. Coll, *J. Mol. Biol.* **238,** 489 (1994).

[93] M. Bruix, J. Pascual, J. Santoro, J. Prieto, L. Serrano, and M. Rico, *Eur. J. Biochem.* **215,** 573 (1993).

[94] F. J. Moy, D. F. Lowry, P. Matsumura, F. W. Dahlquist, J. E. Krywko, and P. J. Domaille, *Biochemistry* **33,** 10731 (1994).

[95] B. F. Volkman, M. J. Nohaile, N. K. Amy, S. Kustu, and D. E. Wemmer, *Biochemistry* **34,** 1413 (1995).

five-stranded parallel β sheet flanked by five α helices. The side chains of the highly conserved aspartates and lysine cluster together at the surface formed by loops between the C-terminal ends of β-strands and the N-termini of subsequent α-helical elements to form an active phosphoacceptor site that binds Mg^{2+} and catalyzes the transfer of phosphoryl groups from various phosphodonors to the central conserved aspartate. The structure has been primarily determined from studies of the single-domain response regulator CheY, which has provided a template for the general analysis of receiver domain structure. No histidine kinase structure has been reported.

Although no one has crystallized an intact multidomain response regulator protein, it has been possible to obtain diffracting crystals of C-terminal effector domains expressed from constructs from which the N-terminal receiver domain has been deleted. The structure of one of these domains, the catalytic domain of CheB, the esterase that mediates adaptive responses in bacterial chemotaxis, indicates an α/β fold with a clearly discernable active site. There is no feature of the structure, however, that provides a clue as to the mechanism of its control by the associated receiver domain in the intact full-length CheB protein. Crystallization of the C-terminal DNA-binding domain of OmpR has also been reported, but no structural data are yet available.[96]

Concluding Remarks

Bacteria possess extensive signal transduction networks that function to regulate transcription. These systems are composed of modular domains that interact with one another to process information and effect appropriate responses to changing environmental conditions. Two superfamilies of signal transduction proteins are involved: protein kinases that autophosphorylate themselves at specific histidine residues, and response regulators that use these phosphoryl groups as well as phosphoryl groups from small-molecule phosphodonors such as acetyl phosphate to autophosphorylate themselves at a specific aspartate residue. Generally, phosphorylation of the response regulator receiver domain is critical for activity of an associated DNA binding–transcriptional regulation domain. The regulatory circuitry that controls the level of phosphorylation of a given response regulator can be quite complex. Histidine kinases clearly play a central role in this process, both by supplying phosphoryl groups and in some cases by facilitat-

[96] H. Kondo, T. Miyaji, M. Suzuki, S. Tate, T. Mizuno, Y. Nishimura, and I. Tanaka, *J. Mol. Biol.* **235,** 780 (1994).

ing the removal of these groups. The sensory mechanisms that control histidine kinase activity are, however, poorly understood. It is generally not possible to interpret spontaneous mutations unambiguously, or to predict the effects of mutations engineered *in vitro*.

Beyond the well-established problems of understanding the molecular mechanisms of distinct modules within these regulatory systems are larger issues associated with the integration of the entire network to effect global regulatory outputs such as sporulation, virulence, and responses to altered nutritional states. As bacterial genomic sequences become available it will be possible to discern the entire complement of regulatory components within a given type of cell. The problem will then be to understand how individual signal transduction modules are integrated. Such research will likely produce new paradigms for understanding global mechanisms that function in the integrated regulation of gene expression that underlies processes of development and differentiation both in bacteria and in higher organisms.

Acknowledgments

We are grateful to Austin Newton, Michael Surrette, and Thorsten Grebe for their helpful discussions. This work was supported by a grant from the National Institutes of Health (AI20980 to J.B.S.).

[25] Mutational Analysis of Structure–Function Relationship of RNA Polymerase in *Escherichia coli*

By DING JUN JIN and YAN NING ZHOU

Bacterial RNA polymerase (RNAP) is a large and complex molecule consisting of multiple subunits. Although its size and complexity preclude detailed physical structure studies of the molecule at present,[1,2] genetic and biochemical studies are readily amenable. Therefore, mutational analysis

[1] H. Heumann, H. Lederer, G. Baer, R. P. May, J. K. Kjems, and H. L. Crespi, *J. Mol. Biol.* **201,** 115 (1988).

[2] S. A. Darst, E. W. Kubalek, and R. D. Kornberg, *Nature* (*London*) **340,** 730 (1989).

has become a feasible and important tool with which to study bacterial RNAP. This chapter describes mutational analysis of *Escherichia coli* RNAP to illustrate how structure and function relationships of RNAP can be studied using a combined approach of genetics, biochemistry, and molecular biology. Extensive sequence similarities between prokaryotes and eukaryotes observed for the subunits of RNAP[3–5] suggest that the basic structure and function of RNAP are conserved throughout evolution. Mutational analyses of eukaryotic RNAP have been described (for review see Ref. 5 and references therein). Owing to the complex nature of these systems, however, a genetic analysis of structure and function relationships of RNAP in eukaryotes is not simple.

Early studies of the *E. coli* RNAP mutants contributed to our knowledge in identifying the *rpo* genes encoding the different subunits of the enzyme on the *E. coli* genetic map, and in assigning function(s) to the individual subunits of RNAP in the transcription process (for review see Ref. 6 and references therein). Development of sophisticated genetic and recombinant DNA techniques, as well as of a variety of biochemical assays designed to characterize defined step(s) of transcription under different conditions, has facilitated systematic mutational analyses of RNAP. Because RNAP is an essential enzyme for cell survival, its mutational analysis is challenging. Normally only those mutations that can sustain cell growth are likely to be isolated and analyzed, although lethal mutations can be studied under some special conditions. It is also important to determine whether an observed *in vivo* phenotype of a given RNAP mutant is directly related to a particular function of RNAP in which one is interested. This is because the phenotype of a given mutant can be manifested in different ways, directly or indirectly. Thus, it is necessary to purify mutant RNAPs and study their properties biochemically to determine whether the altered property of mutant RNAPs is responsible for the observed phenotype *in vivo*. In this way one can correlate a site in the polypeptide chain with a particular function of RNAP. The study of perturbation in transcription caused by mutant RNAPs may also suggest the mechanism by which a particular step in transcription is carried out by wild-type RNAP. This chapter provides examples to illustrate the usefulness (as well as the limitations) of such approaches and concentrates only on the studies of the two largest subunits of RNAP, β and β', which are part of the same operon and encoded by

[3] R. S. Jokerst, J. R. Weeks, W. A. Zehring, and A. L. Greenleaf, *Mol. Gen. Genet.* **215,** 266 (1989).

[4] R. A. Young, *Annu. Rev. Biochem.* **60,** 689 (1991).

[5] J. Archambault and J. Freisen, *Microbiol. Rev.* **57,** 703 (1993).

[6] T. Yura and A. Ishihama, *Annu. Rev. Genet.* **13,** 59 (1979).

the *rpoB* and *rpoC* genes, respectively. For related information on σ and α subunits, see Section III in this volume.

Identification of Sites or Domains of β and β' Subunits Important for Enzyme Function

The β and β' subunits of prokaryotic RNAP were implicated in basic functions of RNAP from early studies.[6] Significant homology is conserved evolutionarily in these two largest subunits of RNAP among different organisms. Nine segments of high homology, termed A through I, have been identified in β, and similarly, eight segments, termed A through H, have been identified in β'[3-5] (Fig. 1). It is likely that some sites within

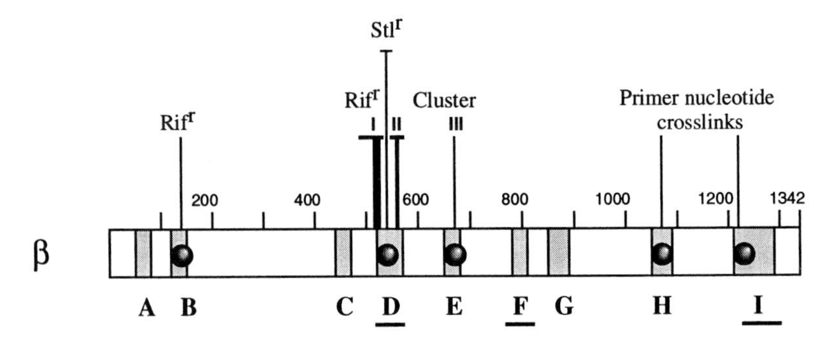

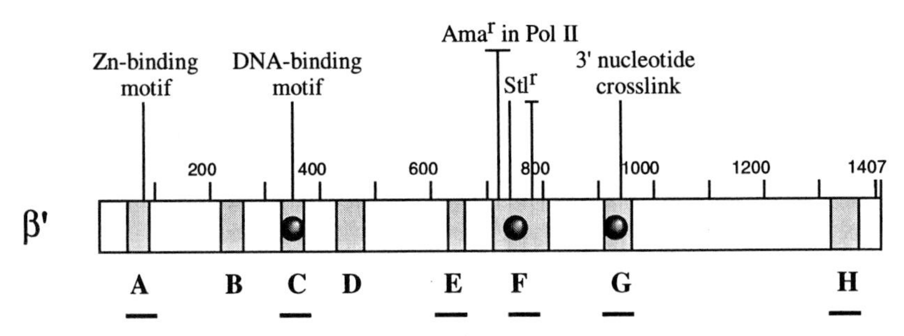

FIG. 1. The β and β' subunits of *E. coli* RNAP showing important sites or domains identified by mutational analyses. Evolutionarily conserved segments in the two largest subunits of RNAP, labeled A to I in β, and A to H in β', are shaded. The mutational changes conferring antibiotic resistance, and the sites cross-linked to primer nucleotide derivatives, are indicated. The major regions at which mutational changes led to altered termination phenotypes are underlined. Other landmarks noted in β' are the Zn-binding motif, the DNA-binding motif of DNA polymerase I, α-amanitin resistance substitutions occurring in Pol II β' homologs, and sites cross-linked to the 3' end of the RNA transcript. The sites or regions that are likely to be in proximity, forming the catalytic center, are indicated by shaded circles. See text for details.

these segments are important for the structure and function of RNAP. The *rpoB* or *rpoC* mutations with altered functions of RNAP can be isolated either on the chromosome or in plasmid-borne genes. The advantage of chromosomal mutations is that their phenotypes can be analyzed *in vivo,* making a correlation between *in vivo* and *in vitro* effects possible, but such studies are limited to viable or conditional lethal mutations (e.g., mutations that make the cell temperature sensitive for growth). Although both lethal and viable mutants can be isolated in the plasmid-borne *rpoB* or *rpoC* gene, it may not be easy to evaluate what their *in vivo* effects would be when present in a single-copy haploid chromosome if mutations are lethal under such a condition.

In general, important sites or domains of RNAP have been successfully identified using mutational analyses by two opposite approaches. One approach is to isolate mutations with changes at defined sites in RNAP, such as antibiotic-binding sites by isolating antibiotic resistance mutants, and then to identify the functions associated with these sites by biochemically characterizing the resulting antibiotic-resistant mutant RNAPs for altered functions. The second approach is to isolate mutations (sometimes conditional lethal) that affect a particular function of RNAP, such as transcription termination function, followed by biochemical dissection of the altered step and determination of the mutational change(s) in the *rpoB* and *rpoC* genes.

Different Antibiotic-Binding Sites and Their Functions

Several antibiotics including rifampicin, sorangicin A, and streptolydigin act on eubacterial RNAP,[7–9] indicating that the binding sites for these antibiotics are highly conserved among bacterial RNAPs. Because these antibiotics inhibit transcription, the binding sites for these antibiotics are likely to be important for RNAP structure and function. Thus, one of the effective ways to probe the important functional domains of RNAP is to analyze these antibiotic-binding sites in the enzyme. Because the steps that are inhibited by different antibiotics might be different, it provides a simple way to probe different functional domains in RNAP.

Rifampicin-Binding Site. Shortly after the discovery of the antibiotic rifampicin, *E. coli* mutants that confer resistance to rifampicin (Rif^r) were isolated and found to have alteration in the β subunit.[10] Because *E. coli* cells are relatively permeable to rifampicin, *rif*^{r} mutants usually can be

[7] G. Hartmann, K. O. Honikel, F. Knusel, and J. Nuesch, *Biochim. Biophys. Acta* **145,** 843 (1967).

[8] H. Irschik, R. Jansen, K. Gerth, G. Hofle, and H. Reichenbach, *J. Antibiotics* **40,** 7 (1987).

[9] R. Schlief, *Nature (London)* **223,** 1068 (1969).

[10] D. Rabussay and W. Zillig, *FEBS Lett.* **5,** 104 (1969).

easily selected by plating out cells on LB or minimal plates containing the antibiotic (≥50 μg/ml) with a frequency of about 1 to 5 × 10^{-8}. Using the Rif^r phenotype as a tool to obtain RNAP mutants and to probe the involvement of RNAP in a variety of physiological processes, a large number of *rif*^{r} mutants affecting different phenotypes have been isolated by different laboratories.[11–17]

To identify all of the rifampicin-sensitive regions, one needs to saturate the *rif*^{r} mutation sites. A large number of chromosomal *rif*^{r} nonlethal mutations that also affect other properties of the enzyme as well were studied. All of these mutations were mapped and sequenced, and found to be located in the middle of the *rpoB* gene.[16,18,19] Alteration at 14 positions in the 1342 amino acid residues of the β subunit of RNAP leads to the Rif^r phenotype.[16] These sites can be divided into three clusters (Fig. 1). The vast majority of these sites are located in cluster I, encompassing amino acid residues 507 to 533, and most of the remainder are located in cluster II, encompassing amino acid residues 563 to 572. A single *rif*^{r} mutation, located at amino acid residue 687, defines cluster III. *rif*^{r} mutants conferring resistance to different levels of rifampicin were found with those imparting resistance to very high levels of the antibiotics (≥1 mg/ml) being located in clusters I and II. The *rif*^{r} mutant in cluster III is resistant only to a very low level of rifampicin (40 μg/ml). Previously, *rif*^{r} mutants capable of growing only at low concentrations of the antibiotic were reported to map around residue 650 of β.[20]

rif^r mutations were also isolated by locally mutagenizing, either chemically or by the use of polymerase chain reaction (PCR) technology, different portions of the *rpoB* gene, followed by cloning of the mutagenized segment into an intact *rpoB* gene under P_{lac} and then transformation of these plasmids into an *rpoB*$^+$ cell.[21,22] *rif*^{r} mutants were selected after the expression

[11] L. Snyder, *Virology* **50,** 396 (1972).
[12] J. P. Lecocq and C. Dambly, *Mol. Gen. Genet.* **145,** 53 (1976).
[13] L. P. Guarente and J. Beckwith, *Proc. Natl. Acad. Sci. U.S.A.* **75,** 294 (1978).
[14] A. Das, C. Merril, and S. Adhya, *Proc. Natl. Acad. Sci. U.S.A.* **75,** 4828 (1978).
[15] C. Yanofsky and V. Horn, *J. Bacteriol.* **145,** 1334 (1981).
[16] D. J. Jin and C. A. Gross, *J. Mol. Biol.* **202,** 45 (1988).
[17] D. J. Jin and C. A. Gross, *J. Bacteriol.* **171,** 5229 (1989).
[18] Y. A. Ovchinnikov, G. S. Monastyrskaya, V. V. Gubanov, V. M. Lipkin, E. D. Sverdlov, I. F. Kiver, I. A. Bass, S. Z. Midlin, O. N. Danilevskaya, and R. B. Khesin, *Mol. Gen. Genet.* **184,** 536 (1981).
[19] Y. A. Ovchinnikov, G. S. Monastyrskaya, S. O. Guriev, N. F. Lalinina, E. D. Sverdlov, A. I. Gragerov, I. A. Bass, I. F. Kiver, E. P. Moiseyeva, V. N. Igumnov, S. Z. Mindlin, V. G. Nikiforov, and R. B. Khesin, *Mol. Gen. Genet.* **190,** 344 (1983).
[20] C. M. Boothroyd, R. M. Malet, V. Nene, and R. Glass, *Mol. Gen. Genet.* **190,** 523 (1983).
[21] R. Landick, J. Stewart, and D. N. Lee, *Genes Dev.* **4,** 1623 (1990).
[22] K. Severinov, M. Soushko, A. Goldfarb, and V. Nikiforov, *J. Biol. Chem.* **268,** 14820 (1993).

of the mutated *rpoB* gene by the induction of isopropyl-β-D-thiogalactopyranoside (IPTG). Owing to the larger number of plasmid copies, the mutations from the plasmid-borne P_{lac}-*rpoB* gene, when expressed, are usually dominant over the chromosomal *rpoB*$^+$ gene. Only mutations in the middle region of the *rpoB* gene showed Rif^r phenotypes. These approaches identified additional residues (S512, S534, and S574) in clusters I and II conferring rifampicin resistance.[21,22] The new sites (amino acid G534 and S574) extended the boundaries for clusters I and II by one and two amino acid residues, respectively. Because the new *rif*^{r} mutations were isolated in the presence of the *rpoB*$^+$ gene in the chromosome, some of them could be lethal mutations when existing in a haploid state. This possibility needs to be tested. A rare *rif*^{r} mutation located outside clusters I, II, and III, affecting amino acid residue 146 of the β subunit of RNAP, has also been identified.[23] Site-directed mutagenesis has confirmed the contribution of this site in producing *rif*^{r} mutation.[24]

Among the *rif*^{r} mutations that are located in the middle of β, there is genetic evidence indicating that amino acid 687 (in cluster III) interacts with that of 529 (in cluster I), even though these two sites are far apart in the linear polypeptide.[25] Cells containing the *rpoB3406* (R687H) allele are temperature sensitive for growth; they cannot grow at either very low temperature (≤25°), or at very high temperature (≥42°). Taking advantage of its growth phenotypes, second site(s) intragenic suppressers of the *rpoB3406* mutation were isolated so that the cells could grow at a very low temperature (≤25°). Interestingly, some of these revertants could also grow at a very high temperature and were resistant to a high level of rifampicin (≥1 mg/ml). Three of the intragenic suppressers of *rpoB3406* were sequenced and found to be identical to two previously characterized *rif*^{r} alleles, *rpoB3401* (R529C) and *rpoB3402* (R529S). Interestingly, both of these *rif*^{r} mutations, when present alone, confer temperature sensitivity. The cells do not grow at ≥42°. This mutational relationship between amino acid residues 529 and 687 of the β subunit of RNAP appears to be allele specific because *rif*^{r} mutations at other positions could not suppress the growth defect of *rpoB3406*. This genetic interaction argues strongly that the residues Arg-529 and Arg-687 are in close proximity in native β.

Furthermore, genetic evidence also showed that part of the protein structure in the region between clusters I and II is dispensable.[22] Mutant RNAPs either missing some of these amino acid residues or having an insertion of extra amino acid residues into this region are still capable of

[23] N. A. Lisitsyn, E. D. Sverdlov, E. P. Moiseyeva, O. N. Danilevskaya, and V. G. Nikiforov, *Mol. Gen. Genet.* **196,** 173 (1984).

[24] K. Severinov, M. Soushko, A. Goldfarb, and V. Nikiforov, *Mol. Gen. Genet.* **244,** 120 (1994).

[25] M. Singer, D. J. Jin, W. A. Walter, and C. A. Gross, *J. Mol. Biol.* **231,** 1 (1993).

carrying out basic functions of the enzyme and are sensitive to rifampicin (but became resistant to streptolydigin; see below). These data indicate that amino acid residues in the region between clusters I and II are likely to form an autonomous domain that probably does not have any essential function. It is likely that the 17 amino acid residues identified by these *rif*^{r} mutations, possibly together with amino acid 146, fold to form the rifampicin-binding site. This model can be tested by the demonstration of cross-linking of riampicin to this region. The importance of this putative rifampicin-binding site is also reflected by the fact that the affected amino acid residues are within or near the evolutionarily conserved segment D of β (see Fig. 1). Not all mutations in this region confer Rif^r phenotypes, indicating that either only a limited number of residues interact with rifampicin, or that the other changes are lethal. *rif*^{s} mutations within this region also affect transcription termination properties, further confirming the importance of this region in RNAP function (see below).

To test whether the *Rif*^{r} mutant RNAPs have indeed altered interaction with rifampicin, their ability to bind to a radioactively labeled rifampicin, as well as their resistance to rifampicin, were determined *in vitro.*[25a] A complete inverse correlation between these two parameters is observed. The higher the resistance to rifampicin, the lower is the ability of the enzyme to bind to rifampicin. Rif^r RNAPs that are resistant to a very high level of rifampicin have lost their ability to bind to rifampicin. These biochemical results support the notion that the sites identified by the *rif*^{r} mutations are likely the binding site for rifampicin.

Functional Role(s) of Rifampicin-Binding Site. The possible functional role(s) of the rifampicin-binding site of RNAP have been studied extensively using defined *rif*^{r} mutants both *in vivo* and *in vitro.* Pleiotropic phenotypes, including altered transcription termination and antitermination functions as well as altered interaction with NusA and mutant σ factor, were associated with some *rif*^{r} mutants *in vivo.*[12–17,26,27] It is interesting to note that mutational changes leading to the same phenotype(s) were often located in cluster.

To analyze the biochemical properties of mutant Rif^r RNAPs, 12 different Rif^r mutant RNAPs, which have changes in the central region of β, were purified and a variety of assays were performed to compare the behavior of the Rif^r mutant RNAPs with that of wild-type RNAP. Interestingly, Rif^r RNAPs affecting different steps of transcription—initiation,

[25a] M. Xu, Y. N. Zhou, B. P. Goldstein, and D. J. Jin, in preparation (1996).

[26] D. J. Jin, W. Walter, and C. A. Gross, *J. Mol. Biol.* **202,** 245 (1988).

[27] D. J. Jin, M. Cashel, D. Friedman, Y. Nakamura, W. Walter, and C. A. Gross, *J. Mol. Biol.* **204,** 247 (1988).

elongation, and termination—have been found, indicating that the corresponding rifampicin-binding sites participate in these essential functions of RNAP.

Rif^r RNAP, RpoB114 (S531F), was found to have reduced transcription initiation at the promoters of the *rpoD* operon encoding ribosomal protein S21, DNA primase, and σ^{70}.[27a] The effect was on a step prior to the promoter clearance by RNAP and appears to be promoter specific. This result indicates that part of the rifampicin-binding site is involved in the early step(s) in transcription initiation.

Rif^r RNAPs also showed altered properties in the transition step between initiation and elongation. During promoter clearance, in addition to making productive initiation products, RNAP sometimes made nonproductive initiation products at several promoters. One such Rif^r RNAP, RpoB3401(R529C), overproduced nonproductive initiation products and reduced productive initiation,[28] whereas another one, RpoB3449(ΔA532) showed a reduced amount of nonproductive synthesis and an increased amount of productive initiation.[29] Interestingly, the effects of these two Rif^r RNAPs on promoter clearance can be influenced by changes in the concentrations of nucleotides, a situation similar to that of other Rif^r RNAPs that are affected in the elongation and termination steps (see below).

The multiple effects of Rif^r RNAPs in transcription initiation described above are consistent with the studies of the mode of action of rifampicin. Rifampicin prevents an initially transcribing complex from entering productive initiation mode. It interferes with the transition between initiation and elongation. In the presence of rifampicin, RNAP makes only very short RNA oligomers (mostly dimers or trimers); the synthesis of longer nascent RNA chain is inhibited.[30] It has been suggested that bound rifampicin lies in the path of nascent RNA emerging from the catalytic center,[31] and either sterically hinders the translocation of RNAP,[30] or destabilizes an initially transcribing complex,[32] or both. When an initially transcribing complex contains an RNA chain longer than a trimer, the RNAP nevertheless becomes immune to rifampicin action; the antibiotic no longer binds to the enzyme at this stage.[32,33] This is presumably because the binding site for

[27a] Y. N. Zhou and D. J. Jin, in preparation (1996).

[28] D. J. Jin and C. L. Turnbourgh, Jr., *J. Mol. Biol.* **236,** 72 (1994).

[29] D. J. Jin, *J. Biol. Chem.,* **271,** 11659 (1996).

[30] R. M. McClure and C. L. Cech, *J. Biol. Chem.* **253,** 8949 (1978).

[31] A. Mustaev, E. Zaychikov, K. Severinov, M. Kashlev, A. Polyakov, V. Nikiforov, and A. Goldfarb, *Proc. Natl. Acad. Sci. U.S.A.* **91,** 12036 (1994).

[32] W. Schulz and W. Zillig, *Nucleic Acids Res.* **9,** 6889 (1981).

[33] E. Eilen and J. Krakow, *Biochem. Biophys. Res. Commun.* **55,** 282 (1973).

rifampicin is masked by a longer (> trimer) nascent RNA molecule. These genetic and biochemical data indicate that the putative rifampicin-binding site is part of the catalytic center of RNAP and is near the 3′ end of a nascent RNA molecule.

Interestingly, about 50% of Rif^r RNAPs show altered elongation properties. Some showed increased pause at some sites leading to a slower elongation rate, whereas others had decreased pause leading to a faster elongation rate compared to wild-type RNAP. In general, the effects of these Rif^r RNAPs on elongation can be influenced by the changes in the concentrations of nucleotides in the reactions. One of these mutant RNAPs, RpoB8(Q513P), displayed a decreased elongation rate on poly[d(AT)] templates, which presumably have no specific pause sites similar to the one on natural DNA templates. This indicates that the Rif^r RNAP does not have an altered recognition of pause signals, but rather has changed its intrinsic catalytic activity, such as K_m for nucleotides.[34]

Those Rif^r RNAPs that had greatly altered elongation rate also had profoundly altered termination function. In general, Rif^r RNAPs that showed a slower elongation rate tend to have increased termination, and vice versa.[34–37] However, there were exceptions to this general rule, indicating that termination is a complex process. Not all Rif^r RNAPs show their *in vivo* effects on termination *in vitro,* suggesting that *in vitro* and *in vivo* conditions are different. Alternatively, the apparent *in vivo* effects of *rif*^{r} mutations could be indirect; the mutations may affect the expression of some other genes leading to altered termination phenotypes.

The effect of Rif^r RNAPs on transcription elongation and termination further argues for the importance of the rifampicin-binding site in RNAP functions, suggesting that the rifampicin-binding site is either near the catalytic center, or has an allosteric effect on the catalytic center of the enzyme.

Sorangicin-Binding Sites. Sorangicin A, a member of a new class of antibiotics structurally unrelated to rifampicin, was reported to inhibit transcription of bacterial RNAP.[8] Some *rif*^{r} mutants were also resistant to antibiotic sorangicin A.[38] The relationship between the binding sites for these two antibiotics was analyzed using the previously characterized mutant Rif^r RNAPs.[25a] Six of 12 purified Rif^r RNAPs were found to be resistant

[34] D. J. Jin and C. A. Gross, *J. Biol. Chem.* **266,** 14478 (1991).

[35] D. J. Jin, R. R. Burger, J. P. Richardson, and C. A. Gross, *Proc. Natl. Acad. Sci. U.S.A.* **89,** 1453 (1992).

[36] J. Greenblatt, M. McLimont, and S. Hanly, *Nature* (*London*) **292,** 215 (1981).

[37] R. F. Fisher and C. Yanofsky, *J. Biol. Chem.* **258,** 8146 (1983).

[38] G. Rommele, G. Wirz, R. Solf, K. Vosbeck, J. Gruner, and W. Wehrli, *J. Antibiotics* **43,** 88 (1990).

to sorangicin A to some degree. Besides, 100% of sorangicin A-resistant mutant cells have acquired the Rif^r phenotype. These results suggest that the binding sites for rifampicin and sorangicin partially overlap, and the binding site for the latter may be a subset of the binding site for the former. Indeed, sorangicin A competitively inhibited the binding of rifampicin to RNAP, confirming that the binding sites for rifampicin and sorangicin A are shared. In the presence of sorangicin A, RNAP synthesized only abortive initiation products, indicating that its mode of action is also similar to that of rifampicin.[25a]

Streptolydigin-Binding Sites. The antibiotic streptolydigin inhibits transcription elongation of bacterial RNAP.[39] Although gram-positive bacteria are permeable to streptolydigin, *E. coli* and other gram-negative bacteria are nonpermeable to this antibiotic. However, *E. coli* mutants permeable to streptolydigin were isolated and were shown to be sensitive to the antibiotic. From such a streptolydigin-permeable *E. coli* strain, streptolydigin-resistant (*stl*^{r}) mutants have been isolated. Several *stl*^{r} mutations of *E. coli* mapped very closely to *rif*^{r} mutations in the β subunit of RNAP.[40,41] One such *stl*^{r} mutation was sequenced and found to be a double mutation affecting amino acid residues 544 and 545 of β.[42] *stl*^{r} mutations alter only one of the four contiguous amino acid residues (543 to 546) in the β subunit.[43] While alteration of each of these positions leads to the Stl^r phenotype, only mutations at position 544 or 545 lead to resistance to high levels of the antibiotic. Because these amino acid residues are within the region between clusters I and II of the rifampicin-resistance site, it suggests that these four positions are involved in the binding of streptolydigin. This could also explain why the mutant RNAPs that have deleted the amino acid residues in this region also become *stl*^{r}.[22]

Reconstitution studies in RNAP from *Bacillus subtilis* found that although a *rif*^{r} mutation is associated with the β subunit, a *stl*^{r} mutation is associated with the β' subunit.[44,45] Because the primary sequences of β and β' are highly conserved between *E. coli* and *B. subtilis,* it was reasonable to presume that additional sites for *stl*^{r} mutation may be found in the β' subunit of *E. coli.* An extensive search for *stl*^{r} mutations in the largest

[39] C. Siddhikol, J. W. Erbstoeszer, and B. Weisblum, *J. Bacteriol.* **99,** 151 (1969).

[40] Y. Iwakura, A. Ishichama, and T. Yura, *Mol. Gen. Genet.* **121,** 181 (1973).

[41] W. Zillig, K. Zechel, D. Rabussay, M. Schlachner, V. S. Sethi, P. Palm, A. Heil, and W. Seifert, *Cold Spring Harbor Symp. Quant. Biol.* **35,** 47 (1970).

[42] N. A. Lisitsyn, E. D. Sverdlov, E. P. Moiseyeva, and V. G. Nikiforov, *Bioorg. Khim.* **11,** 132 (1985).

[43] L. M. Heisler, H. Suzuki, R. Landick, and C. A. Gross, *J. Biol. Chem.* **268,** 25369 (1993).

[44] S. M. Halling, K. C. Burtis, and R. H. Doi, *J. Biol. Chem.* **252,** 9024 (1977).

[45] S. M. Halling, K. C. Burtis, and R. H. Doi, *Nature* (*London*) **272,** 837 (1978).

subunit of *E. coli* RNAP, indeed, showed that changes at three positions in the β' subunit also lead to Stl^r.[46] One (S793F) leads to resistance to high levels of streptolydigin, whereas the other two (M747I or R780H) confer only weak Stl^r. Simultaneously, *stl*^{r} mutations were also identified in the homologous regions (both in β and β') of *B. subtilis* RNAP.[47] The *stl*^{r} mutations of β' in prokaryotes are located in region F, one of the evolutionarily conserved sequence segments of this subunit. In summary, two regions in bacterial RNAP, one in β and the other in β', affect the binding of streptolydigin. It is possible that these two regions of RNAP are in close proximity in the putative binding site for streptolydigin. Alternatively, one of the regions may be the binding site while the other affects the binding of the antibiotic allosterically. Interestingly, the eukaryotic RNAP mutations conferring resistance to α-amanitin, an antibiotic that can be considered as a functional analog of streptolydigin in that both inhibit transcription elongation, are all located in or immediately adjacent to region F.[48–50]

Functional Role of Streptolydigin-Binding Sites. While rifampicin is incapable of inhibiting an elongation complex as described above, streptolydigin is able to do so, probably by interfering with the formation of phosphodiester bonds.[51,52] This suggests that the binding site for streptolydigin is distinct from that of rifampicin. Consistently, streptolydigin did not interfere with the binding of rifampicin to RNAP in a competition assay.[25a]

The effects of *stl*^{r} mutants on transcription have been studied. All of the mutant RNAPs with *stl*^{r} mutations in *rpoB* had no or only a marginal effect on elongation assays *in vitro,* and none of them had effects on termination *in vivo,*[43] suggesting that the alterations are not critical for RNAP function. However, some of the RNAPs containing *stl*^{r} mutations in *rpoC* had altered both elongation and termination functions.[46] Interestingly, an α-amanitin resistance mutant RNAP II from *Drosophila* had a slow rate of elongation.[53] These results suggest that part of this region is important for elongation and termination function of RNAP (also see below).

Sites Cross-Linked to Primer Nucleotide

Nucleotide derivatives with selective affinity near the primer nucleotide-binding site of RNAP were used as probes to locate the catalytic center

[46] K. Severinov, D. Markov, E. Severinova, V. Nikiforov, R. Landick, S. Darst, and A. Goldfarb, *J. Biol. Chem.* **270,** 23926 (1995).
[47] X. Yang and C. W. Price, *J. Biol. Chem.* **270,** 23930 (1995).
[48] M. Bartolomei and J. Corden, *Mol. Cell. Biol.* **7,** 586 (1987).
[49] M. S. Bartolomei and J. L. Corden, *Mol. Gen. Genet.* **246,** 778 (1995).
[50] Y. Chen, J. Weeks, M. A. Mortin, and A. L. Greenleaf, *Mol. Cell. Biol.* **13,** 4214 (1993).
[51] W. R. McClure, *J. Biol. Chem.* **255,** 1610 (1979).
[52] G. Cassani, R. R. Burgess, H. M. Goodman, and L. Gold, *Nature New Biol.* **230,** 197 (1971).
[53] D. E. Coulter and A. L. Greenleaf, *J. Biol. Chem.* **260,** 13190 (1985).

of the enzyme. From such analysis, a lysine residue in the region containing residues 1036 to 1066 and a histidine residue in the region containing residues 1232 to 1243 of β were implicated in the binding of the initial nucleotide.[54] These results are significant because these two regions are located in the evolutionarily conserved segments H and I, respectively. To locate the lysine and histidine residues responsible for the cross-linking, site-directed mutagenesis was performed to change the lysine and histidine residues in or near these two regions, and the resultant mutant RNAPs were assayed for their ability to interact with the cross-linkable substrate analogs. Only the changes, K1065R and H1237A, at two evolutionarily invariant positions, abolished their abilities to cross-link to the nucleotide derivatives.[55] These results indicate that amino acid residues 1065 and 1237 are the targets for the priming nucleotide derivatives. K1065 and H1237 are located about ~3 Å from the α-phosphate of the priming nucleotide; because this is an effective range of the affinity probes used, it indicates that these two sites are close in the enzyme structure. These two residues are essential for RNAP function because plasmid-borne *rpoB* containing either the K1065R or H1237A mutation is unable to complement an amber mutation in chromosomal *rpoB*. The enzymes containing such mutations were purified and the functional role of K1065R and H1237A changes was assessed biochemically.[55,56] Even though K1065 and H1237 of β appear to be in immediate contact with the priming nucleotide derivatives, the two mutant RNAPs were normal in nucleotide binding and forming phosphodiester bonds. These two mutant RNAPs, nevertheless, were defective in promoter clearance. K1065R synthesized mostly abortive initiation products and almost no productive initiation products, whereas the H1237A displayed a less severe defect. Interestingly, the phenotype of the mutant enzyme containing K1065R mimicks the behavior of wild-type RNAP in the presence of rifampicin, whereas the phenotype of the mutant enzyme containing H1237A resembles the Rif^r RNAP (R529C). The latter overproduced abortive initiation products leading to reduced productive initiation.[28] The fact that alterations of RNAP function either in the wild-type enzyme caused by rifampicin or in a mutant enzyme caused by a *rif*^{r} mutation can be simulated by mutational alterations in regions of RNAP that are in close contact to the initiating nucleotide suggests that both the rifampicin-binding site and the sites cross-linked to primer nucleotide participate in a common function and may be in proximity in the enzyme structure.

[54] M. A. Grachev, E. A. Lukhtamov, A. A. Mustaev, E. F. Zaychikov, M. N. Abdukayumov, I. V. Rabinov, V. I. Richert, Y. S. Skoblow, and P. G. Chistyakov, *Eur. J. Biochem.* **180,** 577 (1989).

[55] A. Mustaev, M. Kashlev, J. Lee, A. Polyakov, A. Lebedev, K. Zalenskaya, M. Grachev, A. Goldfarb, and V. Nikiforov, *J. Biol. Chem.* **266,** 23927 (1991).

[56] M. Kashlev, J. Lee, K. Zalenskaya, V. Nikiforov, and A. Goldfarb, *Science* **248,** 1006 (1990).

The importance of the amino residue(s) at or near H1237 of β at conserved segment I is further supported by mutations isolated in this region. A plasmid-borne *rpoB* mutant containing any of the following changes (H1237Y, G1249S, G1266D, and E1272K) was recessive lethal and viable only in the presence of a functional *rpoB* gene in the chromosome. RNAPs from G1249S, G1266D, or E1272K were found to have a slow elongation rate compared to wild-type enzyme.[57a] In addition, other plasmid-borne *rpoB* mutants affecting residue H1237, K1242, or G1271 of β became *trans*-dominant when they were expressed by IPTG induction in the presence of a functional *rpoB* gene in the chromosome,[57b] consistent with the notion that this region of RNAP is critical for enzyme function.

Sites Affecting Termination

As described above, antibiotic-resistant mutants are useful in dissecting the sites in RNAP that are important for termination. Several different approaches have been used to isolate RNAP mutants that have altered termination function.

1. Amber mutations scattered in *rpoB* were isolated and screened for altered termination phenotypes in the presence of different nonsense suppressor mutations.[57c] In this way, several residues of β that, when altered, affect termination were identified. Two of these amber mutations were sequenced and found to be at residues W183 and K844 of β.[57b]

2. More recently, by randomly mutagenizing different portions of the plasmid-borne *rpoB* or *rpoC* genes, followed by screening for mutant RNAPs with either increased or decreased termination function, a large number of mutations have been generated[21,58] (see Fig. 1). Mutations in *rpoB* that affect termination were located predominantly in three regions.[21] Most of the mutations are located between amino acid residues 500 and 575. Some of these mutants were also Rif^r. The latter observation implicates the putative rifampicin-binding site in transcription termination. As indicated above, this region is near or within the highly conserved segment D. A second group of mutations with a termination defect was located between amino acid residues 740 and 840. This region was near or within the highly conserved segment F. A third group maps near the carboxy terminus of β, from amino acids 1225 to 1342, in the highly conserved segment I. Nearly all of the mutant RNAPs analyzed *in vitro* also show altered termination or elongation behavior.

[57a] P. L. Tavormina, R. Landick, and C. A. Gross, *J. Bacteriol.* in press (1996).

[57b] R. E. Glass, personal communication (1995).

[57c] V. Nene and R. E. Glass, *Mol. Gen. Genet.* **194,** 166 (1984).

[58] R. Weilbaecher, C. Hebron, G. Feng, and R. Landick, *Genes Dev.* **8,** 2913 (1994).

For the *rpoC* gene, 55 single substitution mutations were isolated with altered termination phenotypes.[58] Most of these mutations are clustered in several evolutionarily conserved segments including C, E, F, G, and H. In paritcular, mutations in the conserved segment C (amino acid residues 311 to 386) are clustered around a short sequence that is similar to the sequence that forms a DNA-binding cleft in *E. coli* DNA polymerase I.[59] Mutations in the conserved segment F (amino acid residues 718 to 798) correspond to the α-amanitin-resistant mutations located in homologous region in the largest subunit of RNAP II.[48–50] Some of the *stl*^{r} mutations described above also came from this collection.[46] Mutations in the conserved segment G (amino acid residues 933 to 936) occur in a region implicated in close contact to the transcript 3′ end.[60] Mutations in the conserved segment H (amino acid residues 1308 to 1356) are also clustered near the carboxy terminus of β'. Haploid viability assay of 15 β' mutants showed that 8 of them are lethal with the rest being partially lethal, indicating that these sites are critical for RNAP function. The *in vivo* effects of these *rpoC* mutations (in the presence of wild-type *rpoC*$^+$) on termination phenotypes showed mixed results. Ten of them had either increased or decreased termination efficiency, whereas the other 5 exhibited increased termination at one site and decreased termination at another. RNAPs purified from these 15 mutant strains showed altered elongation and termination properties. About half of them showed significantly lower or higher elongation rates. All of the mutant RNAPs had effects on termination assays using six different terminators, but there were no good correlations between *in vivo* and *in vitro* termination phenotypes for some of the mutants.

3. Suppressors of the mutations in the *rho* gene encoding the termination factor Rho protein have been selected. Two *rif*^{r} mutations, *rpoB101* (Q513L) and *rpoB3370* (T563P), were able to suppress the termination defect of the *rho15* allele.[14,26] Purified RNAP Rpo101 had normal elongation property, but RpoB3370 had a slower rate of elongation compared to wild-type RNAP. A third *rif*^{r} mutation, Q513P, suppressed termination defects of several *rho* alleles.[13,35] This mutant RNAP had a reduced rate of elongation due to an increase in K_m for ATP and GTP.[34] Three *rpo* mutations that suppressed the termination defect of *rho201,* but were not *Rif*^{r}, were also isolated and studied.[61,62] Two of them had an identical change, N1072H, in β; the change was close to one (K1065) of the two sites cross-linked to primer nucleotide derivatives in the conserved segment

[59] L. Allison, M. Moyle, M. Shales, and C. Ingles, *Cell* **42,** 599 (1985).
[60] S. Borukhov, J. Lee, and A. Goldfarb, *J. Biol. Chem.* **266,** 23932 (1991).
[61] L. P. Guarente, *J. Mol. Biol.* **129,** 295 (1979).
[62] D. J. Jin and C. A. Gross, *Mol. Gen. Genet.* **216,** 269 (1989).

H. The other had the R352C change in β'. This alteration was within the DNA-binding cleft of DNA polymerase I in conserved segment C. RNAPs from both the β and β' mutants had slower elongation rates compared to the wild-type enzyme. The β' mutant also exhibited reduced stability of the RNAP · DNA complex.[62a]

4. RNAP mutant with altered termination property was isolated using the antitermination system in bacteria phage HK022. Unlike the N-dependent antitermination system in phage λ,[62b] antitermination of early transcription in phage HK022 requires only a *nut* sequence *in cis* and no virus-encoded proteins. In a search for host factors that might affect the antitermination system in phage HK022, *E. coli* mutants that blocked HK022 antitermination were isolated.[63] Interestingly, all of the 14 mutants with such a phenotype mapped in the *rpoC* gene and affected only one of the three amino acid residues in β' (Y75, R77, and E86), which are clustered in the zinc finger domain within the evolutionarily conserved segment A.[64] One of the mutant RNAPs (Y75N) was tested *in vitro* and showed reduced readthrough of a terminator located downstream of the phage P_L promoter and *nutL* site. Furthermore, the mutant RNAP also had a reduced elongation rate compared to the wild-type enzyme on HK022 DNA. These results identified another region in β' that is important for termination and elongation, and probably is involved in the interaction with the *nut* site as well.

Regions That Form Catalytic Center

The mutational analysis described above suggests that the following regions potentially form the interface of the catalytic center of RNAP (see Fig. 1): (1) rifampicin-binding sites (residues at several clusters of β) that presumably form part of the channel of an emerging nascent RNA; (2) sites in β, K1065 and H123, that cross-link to primer nucleotide derivatives and likely interact with or are close to the rifampicin-binding site because the mutant enzymes at these two positions behaved the same way the wild-type RNAP did in the presence of rifampicin; (3) the streptolydigin-binding site (composed of β and β' subunits) which probably is in the proximity of the rifampicin-binding site because some of the *rpoB* mutations conferring Stl^r lie in the region between clusters I and II of the rifampicin-binding site; (4) the amino acid residues near 932 to 1020 of β', which are implicated in cross-linking to the 3′ end of the transcript in the conserved segment G;

[62a] L. M. Heisler, G. Feng, D. J. Jin, C. A. Gross, and R. Landick, *J. Biol. Chem.* in press (1996).
[62b] A. Das, *Methods Enzymol.* **274,** in press (1996).
[63] M. Clerget, D. J. Jin, and R. A. Weisberg, *J. Mol. Biol.* **248,** 768 (1995).
[64] S. Borukhov, K. Severinov, M. Kashley, A. Lebedev, I. Bass, G. Rowland, P. Lim, R. Glass, V. Nikiforov, and A. Goldfarb, *J. Biol. Chem.* **266,** 23921 (1991).

and (5) some amino acid residues located at the conserved segment C of β' that have a sequence resembling the DNA-binding cleft of DNA polymerase I. These conclusions can be tested genetically by studying intragenic suppressors of mutations at these regions.

Dissecting of Mechanisms of Transcription at Defined Steps Using Mutant RNA Polymerase

Mutational analysis of RNAP can be used as a probe to investigate the mechanisms of transcription at different defined steps. The catalytic activity of RNAP in synthesizing RNA is a concerted reaction involving multiple elementary steps. These steps include contacting DNA template covering a 25- to 40-bp region, 12–20 of which may be separated into a single stranded "transcription bubble"; melting DNA double strands at the forward contact and reannealing the single strands at the lagging end; binding an incoming nucleotide; binding a nascent RNA; forming RNA · DNA hybrids of undefined length near the nascent RNA 3′ end; forming a phosphodiester bond; releasing PP_i; and translocation (for reviews, see Refs. 65–67). The dynamic aspect of transcription is reflected in the alternate inchworm and monotonic movement of the elongation complex in a nucleic acid context-dependent manner.[67,68] Theoretically, every elementary step in transcription can be altered or perturbed by a mutational change in RNAP leading to an altered function in transcription. Biochemical analysis of a mutant RNAP that has altered a particular function helps elucidate the mechanism underlying the alteration, which in turn will help dissect the mechanism underlying the normal transcription process.

Promoter Clearance

Promoter clearance is a transition step in which an initially transcribing complex converts to an elongation complex. RNAP of an initially transcribing complex usually makes two kinds of nonproductive initiation products: aborted or stuttered, in a promoter-specific manner. The mechanisms underlying the switch between nonproductive initiation and productive initiation at promoter clearance have been studied with mutant RNAPs that are altered in the switch. Results with some β mutant RNAPs clearly indicated

[65] A. Das, *Annu. Rev. Biochem.* **62,** 893 (1993).
[66] C. Chen and R. Landick, *in* "Transcription: Mechanism and Regulation" (R. Conaway and J. Conaway, eds.), p. 297. Raven, New York, 1994.
[67] M. J. Chamberlin, *in* "The Harvey Lectures," Series 88, p. 1. Wiley-Liss, New York, 1994.
[68] E. Nudler, A. Goldfarb, and M. Kashlev, *Science* **265,** 793 (1994).

that there is a kinetic element in determining the switch between these two alternative pathways during promoter clearance.

One mutant RNAP (R529C) showed reduced transcription initiation from the *pyrBI* promoter *in vivo*.[28] The purified mutant protein overproduced abortive initiation products and reduced productive initiation at the promoter, demonstrating that promoter clearance is biologically significant. Furthermore, the defect of the mutant enzyme on promoter clearance was sensitive to the concentration of UTP. The defect of the mutant was most dramatic when the concentration of UTP was low; conversely, when the concentration of UTP was high, the defect was suppressed. This result suggests that the mutant enzyme is constrained in incorporating the initially transcribing nucleotides (which happen to be a run of three Us) at the *pyrBI* promoter. This result also implicates that a K_m for nucleotides of an initially transcribing RNAP is important for promoter clearance.

Another mutant RNAP (ΔA532) reduced stuttering synthesis at the *galP2* promoter, leading to an increased productive initiation compared to wild-type enzyme.[29] The stuttering synthesis by wild-type enzyme at *galP2* was shown to be UTP concentration dependent.[69] Interestingly, the alteration of the mutant RNAP in stuttering synthesis depends on the presence of CTP. In the absence of CTP, it was capable of stuttering just like wild-type enzyme, but in the presence of CTP the mutant enzyme showed decreased nonproductive synthesis. The initially transcribed sequence of *galP2* is pppAUUUC, C being the nucleotide next to the run of uridine residues that is responsible for stuttering synthesis at *galP2*. Compared to the wild-type enzyme, the mutant RNAP may have a faster rate of incorporating CTP after the run of uridine residues, leading to a reduced stuttering synthesis. Consistently, the stuttering synthesis at *galP2* was sensitive to the concentration of CTP for both the mutant and wild-type enzymes. Thus, the rate of incorporating CTP at the fifth position of the *galP2* promoter after the run of Us is critical in determining the switch between nonproductive stuttering synthesis and productive initiation.

To determine further whether the stuttering synthesis at *galP2* is CTP specific or position specific, the *galP2* template was mutated so that the fifth position became G instead of C. If the stuttering synthesis were CTP specific, one would expect that the stuttering synthesis at the mutated *galP2* promoter would not be sensitive to GTP. Alternatively, if the stuttering synthesis were position specific, the stuttering synthesis would be sensitive to GTP at the mutated *galP2* promoter. The experimental results demonstrated that stuttering synthesis at the mutant *galP2* promoter became sensitive to changes in the concentration of GTP. Thus, this result argues

[69] D. J. Jin, *J. Biol. Chem.* **269,** 17221 (1994).

for a critical role played by the rate of incorporation of the nucleotide after the run of uridine residues in *galP2* for promoter clearance. This result also raises the question of whether a critical position exists in each promoter regardless of the presence of a run of uridine residues and whether the rate of incorporation of a nucleotide at the critical position determines the efficiency of promoter clearance.

Termination

When an elongating RNAP encountered a terminator, termination occurred and RNA transcript was released. There are two kinds of terminators: a Rho-independent terminator and a Rho-dependent terminator (for reviews, see Refs. 65 and 66). Although termination mechanisms at these two terminators may be different, some mutant RNAPs with single amino acid changes had altered termination at both types of terminators,[26] indicating a common step at these different terminators. These mutant RNAPs showed altered elongation rates compared to the wild-type enzyme. The mutant RNAPs that had a faster rate of elongation decreased the efficiency of termination; conversely, the mutant RNAPs that had a slower rate of elongation increased the efficiency of termination at both intrinsic and Rho-dependent terminator.[34–37] Such correlation suggests that termination at both terminators is subject to a kinetic mechanism. This hypothesis was tested by modulating the rates of elongation for the mutant and wild-type RNAPs. Two β mutants, Q513P and H526Y, which affect both termination and elongation were compared for their termination pattern with the pattern of wild-type RNAP at eight well-characterized Rho-independent terminators.[70] Q513P was termination proficient and had a slower elongation rate, whereas the opposite was true for H526Y. Readthrough at these terminators was lower with Q513P, but higher with H526Y than that with wild type. Furthermore, readthrough by H526Y could be raised at each terminator by increasing the concentration of a single nucleotide. The nature of the nucleotide was terminator specific, as if for each terminator there was a critical nucleotide in determining the efficiency of termination. The positions at which different RNAPs terminated at the bacteriophage 21 later terminator, t_{21}, were further analyzed. For Q513P, the population of RNA ends was shifted by 2 nucleotides toward the promoter relative to wild type, whereas for H526Y the population of RNA ends was shifted by 2 nucleotides away from the promoter relative to wild type. When the nucleotide concentration was decreased, the population of RNA ends was shifted toward the promoter for all enzymes, and vice versa. These results

[70] J. C. McDowell, J. W. Roberts, D. J. Jin, and C. Gross, *Science* **266,** 822 (1994).

support the argument that when an elongating RNAP moves near the 3′ end of an intrinsic terminator, it is controlled by the K_m of the enzyme for NTPs, the rate of elongation through this region determines transcript release efficiency. A slower rate would likely increase the probability of RNA release and a faster rate would decrease the probability of RNA release.[71]

To test whether termination at Rho-dependent terminators depends on a kinetic coupling between RNAP and termination factor Rho protein, two β mutants with altered termination property were tested[35]: Q513P, which as described above also terminated efficiently at Rho-dependent terminators and had a slower rate of elongation compared to wild type *in vivo,* and S522F, which behaved oppositely. In an *in vitro* Rho-dependent termination assay, Q513P terminated more efficiently, whereas S522F terminated less efficiently, than wild type. Relative to wild type, the population of RNA ends was shifted toward the promoter with Q513P and away from the promoter with S522F, a situation similar to the intrinsic termination described above. However, the termination profiles of the mutant RNAPs were comparable to that of wild type if the rates of elongation for the three enzymes were comparable (by adjusting the nucleotides in the reactions). Thus, the altered termination phenotypes of the mutant RNAPs can be explained solely by their altered rate of elongation. Evidence for the intimate relationship between the rate of RNAP elongation and the rate of action of Rho protein also came from an analysis of the mechanism by which the *rpoB8* mutation suppresses the termination defect of *rho1* mutation *in vivo.*[35] Rho1 protein acted more slowly in releasing RNA *in vitro* and led to increased readthrough of Rho-dependent terminators *in vivo.* RpoB8(Q513P), which has a slower rate of elongation, compensated for the defect of Rho1 *in vitro.* Slowing down the wild-type RNAP to the rate of RpoB8 also resulted in suppressing the termination defect of Rho1, indicating that the altered elongation rate of the mutant RNAP is responsible for the suppression effect. These results demonstrate directly that termination efficiency at Rho-dependent terminators is governed by a precise kinetic coupling between the rate of elongating RNAP and the rate of action of Rho.

Summary

The structure and function relationships of RNAP can be studied by correlating the change(s) in the *rpo* genes encoding different subunits of RNAP and the altered function(s) of the mutant RNAP both *in vivo* and

[71] P. H. von Hippel and T. D. Yager, *Proc. Natl. Acad. Sci. U.S.A.* **88,** 2307 (1991).

in vitro. Mutational analysis has proven to be a powerful approach to the study of RNAP structure–function and to the elucidation of the mechanisms underlying different steps of transcription. Rapid progress has been made in isolating and characterizing RNAP mutants with altered step(s) of transcription. Analysis of the steps altered in these mutant RNAPs in turn has helped in studying the normal reaction mechanisms.

Acknowledgments

We thank many colleagues for support in writing this chapter, in particular Drs. Robert Glass, Carol Gross, Robert Landick, Chester Price, and Konstantin Severinov for sharing their unpublished results; and Drs. Robert Glass, Sue Garges, and Darren Sledjeski for comments on the manuscript.

[26] Vectors with Bidirectional Reporter Genes for Studying Divergent Promoters

By VINAY K. JAIN

Bidirectional promoters with divergent transcription have been described in both higher and lower organisms and seem to fulfill various functions.[1] In many cases, the divergent transcripts encode proteins. For example, the avian genes for purine nucleotide synthesis, *GPAT* and *AIRC,* are transcribed from a bidirectional promoter.[2] Similarly, the murine and human type IV collagen $\alpha 1$ and $\alpha 2$ genes are organized in a head-to-head configuration with a bidirectional promoter between them.[3,4] The *GAL1–GAL10* genes of the yeast *Saccharomyces cerevisiae* are also arranged in a head-to-head fashion with a divergent promoter separating them.[5] In the plant-infecting bacterium *Rhizobium meliloti,* the *nodA* and the *nodD* genes are also arranged in a head-to-head configuration and similarly have a divergent promoter between them.[6]

Antisense transcription has been found to occur from both murine dihydrofolate reductase and the rat insulin II genes.[7,8] Bidirectional tran-

[1] C. F. Beck and R. A. J. Warren, *Microbiol. Rev.* **52,** 318 (1988).
[2] A. Gavalas, J. E. Dixon, K. A. Brayton, and H. Zalkin, *Mol. Cell. Biol.* **13,** 4784 (1993).
[3] P. D. Burbelo, G. R. Martin, and Y. Yamada, *Proc. Natl. Acad. Sci. U.S.A.* **85,** 9679 (1988).
[4] E. Poschl, R. Pollner, and K. Kuhn, *EMBO J.* **7,** 2687 (1988).
[5] M. Johnston and R. D. Davis, *Mol. Cell. Biol.* **4,** 1440 (1984).
[6] S. K. Parry, S. B. Sharma, and E. A. Terzaghi, *Gene* **150,** 105 (1994).
[7] S. Efrat and D. Hanahan, *Mol. Cell. Biol.* **7,** 192 (1987).
[8] P. J. Farnham, J. M. Abrams, and R. T. Schimke, *Proc. Natl. Acad. Sci. U.S.A.* **82,** 3978 (1985).

scription has also been found in promoters of c-*myc*, a cellular protein critical for both cell growth and apoptosis.[9,10] Although it has been hypothesized that antisense transcripts may fulfill a regulatory function, rigorous evidence for this hypothesis has not been obtained. In other bidirectional promoters, the precise role, if any, of the antisense transcript has not been elucidated. One obvious function for an antisense transcript would be to downregulate the translation of the sense protein.

A number of vectors have been developed with two reporter genes arranged in a head-to-head fashion to analyze simultaneously divergent transcription from the same promoters.[6,11,12] The reporter genes used in these vectors included firefly luciferase (LUX, from *Photinus pyralis*), β-galactosidase (*LacZ*α, from *Escherichia coli*), chloramphenicol acetyltransferase (CAT, from *E. coli*), β-glucuronidase (*gusA*, from *E. coli*), growth hormone (both human and bovine), and secreted alkaline phosphatase (see Alam and Cook for a review of reporter genes[13]). This chapter reviews some of these dual reporter vectors and their potential applications and pitfalls.

Construction of Luciferase/Chloramphenicol Acetyltransferase Vector

The construction[11] is started using CMV-LUX, a pUC9 vector that contains the firefly luciferase gene downstream of the cytomegalovirus (CMV) immediate early promoter.[14] The simian virus 40 (SV40) small t intron and the SV40 early polyadenylation site are present downstream of firefly luciferase coding sequences. The CMV early promoter is removed using *Hin*dIII–*Xho*I digestion. The CAT gene is obtained by *Hin*dIII–*Xho*I digestion of a vector called EBV (Epstein–Barr virus)-CAT.[15] This 1.6-kb *Hin*dIII–*Xho*I fragment contains the CAT gene along with the SV40 small t intron and the SV40 early polyadenylation signal. It is ligated into the *Hin*dIII–*Xho*I-digested CMV-LUX vectors such that the two reporter genes are arranged in a head-to-head orientation. Two unique restriction

[9] F. Calabi and M. S. Neuberger, *EMBO J.* **4,** 667 (1985).

[10] A. Nepveu and K. B. Marcu, *EMBO J.* **5,** 2859 (1986).

[11] V. K. Jain, I. T. Magrath, and T. Shimada, *Biotechniques* **12,** 681 (1992).

[12] M. N. Patel and M. Kurkinen, *Biotechniques* **15,** 615 (1993).

[13] J. Alam and J. L. Cook, *Anal. Biochem.* **188,** 245 (1990).

[14] J. M. Liu, H. Fujii, S. W. G, N. Komatsu, N. S. Young, and T. Shimada, *Virology* **182,** 361 (1991).

[15] M. Woisetschlaeger, J. L. Strominger, and S. H. Speck, *Proc. Natl. Acad. Sci. U.S.A.* **86,** 6498 (1989).

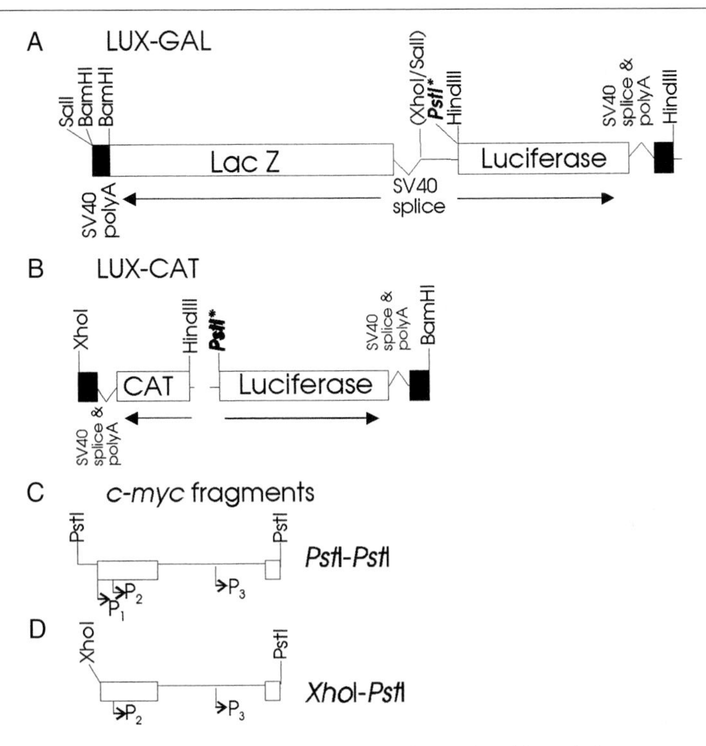

FIG. 1. Structure of the LUX-GAL and LUX-CAT plasmids and the c-*myc* promoter fragments. The unique *Pst*I cloning site in each vector is set in boldface type.

sites, *Hin*dIII and *Pst*I, are used to insert a promoter of interest in this vector (Fig 1B).

Construction of Luciferase/β-Galactosidase Vector

pGEM4Z is used as a backbone for the LUX-GAL vector.[11] The luciferase coding sequence along with SV40 small t intron and the SV40 early polyadenylation signal are excised from CMV-LUX using *Hin*dIII–*Bam*HI digestion.[14] This 2.3-kb fragment is ligated into the *Hin*dIII site of the pGEM4Z using a *Bam*HI–*Hin*dIII adaptor. The *E. coli LacZα* (β-galactosidase) fragment is obtained from an SV40 β-galactosidase vector using *Xho*I–*Sal*I digestion.[16] This yields a 3.7-kb fragment that contains the *E. coli LacZα* gene along with the SV40 16s/19s splice donor/acceptor sequences

[16] G. R. MacGregor and C. T. Caskey, *Nucleic Acids Res.* **17,** 2365 (1989).

upstream of the *LacZα* coding sequences and the SV40 polyadenylation sequence downstream of the *LacZα* coding sequences. This *Xho*I–*Sal*I fragment is ligated into the *Sal*I site of the pGEM4Z vector containing the luciferase gene such that *LacZα* and the luciferase gene are a head-to-head orientation (Fig. 1A). A unique *Pst*I site present between these two reporter genes is used to insert promoter fragments of interest in this vector.

Insertion of c-*myc* Promoter Fragment into LUX-GAL and LUX-CAT Vectors

As c-*myc* has been shown to have bidirectional transcriptional activity, we decided to use this promoter for our experiments.[9,10] A 4638-bp *Pst*I–*Pst*I genomic c-*myc* fragment (which includes c-*myc* upstream regulatory elements, the first exon, first intron, and part of the second exon) is ligated in the *Pst*I site of both the LUX-GAL and LUX-CAT vectors. This fragment contains all three c-*myc* promoters (P_1, P_2, and P_3; Fig. 1C). We have also generated a 2244-bp *Xho*I–*Pst*I fragment of c-*myc* by cutting *Pst*I–*Pst*I with *Hin*dIII and *Xho*I and ligating with a *Hin*dIII–*Xho*I adaptor. This *Xho*I–*Pst*I fragment of c-*myc*, which contained the c-*myc* P_2 and P_3 promoters but not the P_1 promoter, is also inserted into the LUX-GAL and LUX-CAT vectors (Fig. 1D). In each vector, the c-*myc* fragments are placed in both orientations. This enables us to study the effect, if any, of the type of reporter gene on the transcriptional activity from each strand of the promoter. Thus a total of eight plasmids is used in the experiments described below. All plasmids are purified by double banding in cesium chloride.

The plasmids are linearized prior to transfection using the *Bam*HI site for the LUX-CAT vector and *Sal*I for the LUX-GAL vectors. This is done to prevent any possibility that transcripts initiating from one direction of the promoter may read past the polyadenylation sequence and possibly transcribe the opposite strand of the other reporter. This could theoretically affect the level of the mRNA from the other reporter.

Transient Transfection Assays

Plasmids containing various c-*myc* promoter fragments are transiently transfected in both COLO320HR cells and also in a number of Burkitt lymphoma cell lines including P3HRI, BJAB, and AG876. Purified plasmid (10–20 μg) is transfected into 5–7 million cells by electroporation. In the LUX-CAT transfections, 2 μg of CMV–β-galactosidase plasmid (gift of G. MacGregor[16]) is cotransfected as an internal control and all results are standardized against this control to correct for variability in transfection efficiency. Similarly, a plasmid expressing a bacterial luciferase *luxAB* fusion gene under the control of a CMV promoter is used as an internal control for the LUX-GAL transfections.[17]

Luciferase and β-Galactosidase Assays

The cells are in general, harvested 48 hr posttransfection by washing twice with cold phosphate-buffered saline (PBS). The cell pellet is then resuspended in 100–200 μl of lysis buffer [0.1 *M* potassium phosphate with 2 m*M* dithiothreitol (DTT, pH 7.8)]. The cells are lysed by three freeze–thaw cycles, transferred to a microcentrifuge tube, and centrifuged to clarify the supernatant. Ten to 100 μl of the supernatant is used for luciferase and β-galactosidase assays, which are both done using luminescent assays.[18,19]

The luciferase assay is performed as described in the Promega (Madison, WI) Technical Bulletin 101.[20] This assay uses the ability of luciferase to interact with coenzyme A to form luciferyl-CoA. This interaction improves the reaction kinetics of light emission by luciferase. The light, instead of being emitted as a flash of about 300 msec, is emitted over a prolonged period of time (up to 30–40 min). Light emission is constant over the first few minutes and a much larger number of photons is generated by the same amount of luciferase, thus increasing the sensitivity of the assay.

For the luciferase assay 10–100 μl of the lysate is transferred to luminometer cuvettes [LKB 1251 luminometer (Pharmacia, Piscataway, NJ) or ILA912 (Tropix, Bedford, MA)]. The reaction buffer (20 m*M* Tricine, 530 μM ATP, 2.5 m*M* magnesium sulfate, 33.3 m*M* DTT, 270 μM coenzyme A, 470 μM potassium luciferin, and 0.1 m*M* EDTA) is made at room temperature and the pH is adjusted to pH 7.8. Potassium luciferin (Analytical Luminescence, San Diego, CA) and coenzyme A, lithium salt, are added to this buffer after the pH has been adjusted to pH 7.8. The buffer is then stored at $-70°$ in 10-ml aliquots. At the time of the assay, the reaction buffer is brought to room temperature. The cuvettes with the cell extracts are then loaded into either the LKB1251 or the ILA912 Tropix luminometer. The luminometer pump is primed with the reaction buffer. Five hundred microliters of this luciferin-, ATP-, and coenzyme A-containing reaction buffer is then injected into each cuvette, using the automatic injection device attached to the luminometer. After a 5-sec delay postinjection, light emission is quantitated for between 5 and 20 sec.

[17] G. Kirchner, J. L. Roberts, G. D. Gustafson, and T. D. Ingolia, *Gene* **81,** 349 (1989).

[18] K. V. Wood (ed.), "Recent Advances and Prospects for Use of Beetle Luciferases as General Reporter Vectors," pp. 543–553. John Wiley & Sons, Chichester, U.K., 1992; P. Stanley and L. Kricka (eds.), "Bioluminescence and Chemiluminescence: Current Status."

[19] V. K. Jain and I. T. Magrath, *Anal. Biochem.* **199,** 119 (1991).

[20] Anonymous, *Promega Tech. Bull.* **101,** 1 (1990).

β-Galactosidase assays are done primarily as described.[19] The chemiluminescent substrate AMPGD (Tropix) is used to quantitate β-galactosidase activity. Between 5 and 20 μl of the cell extract is placed in the luminometer cuvettes. Five hundred microliters of reaction buffer containing 100 m*M* sodium phosphate, 1 m*M* magnesium chloride, and AMPGD (10 μg/ml) is added to each cuvette. The cuvettes are gently mixed and incubated in a dark room at room temperature for 60 min. They are then loaded into the luminometer. Five hundred microliters of 0.2 *M* sodium hydroxide with 10% (v/v) Emerald enhancer (Tropix) is injected into each cuvette. This injection raised the pH of the reaction to more than pH 12.0, thus terminating the β-galactosidase reaction and also substantially increasing light emission. The Emerald enhancer is a water-soluble polymeric quaternary salt poly(benzylmethylvinylbenzyl)ammonium chloride with 0.1% (v/v) fluorescein. This enhancer increases the quantity of light emitted by over 100-fold. Light is quantitated for 5 sec with a 2-sec delay after injection. A β-galactosidase standard curve is used to confirm the linearity of the assay.

The CAT assays are done using the liquid scintillation counting method.[21] Bacterial luciferase catalyzes the oxidation of reduced $FMNH_2$ by fatty aldehyde, using molecular oxygen. Light is emitted during this reaction. Bacterial luciferase assays are performed by measuring decanal-dependent luciferase activity in the extracts. FMN, NADH, and NAD(P)H : FMN oxidoreductase (Boehringer-Manheim Biochemicals, Indianapolis, IN) are added to provide a source of reduced $FMNH_2$ for the bacterial luciferase reaction.[22]

Sensitivity of Reporter Gene Assays

Firefly luciferase was the most sensitive of the above four reporter proteins in our assays. There was no background luciferase activity and the assay was linear over six orders of magnitude. β-Galactosidase could also be detected sensitively by the AMPGD-linked chemiluminescent assay. Unfortunately, there was a significant amount of endogenous β-galactosidase activity in most cell lines that precluded the detection of small amounts of transfected *LacZα* activity. The endogenous β-galactosidase activity varied significantly between cell lines and also increased with increased density and age of the cell culture. The CAT assays were about three orders less sensitive than the firefly luciferase assays. They required radioactive compounds and also took longer to complete. Bacterial luciferase was the

[21] B. Seed and J.-Y. Sheen, *Gene* **67,** 271 (1988).

[22] C. Koncz, O. Olsson, W. H. R. Langridge, J. Schell, and A. A. Szalay, *Proc. Natl. Acad. Sci. U.S.A.* **84,** 131 (1987).

least sensitive of the four reporters. It could be detected only in the presence of a strong promotor and could be used only to control for transfection efficiency. The reagents for this assay [especially NAD(P)H : FMN oxidoreductase] were quite expensive. It did have the advantage of being a bioluminescent assay. Thus a strategy using firefly luciferase, β-galactosidase, and bacterial luciferase permits the detection of all three reporters with a luminometer and without the need for radioactive compounds.

Bidirectional Transcription from c-*myc* Promoters

We observed bidirectional transcriptional activity from both the *Pst*I–*Pst*I and the *Xho*I–*Pst*I c-*myc* promoter fragments in both the LUX-CAT and the LUX-GAL vectors. The data from the LUX-GAL vectors are discussed here. Although the relative strength of the sense to antisense transcription from each promoter fragment was independent of the reporter gene used, the ratio of the sense to antisense transcript was dependent, to some extent, on the context of c-*myc* promoters with the reporter genes (Fig. 2). There was consistently more sense transcription than antisense transcription from both the c-*myc* promoter fragments. With luciferase as the reporter gene, the ratio of sense to antisense ranged from 1.5 : 1. With β-galactosidase reporters the sense-to-antisense ratio was on the order of 3 : 1. Similarly, with the *Xho*I–*Pst*I c-*myc* promoter fragment, the ratio of sense to antisense ranged from 1.3 : 1 with luciferase and 2.1 : 1 with β-galactosidase (Fig. 2).

It is not possible to compare directly in absolute terms the transcription strength of a promoter using two different reporter proteins. There are intrinsic differences in mRNA half-life, translational efficiency, and protein half-life among various reporter molecules that can significantly alter the total amount of reporter protein measured. Hence, when comparing sense to antisense transcription, these dual reporter vectors allow one only to consider the relative and not the absolute abundance of transcription in each direction. It would require reporter genes that are very similar to each other to be able to correlate reporter protein level directly with mRNA transcription. One possible pair of such reporter genes may be the green fluorescent protein (GFP) and its red-shifted mutant (see below).[24,25]

[23] B. Sugden and N. Warren, *Mol. Biol. Med.* **5,** 85 (1988).

[24] M. Chalfie, Y. Tu, G. Euskirchen, W. W. Ward, and D. C. Prasher, *Science* **263,** 802 (1994).

[25] S. Delagrave, R. E. Hawlin, C. M. Silva, M. M. Yang, and D. C. Youvan, *Biotechnology* **13,** 151 (1995).

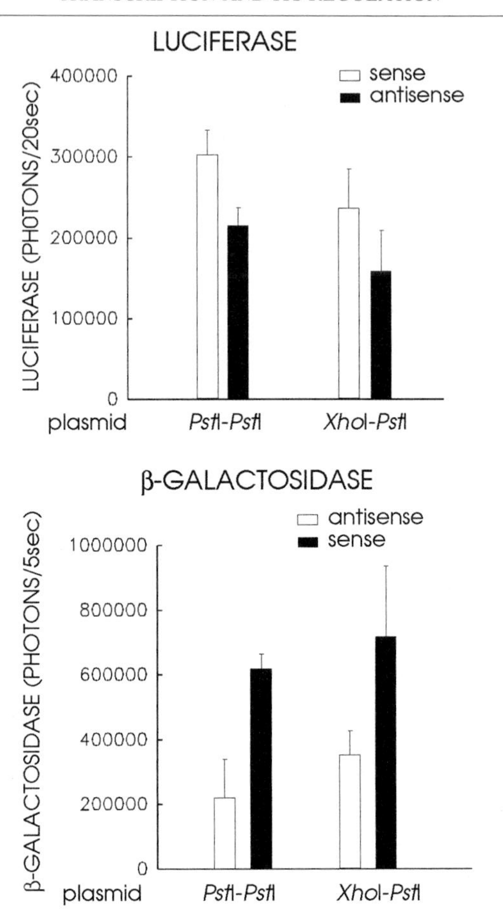

FIG. 2. Results of transient transfection experiments using c-*myc* promoter fragments. There is more sense to antisense transcription from both fragments with both the reporter genes. Experiments were done in duplicate. Means ± SEMs are shown for a representative experiment.

Stable Transfectants with Double Reporter Vectors

Transient transfection experiments have numerous potential problems that can result in variability in promoter activity as measured by reporter gene assays. These include variability in transfection efficiency and differences in the growth state of the transfected cell or its external milieu, which could influence the transcriptional activity of the promoter of interest. In addition, the transient transfection procedure itself exposes the cells to unnatural stimuli, which may include calcium phosphate, various cationic lipids (e.g., Lipofectin), or an electric shock for electroporation that may alter the transcriptional activity of the promoter of interest. It is difficult to control all of these variables precisely in different experiments, resulting in an increase in interassay variability.

Stable transfections of dual reporter vectors containing promoter fragments of interest overcome many of these problems. Although they take longer to establish, they then provide a consistent and reliable source with which to address questions related to the effect of various external agents (including various cytokines, drugs, etc.) on the bidirectional activity of these promoters.

To establish stable cell lines expressing the LUX-GAL vector driven by c-*myc* (*Xho*I–*Pst*I) promoter fragments, 10 μg of each plasmid is cotransfected with pHEBO,[23] a hygromycin resistance-encoding plasmid, into COLO320 (a colon cancer cell line) by electroporation. Forty-eight hours posttransfection, the cells are diluted to a final concentration of 10^4 cells/ml in RPMI 1640 medium containing hygromycin (400 μg/ml) and plated in 24-well plates. Stable clones are observed 10–14 days after transfection and are expanded in hygromycin (400 μg/ml)-containing medium. The activity of the c-*myc* promoter in each stably transfected clone is determined by lysing 10^5 cells in 100 m*M* potassium phosphate (pH 7.8) buffer with 1% (v/v) Triton X-100 and doing luciferase assays. Each clone is assayed at least two times. β-Galactosidase assays are done using a chemiluminescent assay as described above.

Again, we observed a significantly greater transcription in the sense direction as compared to the antisense direction for the *Xho*I–*Pst*I c-*myc* promoter fragment. This ratio of sense to antisense transcription was present regardless of the choice of the reporter (firefly luciferase or β-galactosidase; Fig. 3). This confirmed the findings of the transient transfection experiments. In addition, these cell lines were then used to study the effect of agents such as sodium butyrate, dimethyl sulfoxide (DMSO), and various cytokines on the ratio of sense to antisense transcription from these *myc* promoter fragments (data not shown).

Vectors Containing Reporters with Secreted Protein Products

The traditional reporter gene vectors, such as the ones described above, express reporter proteins that are localized intracellularly. This requires the cells to be lysed before the reporter gene assays can be performed. This makes it difficult to do kinetic studies in which reporter activity is measured sequentially over a period of time. A number of reporter genes

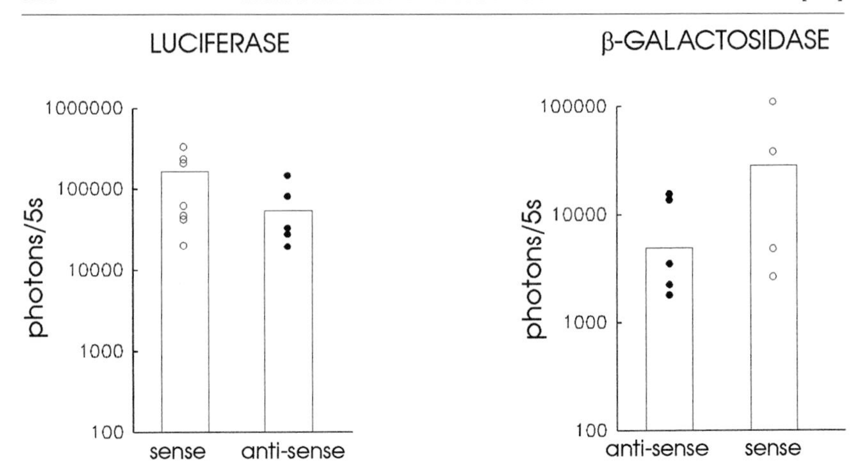

FIG. 3. Result of stable transfection of LUX-GAL driven by *Xho*I–*Pst*I into COLO320 cells. Stable clones were analyzed for both luciferase and β-galactosidase activity. There is more sense to antisense transcription in each clone analyzed.

have been described that encode proteins that are secreted into the culture medium and can be assayed without the need for lysing the transfected cells. These include a secreted version of alkaline phosphatase (SEAP), human growth hormone (HGH), and bovine growth hormones (BGH). Patel and Kurkinen have described two dual reporter vectors that contain HGH and BGH or HGH and SEAP in head-to-head orientations.[12] Promoter fragments can be cloned between these reporter genes, using unique *Bam*HI and *Xba*I sites for the HGH–BGH and HGH–SEAP vectors, respectively. In addition, a polylinker has been provided in the HGH–BGH dual reporter vectors, making additional sites (including *Sca*I, *Not*I, *Sfi*I, and *Spe*I) available for promoter cloning in this vector.

Both these vectors have been tested by transfection into HepG2 and Cos-1 cell lines, respectively. Both HGH and BGH can be assayed by commercially available radioimmunoassay or ELISA techniques. Secreted alkaline phosphatase can be assayed sensitively by commercially available chemiluminescent technology (using dioxetane substrates such as AMPPD, similar to the technique used for β-galactosidase). These vectors permit repeated reporter gene quantitation to be done from the same transfected cells, thus providing kinetic information about the rate of transcription in each direction. In addition, the cells can still be used for additional tests that require RNA or DNA extraction.

Dual Reporter Vectors for Use in Bacteria

A number of bacterial genes have been found to have bidirectionally active promoters. These include the promoters of the *nodA* and *nodD* genes of the *Rhizobium* species, which overlap each other and are transcribed in opposite directions. Parry *et al.* have developed a dual reporter vector, pSPV4, utilizing two reporters [β-galactosidase (*LacZ*α) and β-glucuronidase (*Gus-A*) genes] that are expressed well in plant tissues, in a head-to-head orientation.[6] pSPV4 is constructed on a pUC21 backbone and contains a portable ribosome-binding site derived from the *E. coli cat* gene 5′ of each reporter gene. There is a multiple cloning site with five unique restriction sites (*Hin*dIII, *Sph*I, *Kpn*I, *Eco*RI, and *Xba*I) that can be used for cloning of promoter fragments of interest.

Various fragments of the *nod* gene promoter were ligated into pSPV4 and the resultant vectors were electrotransformed into the *Rhizobium* species and stable transfectants were made. Both β-galactosidase and β-glucuronidase activities could be measured in the transformed cells, confirming bidirectional transcription from the *nod* gene promoters. pSPV4 vector has a broad host range (including numerous gram-negative bacteria such as *Pseudomonas, Agrobacterium,* and *Rhizobium*). Thus this vector may be used to detect gene expression in plant species that can be infected by the above-cited bacteria.

Potential Problems with Dual Reporter Vectors

There are a number of theoretical and practical problems with dual reporter vectors that are intrinsic to the choice of the reporter genes. Some of these can result in significant differences in the levels of the two reporter proteins, which may be unrelated to the rate of transcription from the promoter and can potentially give fallacious results. A few of these include (1) variability in messenger RNA half-life between the two reporter genes present in the same vector, (2) variability in translational efficiency between the two reporter genes, (3) differences in the rate of degradation and half-life of the two reporter genes, and (4) differences in sensitivities of the assays available for each reporter gene.

As one is dealing with two different reporter molecules with different mRNA and protein half-lives and different posttranslational processing and degradation, it is not possible to compare in absolute terms the relative abundance of the sense and antisense transcripts from reporter gene assays. One may need to do nuclear runon experiments with appropriate probes to actually estimate the transcriptional activity of the promoter in each direction.

Also, there might be interaction between the reporter gene itself and the promoter of interest. These include possible inhibitory sequences within the reporter gene coding region that may affect the activity of the promoter, resulting in certain promoters working better with certain reporter genes and not with others.

Ideally, one should be able to quantitate both reporter molecules used in a dual reporter vector at similar levels of sensitivity. As has already been discussed, this in general is not the case. Also, there may exist endogenous background activity for certain reporter genes such as alkaline phosphatase or β-galactosidase because of cellular synthesis of these enzymes. Because of this background activity, it may become difficult to detect low levels of reporter gene activity. This can confound the analysis of weak promoters, transcription from which may be detectable in one direction by the more sensitive reporter gene in a dual reporter vector (e.g., firefly luciferase) but not in the other direction owing to a less sensitive reporter (e.g., CAT).

To obviate many of these pitfalls, it may be prudent to insert promoter of interest in both orientations in a dual reporter vector. One should see similar ratios of sense to antisense transcription regardless of the choice of reporter. Another approach to improving the reliability and utility of dual reporter vectors would be to use reporter gene constructs that do not suffer from many of the drawbacks listed above. One such potential reporter gene is discussed below.

Use of Green Fluorescent Protein Mutants in Dual Reporter Vectors

Green fluorescent protein (GFP) is a small protein that is present in numerous fluorescent marine invertebrates such as *Aequorea victoria* and *Renilla reniformis*. The GFP from *A. victoria* has been cloned and has been expressed in bacterial, nematode, insect, and mammalian cells under heterologous promoters.[24] The GFP has a unique chromophore made of a covalently attached tripeptide that apparently can be synthesized in most cell types. Hence introduction of vector containing GFP reporter allows the synthesis of a fluorescent protein in most cell types and permits monitoring of reporter gene activity without the need for addition of any external agent. All that is required is a device to detect fluorescence (e.g., a fluorescent microscope, flow cytometers, or a digital imaging spectroscopy system). This reporter is unique in that it can be measured in living cells on a sequential basis. It does not appear to damage cells in which it is expressed and can be inserted in embryonic cells to allow for determination of cell fate. There is essentially no cellular background for this fluorescent reporter.

Red-shifted excitation mutants of the GFP have been synthesized.[25] The GFP has an excitation peak of 395 nm (blue light) and an emission peak at 510 nm (green light). The red-shifted GFP mutants have an excitation peak of 490 nm, with an emission peak of 505 nm. Thus, it should be possible to distinguish these two GFPs in an individual cell by using a fluorescent microscope or a cytofluorometer and adjusting the excitation filter for each reporter.

Although dual reporter vectors utilizing GFP mutants have not been published, it is only a matter of time before they become available. These two reporter genes would differ only by a few amino acids and would otherwise be identical.[25] This would obviate most of the problems with dual reporter vectors listed above. These reporters would have similar mRNA and protein half-lives, similar sensitivities and background, and most probably similar interaction with different promoters. The reporters could be detected noninvasively, without the addition of any exogenous agent, on living cells. Sequential and kinetic analysis of gene transcription from promoters of interest could be done. The effect of drugs and cytokines could be assessed *in vivo* in cells stably transfected with GFP-expressing vectors under desired promoters.

Conclusion

Bidirectional transcription has been found to occur from both eukaryotic and prokaryotic promoters. Dual reporter vectors provide a versatile way to analyze the divergent transcription seen from these promoters. A number of dual reporter vectors have been described using currently available reporter genes. These vectors, although useful, do suffer from some systemic deficiencies. The availability of newer reporter proteins such as the GFP and its mutants may allow new vectors to be developed that overcome most of these defects.

Acknowledgments

I thank Dr. Ian Magrath, who made this work possible, and numerous colleagues, including Tom Shima, Michael Chirigos, Thuan Nguyen, Anuj Gupta, and Niti Bhatia, who assisted with this work.

[27] Gene Identification Using the Yeast Two-Hybrid System

By CHANG BAI and STEPHEN J. ELLEDGE

Protein–protein interactions have attracted much attention because they form the basis of a wide variety of biochemical reactions. The identification of proteins that interact with a known protein is an essential aspect of the elucidation of the regulation and function of that protein. This interest has stimulated the development of a number of biochemical and genetic approaches to identify and clone genes encoding interacting proteins, including coimmunoprecipitation, copurification, cross-linking, and direct expression library screening using proteins as probes. However, the development of the yeast two-hybrid system appears to have had the greatest impact on interaction cloning methodology.

The yeast two-hybrid system was devised to identify genes encoding proteins that physically associate with a given protein *in vivo*. This is a versatile and powerful method that is applicable to most, if not all, proteins once their genes have been isolated. In contrast to biochemical methods detecting protein–protein interaction, this system is based on a yeast genetic assay in which the interaction of two proteins is measured by the reconstitution of a functional transcription activator in yeast.[1,2] This method not only allows identification of proteins that interact, but also can be used to define and/or test the domains/residues necessary for the interaction of two proteins.[3] In the last few years, a large number of genes from a variety of studies have been identified using this method,[4–6] including many cell cycle regulators that have contributed significantly to our understanding of the eukaryotic cell cycle.

This chapter provides a practical guide to the yeast two-hybrid system, emphasizing recent improvements of this method. There are several different designs for the actual hybrid proteins and selection utilized in two-hybrid screens. This chapter focuses on derivatives of the *GAL4*-based system (for a description of LexA-based systems see Ref. 7). Owing to the limitation of space, general methods of yeast genetics, for which several comprehensive manuals are available,[8,9] are not covered in detail in this chapter. As with all protocols, the one presented here is not completely optimized and readers are encouraged to try alternative methods to improve performance.

The basis of the two-hybrid system relies on the structure of particular transcription factors that have two physically separable domains: a DNA-binding domain and a transcription activation domain. The DNA-binding domain serves to target the transcription factor to specific promoter sequences (UAS, upstream activating sequence) whereas the activation do-

[1] S. Fields and O. Song, *Nature (London)* **340,** 245 (1989).

[2] C.-T. Chien, P. L. Bartel, R. Sternglanz, and S. Fields, *Proc. Natl. Acad. Sci. U.S.A.* **88,** 9578 (1991).

[3] B. Li and S. Fields, *FASEB J.* **7,** 957 (1993).

[4] J. W. Harper, G. R. Adami, N. Wei, K. Keyomarsi, and S. J. Elledge, *Cell* **75,** 805 (1993).

[5] H. Toyoshima and T. Hunter, *Cell* **78,** 67 (1994).

[6] S. Fields, *Trends Genet.* **10,** 286 (1994).

[7] J. Gyuris, E. Golemis, H. Chertkov, and R. Brent, *Cell* **75,** 791 (1993).

[8] F. M. Ausubel, R. Brent, R. E. Kingston, D. D. Moore, J. G. Seidman, J. A. Smith, and K. Struhl (eds.), "Current Protocols in Molecular Biology." Wiley, New York, 1994.

[9] C. Guthrie and G. R. Fink, *Methods Enzymol.* **194,** (1991).

main serves to facilitate assembly of the transcription complex allowing the initiation of transcription. The fact that a functional transcription factor can be reconstituted through noncovalent interaction of two independent hybrid proteins containing either a DNA-binding domain or an activation domain constitutes the basis of the two-hybrid approach.[1] The hybrid proteins are normally transcriptionally inactive alone or when coexpressed with a noninteracting hybrid protein. However, if when coexpressed they associate via the interaction between the two fusion protein partners, they become active, causing the expression of a reporter gene driven by the specific UAS for the DNA-binding domain.[1] A schematic representation of the two-hybrid genetic selection by activation of transcription of *HIS3* and *lacZ* genes is shown in Fig. 1. The plasmid encoding the fusion protein with the *GAL4* DNA-binding domain is often referred to as the bait plasmid.

Since the emergence of the two-hybrid approach in 1989, a number of improvements have been incorporated into this system that have greatly increased its usefulness; several are listed below. The most significant im-

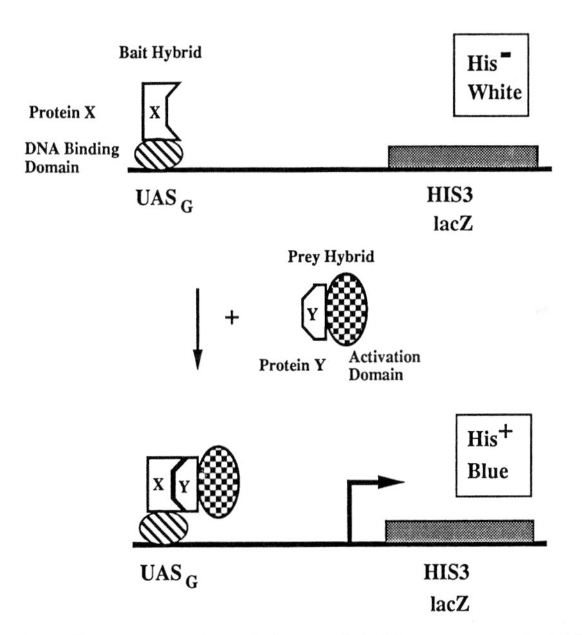

FIG. 1. A schematic representation of the two-hybrid system: two hybrid proteins are generated. The bait hybrid consists of protein X fused to a DNA-binding domain while the prey hybrid consists of protein Y fused to an activation domain. Neither of these alone are able to activate transcription of *lacZ* or *HIS3*. However, if interaction of protein X and Y occurs, a functional transcription activator is generated and results in the transcription of reporter genes that confer the His^+ and blue phenotype on the host cells.

provement has been the conversion of the assay from a color-based screen to a nutritional selection via inclusion of the *HIS3* gene as a transcription reporter in addition to *lacZ*.[10] This allows much larger libraries to be efficiently screened on many fewer plates. New methods for the elimination of false positives have also had a large impact on the efficacy of the system. The mechanistic basis for false positives is not understood, but they are operationally defined as clones that activate multiple unrelated bait proteins. This has served as the basis of a mating assay to eliminate these clones (described below).[4] Improvements in the specificity were also accomplished by the simultaneous use of two different reporters that share minimal sequence overlap in their promoters.[10,11] Finally, advances have been made in the generation of large representative cDNA libraries by the development of highly efficient λ cloning vectors, λACT[10] and λACT2, that also allow direct conversion into plasmid via cre-lox site-specific recombination. λACT2 (S. Elledge, unpublished results, 1995) has an epitope tag and allows directional libraries using unique *Eco*RI and *Xho*I restriction sites. A commercially available kit exists for preparing *Eco*RI–*Xho*I cDNA (Stratagene, La Jolla, CA).

Key Reagents

A flowchart for cloning by the two-hybrid system is outlined in Fig. 2. To search for interactions with a known protein, a yeast reporter strain, a DNA-binding domain vector, and a cDNA library in an activation domain vector are essential, as are yeast media, basic yeast manipulation techniques, and some patience.

Vectors

A number of different DNA-binding domain and transcription activation domain vectors have been successfully employed in this system.[12] The most extensively used vectors are *GAL4* based. Examples are the *GAL4* DNA-binding domain vector pAS1 and pAS2 (see restriction map in Fig. 3) and the *GAL4* activation domain vector pACT1 and pACT2 (Fig. 3). However, it is possible to combine vectors and strains from different source as long as they are compatible with each other with regard to the selectable markers and the UAS for the reporters.

[10] T. Durfee, K. Becherer, P. L. Chen, S. H. Yeh, Y. Yang, A. E. Kilburn, W. H. Lee, and S. J. Elledge, *Genes Dev.* **7,** 555 (1993).

[11] P. Bartel, C.-T. Chien, R. Sternglanz, and S. Fields, *Biotechniques* **14,** 920 (1993).

[12] S. Fields, *Methods* **5,** 116 (1993).

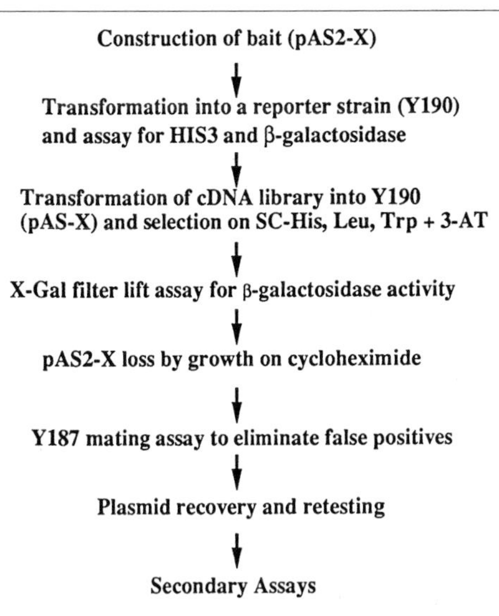

FIG. 2. A flowchart describing the steps employed in a typical two-hybrid screen and false-positive detection assays.

Strains

The reporter strain must contain a reporter gene that is under the control of the UAS corresponding to the DNA-binding domain vector. It should also carry nonreverting mutations in the genes corresponding to those present on the two-hybrid vectors. A number of reporter strains are available in which bacterial gene *lacZ* is the reporter. Some strains (e.g., Y190) also carry the yeast biosynthetic gene *HIS3* as a selectable marker. Please refer to other reviews for other vectors and reporter strains.[12]

Strain Y190 is *MAT***a** *gal4 gal80 his3 trp1-901 ade2-101 ura3-52 leu2-3,-112 + URA3::GAL-lacZ, LYS2::GAL(UAS)-HIS3 cyh*^{r}.

cDNA Libraries

The most commonly used two-hybrid cDNA libraries are made as the *GAL4* transcription activation domain fusion libraries. A large number of libraries from different tissues and organisms have been published and several more are available commercially, but it may be necessary to construct your own library. A description of cDNA library construction is beyond the scope of this chapter, but has been covered previously.[13]

[13] J. T. Mulligan and S. J. Elledge, *in* "Molecular Genetics of Yeast: A Practical Approach," p. 65. Oxford University Press, Oxford, 1994.

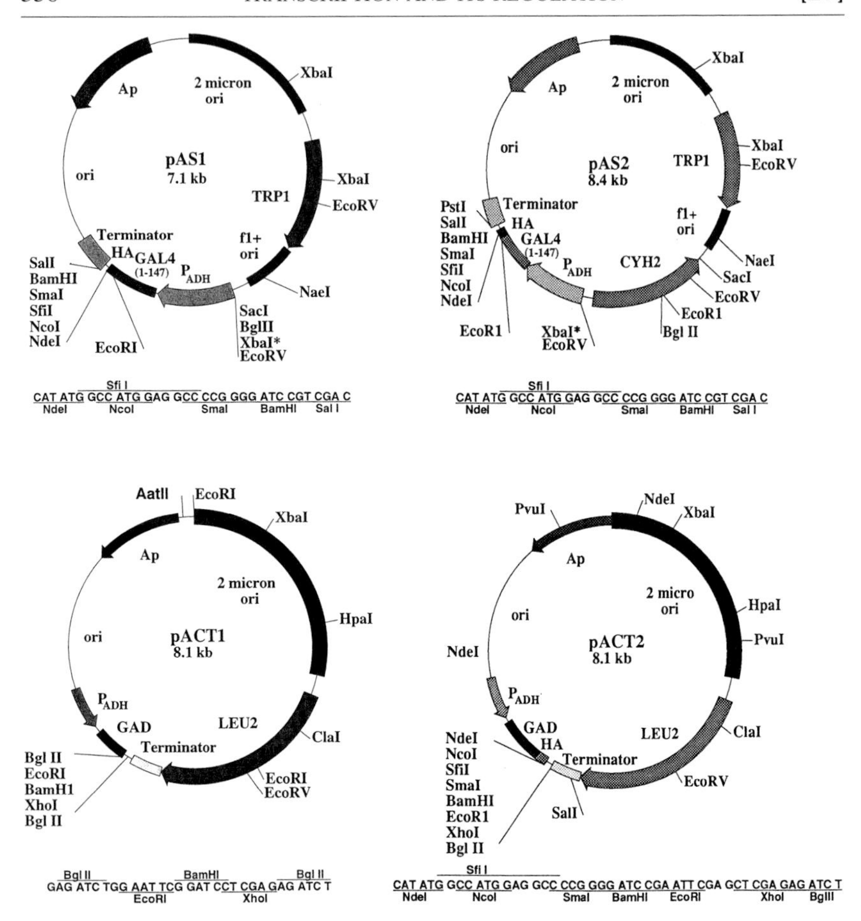

FIG. 3. The restriction maps of DNA-binding domain vectors pAS1 and pAS2, and activation domain vectors pACT1 and pACT2. The reading frames of the polylinkers are shown in coding triplets. *, Dam methylation, which blocks cutting.

Establishing Reporter Strain Expressing *GAL4* DNA-Binding Domain Fused to Gene of Interest

First, the gene to be examined must be subcloned into pAS2 or a similar *GAL4* DNA-binding domain fusion vector. Care must be taken to maintain the proper reading frame to create a hybrid protein. DNA sequencing or Western analysis can be used to verify that the proper fusion has been accomplished. The sequencing primer used for pAS is 5′ GAA TTC ATG GCT TAC CCA TAC. The pAS vectors also encodes a hemagglutinin (HA) epitope tag that allows the protein to be visualized with commercially available antibodies if present in sufficient amounts. This construct (pAS-X) must be introduced into a reporter strain such as Y190 by transformation and nutritional selection using synthetic complete (SC) medium lacking the appropriate amino acid. Minipreparation plasmid DNA is sufficient for this purpose.

Reagents and Media

LiAcTE: 100 m*M* lithium acetate, 10 m*M* Tris (pH 8), 1 m*M* EDTA; autoclave

LiSORB: 100 m*M* lithium acetate, 10 m*M* Tris (pH 8), 1 m*M* EDTA, 1 *M* sorbitol; autoclave

PEG (70%): 70% Polyethylene glycol (M_r 3350) in water; store in a tightly sealed container to prevent evaporation after autoclaving

YEPD: 10 g of yeast extract, 20 g of peptone, 20 g of dextrose per liter; autoclave

SC: Synthetic medium used for selection of yeast containing plasmids with specific nutritional markers. SC lacking a particular amino acid is referred to as *dropout medium*

10× YNB: 67 g of yeast nitrogen base without amino acids in 1 liter of water, filter sterilized; store in the dark

Dropout mixture components:

Component	Amount (mg)	Component	Amount (mg)
Adenine	800	Arginine	800
Aspartic acid	4000	Histidine	800
Leucine	2400	Lysine	1200
Methionine	800	Phenylalanine	2000
Threonine	8000	Tryptophan	800
Tyrosine	1200	Uracil	800

To make a dropout mixture, simply weigh out the different components, leaving out the ones that are going to be selected for, combine them, and grind into a fine powder with a mortar and pestle

SC-Trp plates: 870 mg of dropout mixture (minus tryptophan), 20 g of dextrose, 1 ml of 1 *N* NaOH, 20 g of agar, add water to 900 ml, and autoclave. Add 100 ml of 10× YNB just prior to pouring plates

Procedure for Small-Scale Yeast Transformation

1. Inoculate Y190 in 10 ml of YPED and grow to OD_{600} 1.
2. Wash with water and resuspend in 200 μl of LiAcTE (30° for 1 hr).

3. Add ~1 μg of pAS-X DNA to 100 μl of cells and 100 μl of 70% PEG[13a]; incubate at 30° for 1 hr.
4. Heat shock at 42° for 5 min, add 1.2 ml of water, and pellet by spinning for 5 sec in an Eppendorf centrifuge at room temperature.
5. Resuspend in 100 μl of water and plate on an SC-Trp plate and incubate at 30° for 3 days.

The transformation mixture should be plated on SC-Trp plates in the case of pAS2. Colonies should be seen after about 48 hr of incubation at 30°. It is always advisable to include a "no DNA" control during the transformation to test the reversion frequency of the yeast-selectable marker or the presence of contaminants. It should be noted that a few independent transformants need to be picked and the expression of the fusion protein should be verified by Western blotting with anti-HA antibodies available from BABCO (Berkeley, CA) if specific antisera are not available for the protein, because reversion and gene conversions do occur in the process of transformation. Occasionally, a functional fusion is made and cannot be detected with anti-HA. In this case, verify the presence of the plasmid in the strain by recovery into *Escherichia coli*, Southern blotting, or the polymerase chain reaction (PCR) with primers specific to the gene of interest.

Checking Feasibility of Using Two-Hybrid Screen for Particular Protein

Some proteins that are not transcription factors themselves activate transcription when fused to the *GAL4* DNA-binding domain. This will severely interfere with the two-hybrid system. Prior to beginning a screen, it is important to check the bait protein for *lacZ* activation and its growth properties on SC-His plates containing differing concentrations of 3-AT (3-amino-1,2,4-triazole; Sigma, St. Louis, MO). 3-AT, an inhibitor of *HIS3*-encoded IGP-dehydratase, is used because the basal level of *HIS3* expression from the reporter construct, in which the *HIS3* UAS sequences have been replaced by the GAL1-10 UAS, is sufficient to allow growth on SC-His medium. We have found that 3-AT concentrations of 25 to 50 m*M* are typically sufficient to select against the growth of strains bearing pAS2 subclones that fail to activate transcription on their own. If the construct fails to activate transcription, like most fusions, proceed to the library transformation step. If it activates transcription alone [turns blue in the 5-bromo-4-chloro-3-indolyl-β-D-galactopyranoside (X-Gal) assay or grows on

[13a] PEG can be briefly heated in a microwave oven to help it go into solution if necessary.

SC-His plus 50 m*M* 3-AT], it cannot be used in this version of the two-hybrid system. Usually the X-Gal assay is the most sensitive.

Reagents and Media

SC-His, Trp + 3-AT plates: 3-AT is dissolved in sterile water as a 2.5 *M* stock, heated to 50° to solubilize if necessary, filter sterilized, and added to SC-His, Trp medium just prior to pouring plates at a concentration of 50 m*M*

Z buffer: Combine $Na_2HPO_4 \cdot 7H_2O$ (16.1 g), $NaH_2PO_4 \cdot H_2O$ (5.5 g), KCl (0.75 g), $MgSO_4 \cdot 7H_2O$ (0.25 g), adjust to pH 7.0, add water to 1 liter, and autoclave. 2-mercaptoethanol (2.7 ml) should be added just before use

X-Gal: 5-Bromo-4-chloro-3-indolyl-β-D-galactoside dissolved in DMF (*N,N*-dimethylformamide) at 100 mg/ml as a stock. Store at −20° in the dark

3-Amino-1,2,4-triazole Histidine Prototrophy Assay

Y190 (pAS-X) colonies are streaked onto SC-Trp, His + 3-AT plates and incubated at 30° for 72 hr to examine their ability to form colonies. It is advisable to include both positive and negative controls at this point.

X-Gal Colony Filter Assay to Detect β-Galactosidase Activity

This procedure is adapted from Ref. 14:

1. Label Schleicher & Schuell (Keene, NH) BA85 45-μm pore size circular nitrocellulose filters or Nytran filters[14a] with a ballpoint pen.
2. Lay the filter onto the plate of yeast colonies[14b] and allow it to wet completely. Place asymmetric orientation markers on filters and plates with India ink and a needle.
3. Lift the filter off of the plate carefully to avoid smearing the colonies and place the filter in liquid nitrogen to permeabilize the cells. Five to 10 sec is sufficient. Filters can either be submerged in the liquid nitrogen alone or placed on an aluminum foil float and submerged, which minimizes handling.
4. Carefully remove the filters from the liquid nitrogen (frozen nitrocellulose filters become brittle), allow to thaw, and place cell side up in a petri

[14] L. Breeden and K. Nasmyth, Cold Spring Harbor Symp. Quant. Biol. **50,** 643 (1985).

[14a] We observe that nitrocellulose filter becomes more brittle in liquid nitrogen than do nytran filters. However, nitrocellulose filters adhere to colonies slightly better than does nytran. Filters can be thoroughly washed with deionized water and reused.

[14b] It is best to have at least medium-size colonies with which to perform this assay.

dish that contains Whatman (Clifton, NJ) 3MM chromatography paper circles soaked with Z buffer (0.3 ml/in^2) containing X-Gal (1 mg/ml).

5. Incubate at 30° for 30 minutes to overnight for the development of color.

Library Transformation and Selection

If the Y190 reporter strain expressing the protein of interest does not activate *HIS3* or *lacZ*, this strain can now be used for screening an activation domain cDNA library.

Reagents

LiAcTE
LiSORB
LiAcPEG: LiAcTE containing 40% polyethylene glycol (M_r 3350)
Carrier ssDNA or total yeast RNA at 20 mg/ml[15,16]
SC-Trp liquid medium
YEPD liquid medium
Sc-Trp, Leu, His liquid medium
SC-Trp, Leu, His + 3-AT: 15-cm dropout plates (minus tryptophan, leucine, and histidine) with the addition of 25 or 50 m*M* 3-AT

Procedure

1. Grow the recipient strain, Y190 (pAS-X), to mid-log phase (1×10^7 cells/ml) in SC-Trp.

2. Determine the OD_{600} of the above culture and inoculate 1 liter of YPED such that in two generations the cell density becomes 1×10^7 cells/ml (OD_{600} = 0.5 to 0.8).[16a]

3. Pellet the cells and resuspend the pellets in LiAcTE: the volume is not critical because this is a wash step.

4. Pellet the cells once more, and this time resuspend the cells in 5 ml of LiSORB for every 200 ml of starting culture.

5. Incubate the cells for 30 min at 30° with shaking; the incubation should not last longer than 1 hour, at which point the cells begin to starve, thus decreasing their survival.

6. Pellet the cells as described above and resuspend in 500 μl of

[15] R. D. Gietz and R. A. Woods, *in* "Molecular Genetics of Yeast: A Practical Approach." Oxford University Press, Oxford, 1994.

[16] K. Kohrer and H. Domdey, *Methods Enzymol.* **194,** 398 (1991).

[16a] SC-Trp medium is used to select for pAS-X but YPED gives the best transformation efficiencies. The doubling time for Y190 (pAS-X) in YPED varies between 2.0 to 3.5 hr depending on the plasmid it carries.

LiSORB per 200 ml of culture. (We usually do large-scale transformations with 1 liter of cells, so that the final volume is 2.5 ml.)

7. After removing 100 μl of cells for a negative control, add 2 μg of pACT library DNA and 200 μg of yeast total RNA or single-stranded DNA (ssDNA)[16b] for every 100 μl of cells remaining (50 μg of DNA and 5 mg of total RNA carrier for 1 liter's worth of cells).

8. Mix well, then incubate for 10 min at 30° without shaking.

9. Add 900 μl of LiAcPEG for each 100 μl of cells and mix well (22.5 ml for the large-scale preparation). We usually do this in 125- or 250-ml flasks. This aids the heat shock step (see below).

10. Heat shock the tubes or the flask in a 42° water bath for 12 min.

11. At this point cells can be plated out to check the transformation frequency. Five microliters on an SC-Trp, Leu plate should give 1000 or more transformants with a good transformation. However, to plate a large number of cells per plate, we generally take our transformation mixture and add it to 500 ml of SC-Trp, Leu, His medium for an initial 1-liter culture and allow it to recover at 30° for 4 hr.[16c] At this point cells have been established as transformants. They can be pelleted, resuspended in 6 ml of SC-His,Trp, Leu liquid medium, and plated [300 μl/15-cm plate (SC-His, Trp, Leu + 3-AT)] or frozen at −70° with the addition of dimethyl sulfoxide (DMSO) to 9% (v/v) final concentration and stored for future use. Stored, small aliquots may be thawed and plated out later at an appropriate density. It is important not to have extra histidine around during plating, because it interferes with the 3-AT selection.

Screening Using X-Gal β-Galactosidase Assay

Colonies that grow after 4 to 5 days (or longer; we have waited up to 1 week in some cases) are then tested for β-galactosidase activity using the X-Gal colony filter assay (see above). Blue colonies are taken for further study and they can often be recovered directly from the filters. However, when this does not work, they can be recovered from the original plate. All positives should be struck out to single colonies and retested for β-galactosidase activity. If only a few colonies formed on SC-His, Trp, Leu + 3-AT plates, pick all these colonies and streak them on a single

[16b] We have found that total yeast RNA works more reproducibly as a carrier but more work is required to prepare than DNA. Protocols for preparing ssDNA and yeast RNA for transformation can be found in Refs. 15 and 16, respectively.

[16c] Cells in PEG are more fragile and often die when pelleted; thus the recovery step is useful. Plating directly from the PEG without the recovery step also works, but is more messy. Cells lose less than half their viability when frozen and can be stored indefinitely. This is useful because they can be thawed and plated at a optimal density for screening/selection at leisure.

plate before performing the X-Gal assay, to save filters and labor. If a larger number of plates is used, filters can be washed and reused.

The *HIS3*/3-AT selection sometimes functions as a tight selection but more often like an enrichment, depending on the bait employed. We often see many microcolonies on the original selection plates, although it has been reported that inclusion of 3-AT in the recover step eliminates these. Occasionally these microcolonies are 1% of the total Leu^+Trp^+ colonies. In most cases true positives continue to grow into large colonies while the microcolonies stop growing. The secondary X-Gal screen eliminates the vast majority of the microcolonies. The majority of blue colonies are the large His^+ colonies that grow out. An enrichment of 100-fold is useful because it allows one to screen 100 times as many colonies on a single plate, so that large libraries can be screened in only 20 large plates. We have also developed a *GAL→URA3* selection system that requires higher levels of two-hybrid activated transcription than the His selection. That strain, Y166, is available on request.

We typically use pSE1111 (*SNF4* fused to the activation domain in pACT) and pSE1112 (*SNF1* fused to the DNA-binding domain of *GAL4* in pAS1) as a positive control for our X-Gal and 3-AT resistance assays. It should be noted that pAS2 alone can activate *lacZ* weakly and is not a good negative control. pSE1112 is a better negative control. The weak activation observed with pAS2 may be due to residual transcriptional activation activity of amino acids 1–147 of the Gal4 protein used in pAS1 and pAS2 (S. Johnston, personal communication, 1995). This weak activation capacity appears to be eliminated when genes are cloned into pAS1 and pAS2.

Eliminating False Positives

The mechanistic basis of false positives is not well understood. It has been observed that the use of two different *GAL4*-dependent promoters and reporters, *lacZ* and *HIS3*, with little sequence overlap decreases the frequency of false positives. This has led to the idea that some false positives might be dependent on the sequences of a particular promoter context. The best way to determine whether a particular library plasmid encodes a bona fide positive is to test its ability to activate the original bait and several other unrelated bait plasmids as a negative control. This can be cumbersome, requiring the recovery of many plasmids and transformation into several strains. A genetic assay has been developed to facilitate this process.[4] The basis of this assay is the loss of the original bait plasmid, leaving a strain containing only the library plasmid. The plasmids used in the two-hybrid system are lost at a low frequency during growth in the absence of nutritional selection. The mating strain can be made by screening for loss of the pAS derivative by replica plating for inability to grow on SC-Trp, or by the use of a negative selectable marker incorporated into pAS2. The strain Y190 is resistant to cycloheximide (2.5 μg/ml) due to a mutation in the *CYH2* gene. This is a recessive drug resistance. When a plasmid, (e.g., pAS2) carrying the wild-type *CYH2* gene is in the strain, cells become sensitive to cycloheximide. If one begins with pAS2-X, it is a good idea to streak the colonies out first on SC-Leu before streaking on cycloheximide medium, to allow plasmid loss and dilution of the *CYH2* gene product. Loss of the pAS2 derivative can be achieved by streaking colonies onto SC-Leu plates containing cycloheximide (2.5 μg/ml). The colonies that grow should be Trp^-, but they should be checked for loss of the *TRP* marker, just to be safe and to avoid *CYH2* gene conversion events. This plasmid loss allows one to check for plasmid dependency of *lacZ* activation as well as generating a strain that contains only the library plasmid, facilitating plasmid recovery into bacteria (see below). This strain is mated to strains of the opposite mating type (Y187) that harbor pAS2 clones containing unrelated fusions. Because the library plasmid containing strain is Leu^+Trp^- and the tester strains are Leu^-Trp^+, the diploids resulting from the mating can be selected for on SC-Trp, Leu plates. The genotype of Y187 is *MAT*α*gal4 gal80 his3 trpl-901 ade2-101 ura3-52 leu2-3,-112 URA3::GAL-lacZ*. The unrelated fusions we generally use are in pAS1 or pAS2(Leu^-Trp^+) and include *CDK2*, *SNF1*(pSE1112), lamin, and p53 (a gift of S. Fields, University of Washington, Seattle, WA), although other sets of unrelated fusions may be desirable. The resulting diploids, selected by growth on SC-Trp, Leu, can be immediately tested for β-galactosidase activity in the filter screen assay. Colonies that activate *lacZ* expression significantly above background levels (pSE1112 alone) are likely to contain library plasmids encoding false positives that nonspecifically activate fusion and should therefore be discarded.

Media and Reagents

SC-Leu plates and liquid medium
SC-Leu*Cyh*: SC-Leu plus cycloheximide (2.5 μg/ml) plates. Cycloheximide (dissolved in water and filter sterilized) should be added just prior to pouring plates
SC-Trp plates
YEPD plates
SC-Trp, Leu plates
Replica-plating velvets

Procedure for Selecting Loss of DNA-Binding Domain Plasmid

1. Streak the positive clones on an SC-Leu plate and incubate for 48 hr (optional).
2. Pick colonies from the above plates and streak them on SC-Leu containing cycloheximide (2.5 μg/ml); incubate until colonies appear.
3. Replica the plate to SC-Trp and SC-Leu plates and incubate for 48 hr (optional).
4. Colonies growing on SC-Leu only are to be picked for further analysis. Clones that activate *lacZ* by itself are false positives and should be eliminated.

Procedure for Mating Assay to Eliminate False Positives

1. Make a series of wide parallel streaks of individual Y190 derivatives from above in a straight line crossing the length of a SC-Leu plate. Six evenly spaced streaks can be accommodated on a single plate.
2. Do the same for the false-positive detector Y187 derivatives on an SC-Trp plate.
3. Replica plate the above two plates to a common YEPD plate at 90° angles so that cells from the SC-Leu plate cross over with cells from the SC-Trp plate; incubate for 4 hr to overnight to allow mating.
4. Replica plate to an SC-Trp, Leu plate and incubate for 2–3 days. Colonies of diploid cells will grow in patches at the intersection of the two streaks. Care must be taken to label the mating plates properly to ensure that there is no confusion in the identity of the streak. Replica plating can be confusing.
5. Perform the X-Gal filter assay as described earlier.

The more specificity fusions with which to test the positives, the more trustworthy the significance of the positives. Gather bait constructs from others as false-positive detectors. The above strains eliminate many but not all false positives.

Recovery of Plasmids from Yeast into *Escherichia coli*

Once putative positive two-hybrid clones are obtained, the library plasmid can be recovered through bacterial transformation using DNA isolated from these clones.

Reagents

Yeast lysis solution: 300 m*M* NaCl, 10 m*M* Tris, 1 m*M* EDTA (pH 8), 0.1% (w/v) SDS

PCI: 49% (v/v) phenol plus 49% (v/v) $CHCl_3$ plus 2% (v/v) isoamyl alcohol. Phenol is saturated with 10 m*M* Tris, 100 m*M* NaCl, 1 m*M* EDTA (pH 8)
$CHCl_3$
TE: 10 m*M* Tris (pH 8), 1 m*M* EDTA
Acid-washed glass beads, 0.45 μm (Sigma)

Procedure for Small-Scale DNA Isolation from Yeast

1. Spin down a 1-ml culture of Y190 (containing library plasmids) grown in SC-Leu or scrape a few colonies from an SC-Leu plate into an Eppendorf tube.
2. Add 200 μl of yeast lysis solution; vortex to resuspend cells.
3. Add glass beads until the total volume reaches 400 μl; vortex vigorously for 1 min.
4. Add 200 μl of PCI and vortex vigorously for 1 min.
5. Spin for 5 min in an Eppendorf centrifuge.
6. Recover the aqueous phase and repeat steps 4 and 5 for the aqueous phase.
7. Extract once more with 200 μl of $CHCl_3$.
8. Precipitate the DNA by addition of 2 vol of ethanol and incubation at −20° for 1 hr.
9. Spin for 10 min in an Eppendorf centrifuge at room temperature and wash the DNA pellet with 1 ml of 70% ethanol.
10. Dissolve the DNA in 50 μl of TE.

The plasmid can be easily recovered from several of the common *E. coli* strains with ampicillin selection. Electroporation gives us consistently high efficiency of transformation. Five microliters of the above miniprep DNA should give hundreds of colonies. If one is attempting to recover selectively the library plasmid that also has the bait vector (i.e., the pAS2 cycloheximide selection for plasmid loss is not being used) transform into a *leuB*$^-$ *E. coli* strain such as JA226. The yeast *LEU2* gene can complement the *leuB* mutation in *E. coli*. The ampicillin selection must then be performed on *E. coli* minimal medium lacking leucine.

It is important to note that the library transformation into yeast often places more than one plasmid into a single yeast cell; as many as 10% are doubly transformed. Thus, several transformants must be recovered into *E. coli* and checked for inserts and retested for transcriptional activation. Plasmid DNAs recovered into *E. coli* should be retransformed into *Y190* (pAS-X) followed by a secondary X-Gal assay to confirm the interaction. Plasmids that give a positive signal can then have their 5′ ends sequenced to determine if they are previously known genes. The sequencing primers we use for identifying the sequences for the 5′ and 3′ ends of the cDNA in pACT libraries are as follows:

pACT forward 5′: 5′ CTATCTATTCGATGATGAAG
pACT reverse 3′: 5′ ACAGTTGAAGTGAACTTGCG

Characterization of Two-Hybrid Clones

It is always important to have an independent test for the interaction of the two proteins once genetically interacting clones are obtained, because false positives do occur even if clones meet all the criteria described above and secondary evidence of physical association is usually required for publication. Always have some independent biochemical assay for the interaction between the proteins identified and the bait. We have used several different tests to detect *in vitro* binding of positives. One is to make a PCR primer to the library plasmid that has a T7 promoter placed in an appropriate position to place the insert of the library plasmid under T7 control. The PCR-derived fragments can then be directly added to a coupled transcription–translation system (TnT; Promega, Madison, WI) and radiolabeled protein prepared. We usually add 6 μl of a robust 30-cycle PCR reaction to 25 μl of the translation mix (TnT). Five microliters of this reaction mixture run on an SDS gel gives a readily detectable signal after an overnight exposure. This reveals the size of the fused protein and can be used to detect interaction *in vitro* with a column of the bait protein, usually as a glutathione *S*-transferase (GST) fusion.

The sequences of the PCR primers we have used successfully are as follows:

1. TAA TAC GAC TCA CTA TAG GGA GAC CAC ATG GAT GAT GTA TAT AAC TAT CTA TTC
   T7 Promoter — Met — *GAL4* activation domain
   21 aa before the *Bgl*II site in pACT1
2. CTA CCA GAA TTC GGC ATG CCG GTA GAG GTG TGG TCA
   In the ADH terminator

A second method that is useful if antibodies to the bait protein are available is to immunoprecipitate the protein out of yeast extracts prepared from cells containing both hybrid plasmids, and then use antibodies to the activation domain of *GAL4* to detect binding of the fusion protein. Alternatively, immunoprecipitate ^{35}S-labeled yeast extracts to detect their coprecipitation.[17] A new band should appear in extracts from cells containing the library plasmid that is not present in extracts from cells containing control plasmids. Because the two-hybrid system can detect interac-

[17] T. H. Lee, S. J. Elledge, and J. S. Butel, *J. Virol.* **69,** 1107 (1995).

tions that are below the affinity detectable by immunoprecipitation, a negative result in these assays is not definitive.

A third method is to switch the bait and prey in their respective plasmids, i.e., take the library insert out of pACT and insert it inframe into pAS2, and place the original pAS2 insert into pACT. The majority of false positives will not interact in this test. It should be noted that some true positives may not activate for structural reasons; thus only a positive result can be trusted. We have placed the pAS2 polylinker into pACT, creating pACT2, to facilitate this transfer.

Author Index

Numbers in parentheses are footnote reference numbers and indicate that an author's work is referred to although the name is not cited in the text.

C

D

E

H

L

M

N

O

S

T

U

V

W

X

Y

Subject Index

F

G

H

L

M

N

O

P

U

ISBN 0-12-182174-9